Lecture Notes in Mathematic

Edited by A. Dold and B. Eck

765

Padé Approximation and its Applications

Proceedings of a Conference held in Antwerp, Belgium, 1979

Edited by L. Wuytack

Springer-Verlag
Berlin Heidelberg New York 1979

Editor

Luc Wuytack
Department of Mathematics
University of Antwerp
Universiteitsplein 1
B-2610 Wilrijk
Belgium

AMS Subject Classifications (1980): 41A17, 42A16, 65D10, 65D20

ISBN 3-540-09717-1 Springer-Verlag Berlin Heidelberg New York
ISBN 0-387-09717-1 Springer-Verlag New York Heidelberg Berlin

Printing and binding: Beltz Offsetdruck, Hemsbach/Bergstr.
2141/3140-543210

PREFACE

This publication represents the Proceedings of a "Conference on Padé Approximation and Its Applications" which was held at the Department of Mathematics of the University of Antwerp (Wilrijk, Belgium) on April 4-6, 1979. Every speaker at the Conference was so kind to submit a manuscript for the Proceedings. Because of its importance we also included a not yet published paper by F. Cordellier entitled "Démonstration algébrique de l'extension de l'identité de Wynn aux tables de Padé non-normales", which was presented at a similar conference in Lille last year. These Proceedings also contain two bibliographies. The first is a complement to Claude Brezinski's bibliographies on Padé Approximation and related topics and includes items published in 1978 and 1979. The second is a commented list of publications on algorithms for computing Padé approximants.

Padé approximants are nonlinear approximations (in the form of a rational function) for a given function (in the form of a power series) having various interesting properties and applications. The importance of Padé approximation is somewhat reflected in the Proceedings of other conferences on this topic. These conferences were held at the University of Colorado (Boulder, June 1972), the University of Kent (Canterbury, July 1972), the University of Toulon (Toulon, May 1975), the University of South Florida (Tampa, December 1976).

At the Conference in Antwerp several mathematical aspects of Padé approximation and its generalizations were treated. An important part of the talks was on applications of this technique to different fields of Numerical Analysis. Several talks were concerned with algorithms for computing Padé approximants. The conference was a continuation of two conferences on Padé approximation, organized by Claude Brezinski at the University of Lille (France) in 1977 and 1978. The number of participants and talks indicates a still growing interest in the field.

We wish to express our appreciation to the University of Antwerp and the N.F.W.O. (Belgian National Science Foundation) for having provided financial support for the conference. We also express our gratitude to the Department of Mathematics for providing the necessary facilities and an informal and cordial atmosphere. At last we thank Professor A. Dold and the Springer-Verlag for the speedy publishing of this volume.

Antwerp, August 10, 1979
L. Wuytack.

CONTENTS

C. Applications of Padé approximation

D. Bibliographies

Padé-Chebychev Approximants

J. S. R. Chisholm and A. K. Common
University of Kent, Canterbury, England.

I. INTRODUCTION

Expansions of functions in terms of Chebychev polynomials have proved most useful when one needs to approximate such functions in the neighbourhood of an interval of a line rather than in the neighbourhood of a point. Without loss of generality we may take this interval to be [-1,+1] and if f is a function which is homomorphic in an ellipse with foci at ±1, then the Chebychev series expansion to $f(z)$ converges uniformly on [-1,+1]. We may then write, for real x in [-1,+1] ,

$$f(x) = \sum_{r=0}^{\infty}{}' c_r T_r(x) \equiv \tfrac{1}{2} c_0 + \sum_{r=1}^{\infty} c_r T_r(x) \tag{1.1a}$$

$$= \sum_{0}^{\infty}{}' c_r \cos r\theta \tag{1.1b}$$

$$= \tfrac{1}{2} \sum_{-\infty}^{\infty} c_r z^r \quad , \tag{1.1c}$$

where $x=\cos\theta$, $z=e^{i\theta}$ and $c_r=c_{-r}$. The three expressions for f(x), show that Chebychev series are linked with Fourier and Laurent series.

The general Fourier series

$$\sum_{r=0}^{\infty} f_r \cos r\theta + \sum_{r=1}^{\infty} g_r \sin r\theta = \sum_{r=0}^{\infty} f_r T_r(x) +$$

$$+ \sqrt{1-x^2} \sum_{r=1}^{\infty} g_r U_{r-1}(x) = \tfrac{1}{2} \sum_{-\infty}^{\infty} c_r z^r \quad (1.2)$$

where $U_r(x)$ are Chebychev polynomials of the second kind and $c_0=f_0$, $c_{\pm r}=f_r \mp ig_r$, $r>0$. Once again one can see that the three types of series are closely related.

To accelerate convergence of the Chebychev series or continue it outside its domain of convergence one may define Padé-Chebychev approximants from its coefficients, analagous to the Padé approximants to power series. The first method suggested for approximating the Chebychev series in this way was due to Maehly[8]. He proposed approximants of the form

$$f^M_{m/n}(x) = \frac{\sum'^{m}_{r=0} a_r T_r(x)}{\sum'^{n}_{s=0} b_s T_s(x)} , \quad (1.3)$$

where the coefficients in the numerator and denominator are determined from the c_r's in (1.1a) by the formal identity,

$$\sum_{s=0}^{n}{}' \; b_s T_s(x) \; [f^M_{m/n}(x) - f(x)] = O[T_{m+n+1}] \quad . \quad (1.4)$$

However as

$$2T_i(x)\, T_j(x) = T_{i+j}(x) + T_{|i-j|}(x) \quad , \quad (1.5)$$

one needs to know the set of coefficients $\{ c_r \; ; \; r=0,1,\ldots,m+2n \}$ whilst the [m/n] Padé approximants to power series are determined by only the first m+n+1 coefficients. The Chebychev-Padé approximants defined in this way therefore do not make the most economic use of a given number of coefficients and as a consequence tend not to converge as quickly as approximants described later in this work.

When the coefficients in (1.1) are real one may define Chebychev-Padé approximants which make full use of a given number of coefficients in the following manner[5]. Write

$$f(x) = \sum_{r=0}^{\infty}{}' c_r T_r(x) \equiv \mathrm{Re}\,[g(z)] \quad ,$$

where

$$g(z) = \sum_{r=0}^{\infty}{}' c_r z^r \quad . \tag{1.6}$$

Then define the approximant

$$f^T_{m/n}(x) = \mathrm{Re}\,[g_{m/n}(z)] \quad , \tag{1.7}$$

where $g_{m/n}(z)$ is the m/n Padé approximant to $g(z)$. It is easy to show that, for $m \geq n$, $f^T_{m/n}$ is of the form

$$f^T_{m/n}(x) = \frac{\sum_{r=0}^{m} \alpha_r T_r(x)}{\sum_{s=0}^{n} \beta_s T_s(x)} \tag{1.8}$$

and properties of this type of approximant to real Chebychev series have been discussed by Field[4] .

As it stands this method cannot be used when the c_r's are complex. In the next section we will show how this definition may be generalised to complex series, using the concept[1] of "*JI-NUMBERS*". Then in Section 3 we will compare these approximants with other types of Chebychev-Padé approximants and in Section 4 present our conclusions.

II. CHEBYCHEV-PADE APPROXIMANTS AND *JI-NUMBERS*

Let us define the following matrices,

$$\mathbf{I} = \begin{pmatrix} 1 & 0 \\ 0 & 1 \end{pmatrix} \quad ; \quad \mathbf{J} = \begin{pmatrix} 0 & 1 \\ -1 & 0 \end{pmatrix} \quad . \tag{2.1}$$

They satisfy relations

$$\mathbf{I}^2 = -\mathbf{J}^2 = \mathbf{I} \quad \text{and} \quad \mathbf{IJ} = \mathbf{JI} = \mathbf{J} \quad . \tag{2.2}$$

Therefore ($\mathbf{I}$, $\mathbf{J}$) is isomorphic to (1,i) and can be used to represent complex numbers. If z_1, z_2 are ordinary complex numbers we define

$$\mathbf{Z} = z_1\mathbf{I} + z_2\mathbf{J} \quad , \tag{2.3}$$

to be a "*JI-NUMBER*".
The "*JI-CONJUGATE NUMBER*" is

$$\bar{\mathbf{Z}} = z_1\mathbf{I} - z_2\mathbf{J} \quad , \tag{2.4}$$

while the ordinary complex conjugate is

$$\mathbf{Z}^* = z_1^*\mathbf{I} + z_2^*\mathbf{J} \quad . \tag{2.5}$$

The operations of conplex conjugation and JI-conjugation commute. We define,

$$\mathit{JIMAG}\,\mathbf{Z} \equiv z_2 \quad , \quad \mathit{IRE}\,\mathbf{Z} \equiv z_1 \tag{2.6}$$

Now

$$e^{\mathbf{J}\theta} = \mathbf{I}\cos\theta + \mathbf{J}\sin\theta \quad . \tag{2.7}$$

Therefore we can write

$$\sum_{r=0}^{\infty}{}' \; c_r \cos r\theta \;=\; \sum_{r=0}^{\infty}{}' \; c_r T_r(x) \;=\; IRE \sum_{r=0}^{\infty}{}' \; c_r e^{r\mathbf{J}\theta}$$

$$= IRE \sum_{r=0}^{\infty}{}' \; c_r \mathbf{z}^r \;, \qquad (2.8)$$

with $\mathbf{z}=e^{\mathbf{J}\theta}$.

Definition The [m/n] Chebychev-Padé approximant to the series $\sum_{r=0}^{\infty}{}' c_r T_r(x)$ is

$$f^T_{M/N}(x) \;\equiv\; IRE \left\{ \frac{\sum_{r=0}^{m} a_r \mathbf{z}^r}{\sum_{s=0}^{n} b_s \mathbf{z}^s} \right\} \qquad (2.9)$$

where the a's and b's are the corresponding coefficients in the numerator and denominator of the [m/n] Padé approximant to $\sum_{r=0}^{\infty}{}' c_r x^r$.

To calculate the approximant, calculate coefficients a_r, b_s in the usual way. Then write the approximant as

$$f^T_{m/n}(x) \;=\; IRE \; \frac{\left[\left[\sum_{r=0}^{m} a_r \mathbf{z}^r\right]\left[\sum_{s=0}^{n} b_s \mathbf{z}^{-s}\right]\right]}{\left[\sum_{s=0}^{n} b_s \mathbf{z}^{-s}\right]\left[\sum_{t=0}^{n} b_t \mathbf{z}^{t}\right]}$$

$$= \frac{\sum_{r=0}^{m}\sum_{s=0}^{n} a_r b_s \cos (r-s)\theta}{\sum_{t=0}^{n} b_t^{\,2} + 2 \sum_{t=1}^{n} \sum_{\substack{s=0\\ s<t}}^{n} b_s b_t \cos(t-s)\theta} \;. \qquad (2.10)$$

Series of Chebychev polynomials of the second kind are equivalent to Fourier sine series, since

$$\sqrt{1-x^2}\ \sum_{r=1}^{\infty} c_r\, U_{r-1}(x) \;=\; \sum_{r=1}^{\infty} c_r \sin r\theta$$

$$= \mathit{JIMAG}\left[e^{\mathbf{J}\theta} \sum_{r=0}^{\infty} c_{r+1}\, e^{r\mathbf{J}\theta} \right] \qquad (2.11)$$

If $[\sum_{r=0}^{m} a_r z^r] / [\sum_{s=0}^{n} b_s z^s]$ is the [m/n] Padé approximant to $\sum_{r=0}^{\infty} c_{r+1} e^{ir\theta}$,

we define approximants corresponding to (2.11) by

$$f^{u}_{m/n}(x) \;=\; \mathit{JIMAG}\left\{ \frac{e^{\mathbf{J}\theta} \sum_{r=0}^{m} a_r\, e^{r\mathbf{J}\theta}}{\sum_{s=0}^{n} b_s\, e^{s\mathbf{J}\theta}} \right\}$$

$$= \frac{\sum_{r=0}^{m} \sum_{s=0}^{n} a_r\, b_s \sin(r-s+1)\theta}{\sum_{t=0}^{n} b_t^{\,2} + 2 \sum_{t=1}^{n} \sum_{\substack{s=0\\ s<t}}^{n} b_s\, b_t \cos(t-s)\theta} \qquad (2.12)$$

Finally we consider the general Fourier series

$$s(\theta) \;=\; \sum_{r=0}^{\infty} [f_r \cos r\theta + g_r \sin r\theta] \quad , \qquad f_r, g_r \text{ complex}$$

$$= \mathit{IRE}\ \{ \mathbf{I} \sum_{r=0}^{\infty} f_r\, e^{r\mathbf{J}\theta} - \mathbf{J} \sum_{r=0}^{\infty} g_r\, e^{r\mathbf{J}\theta} \}$$

$$= \mathit{IRE}\ \{ \sum_{r=0}^{\infty} \mathbf{c_r}\, e^{r\mathbf{J}\theta} \} \quad , \qquad (2.13)$$

where

$$\mathbf{c_r} \;=\; f_r \mathbf{I} - g_r \mathbf{J} \quad .$$

We can then define the [m/n] Fourier-Padé approximant to $s(\theta)$ to be

$$s^F_{m/n}(\theta) = IRE \left\{ \frac{\sum_{r=0}^{m} \mathbf{A_r}\; e^{r\mathbf{J}\theta}}{\sum_{s=0}^{n} \mathbf{B_s}\; e^{s\mathbf{J}\theta}} \right\} \tag{2.14}$$

where the quotient in (2.14) is the [m/n] matrix Padé approximant to $\sum_{r=0}^{\infty} \mathbf{c_r}\; e^{r\mathbf{J}\theta}$. The matrix coefficients $\mathbf{A_r},\mathbf{B_s}$ are now *JI*-numbers determined by the usual conditions

$$\sum_{s=0}^{\mathrm{Min}(n,r)} \mathbf{B_s}\, \mathbf{c_{r-s}} = \begin{cases} \mathbf{A_s} & r=0,\ldots,m \\ \mathbf{O} & r=m+1,\ldots,m+n \end{cases} \tag{2.15}$$

Let

$$\mathbf{B_O} = \mathbf{I} = \begin{pmatrix} 1 & 0 \\ 0 & 1 \end{pmatrix}, \quad \mathbf{B_s} = \begin{pmatrix} \beta_{1s} & \beta_{2s} \\ -\beta_{2s} & \beta_{1s} \end{pmatrix} ,$$

and put

$$\mathbf{c_r} = \begin{pmatrix} f_r & -g_r \\ g_r & f_r \end{pmatrix} .$$

We have to solve for the 2n unknowns $\beta_{11},\beta_{21},\beta_{12},\beta_{22},\ldots,\beta_{1n},\beta_{2n}$ from equations

$$\begin{bmatrix} f_{m+1} & g_{m+1} & f_m & g_m & \cdots & f_{m+1-n} & g_{m+1-n} \\ -g_{m+1} & f_{m+1} & -g_m & f_m & \cdots & -g_{m+1-n} & f_{m+1-n} \\ f_{m+2} & g_{m+2} & f_{m+1} & g_{m+1} & \cdots & f_{m+2-n} & g_{m+2-n} \\ -g_{m+2} & f_{m+2} & -g_{m+1} & f_{m+1} & \cdots & -g_{m+2-n} & f_{m+2-n} \\ \vdots & \vdots & & & & & \\ f_{m+n} & g_{m+n} & & & & & \\ -g_{m+n} & f_{m+n} & & & & & \end{bmatrix} \begin{bmatrix} 1 \\ 0 \\ \beta_{11} \\ \beta_{21} \\ \vdots \\ \beta_{1n} \\ \beta_{2n} \end{bmatrix} = 0 \tag{2.16}$$

where $f_i=g_i=0$ if $i<0$.

Having obtained $\{\mathbf{B}_r\}$ in this way, we then solve (2.15) for the matrices

$$\mathbf{A}_r = \begin{pmatrix} \alpha_{1r} & \alpha_{2r} \\ -\alpha_{2r} & \alpha_{1r} \end{pmatrix} .$$

Then substituting in (2.14) we obtain the following expression for the approximants to the Fourier series,

$$s^F_{m/n}(\theta) = \frac{\sum\limits_{s=0}^{n} \sum\limits_{r=0}^{m} \{[\alpha_{1r}\beta_{1s} + \alpha_{2r}\beta_{2s}]\cos(r-s)\theta + [\alpha_{1r}\beta_{2s} - \alpha_{2r}\beta_{1s}]\sin(r-s)\theta\}}{\sum\limits_{s=0}^{n} \sum\limits_{t=0}^{n} \{[\beta_{1s}\beta_{1t} + \beta_{2s}\beta_{2t}]\cos(s-t)\theta + [\beta_{1s}\beta_{2t} - \beta_{2s}\beta_{1t}]\sin(s-t)\theta\}} \tag{2.17}$$

We note:-

(i) If $m>n$, then largest angle in numerator is $m\theta$ and in denominator is $n\theta$.

(ii) $$s(\theta) - s^F_{m/n}(\theta) = IRE\left\{ \sum_{r=0}^{\infty} \mathbf{c}_r\, e^{r\mathbf{J}\theta} - \frac{\sum\limits_{r=0}^{m} \mathbf{A}_r\, e^{r\mathbf{J}\theta}}{\sum\limits_{s=0}^{n} \mathbf{B}_s\, e^{s\mathbf{J}\theta}} \right\}$$

$$= IRE\ \{O\,[e^{(m+n+1)\mathbf{J}\theta}]\}$$

$$= O\,[\sin(m+n+1)\theta\ ,\ \cos(m+n+1)\theta] \ . \tag{2.18}$$

As an alternative to $s^F_{m/n}(\theta)$ we can approximate the sine series and cosine series separately using (2.12) and (2.10) respectively. In the table we compare the two ways of approximating the function

$$s(\theta) = e^{(e^{i\theta})} + e^{2(e^{-i\theta})} + e^{3i(e^{i\theta})} + e^{4i(e^{-i\theta})}$$

at the level of the [5/5] approximants.

TABLE

θ (radians)	Re $s(\theta)$	Im $s(\theta)$	$\lvert s(\theta)-s^{F}_{5/5}(\theta)\rvert$	$\lvert s(\theta)-f^{T}_{5/5}(\theta)-f^{u}_{5/5}(\theta)\rvert$
0.000	8.464	- 0.616	0.0131	0.0101
0.628	- 6.770	- 4.287	0.0441	0.0137
1.257	14.969	41.800	0.1514	0.0552
1.885	15.033	-42.357	0.1370	0.0572
2.513	-10.133	0.942	0.0275	0.0131
3.142	- 1.140	0.616	0.0041	0.0035
3.772	- 4.053	- 3.876	0.0016	0.0095
4.398	10.667	-13.981	0.0020	0.0479
5.026	10.603	14.538	0.0058	0.0595
5.655	-0.690	7.222	0.0055	0.0236

It can be seen from these results, that sometimes approximating the sine and cosine series separately gives the better results while sometimes approximating them simultaneously is better. Also the error between $s(\theta)$ and $s^{F}_{5/5}(\theta)$ appears to be least for larger values of θ .

III. COMPARISON WITH OTHER TYPES OF PADE-CHEBYCHEV APPROXIMANTS

We compare here the approximants described in Section 2 with other methods of defining Padé-Chebychev approximants and will show that for m≥n they are all identical.

1. Generating Function Method

Consider the [m/n] Padé approximant to $f(x) = \sum_{r=0}^{\infty}{}' c_r x^r$ which has the partial fraction expansion

$$f_{m/n}(x) = \sum_{j=0}^{m-n} \beta_j x^j + \sum_{s=1}^{n} \frac{\alpha_s}{1-\sigma_s x} \equiv \sum_{r=0}{}' c_r^A x^r \qquad (3.1)$$

where the first sum is absent if $m<n$. Therefore when we take the [m/n] approximant to f(x) we are approximating c_r by

$$c_r^A = \beta_r + \sum_{s=1}^{n} \alpha_s \sigma_s^r \qquad (r=1,2,\ldots)$$
$$\tfrac{1}{2} c_0^A = \beta_0 + \sum_{s=1}^{n} \alpha_s \qquad . \qquad (3.2)$$

Then we can define an [m/n] Chebychev-Padé approximant to f(x) by the formal summation

$$f_{m/n}^T(x) = \sum_{r=0}{}' c_r^A T_r(x) = \sum_{r=0}^{\infty} \sum_{s=1}^{n} \alpha_s \sigma_s^r T_r(x) + \sum_{j=0}^{m-n} \beta_j T_j(x)$$
$$= \sum_{s=1}^{n} \alpha_s \left[\sum_{r=0}^{\infty} \sigma_s^r T_r(x)\right] + \sum_{j=0}^{m-n} \beta_j T_j(x)$$
$$= \sum_{s=1}^{n} \frac{\alpha_s(1-\sigma_s x)}{1-2\sigma_s x+\sigma_s^2} + \sum_{j=0}^{m-n} \beta_j T_j(x) \qquad (3.3)$$

Approximants corresponding to (3.3) have been defined previously by Common[3] for Legendre series and for general series of orthogonal polynomials by Garibottiand Grinstein[6] .

We will now show that the approximants defined in (3.3) are identical for all $m,n\geq 0$ to those defined in (2.9), which are

$$f^{T}_{m/n}(x) = \mathcal{R}e \left\{ \frac{\sum_{r=0}^{m} a_r e^{r\mathbf{J}\theta}}{\sum_{s=0}^{n} b_s e^{s\mathbf{J}\theta}} \right\}$$

$$= \mathcal{R}e \left\{ \sum_{s=1}^{n} \frac{\alpha_s}{1-\sigma_s e^{\mathbf{J}\theta}} + \sum_{j=0}^{m-n} \beta_j e^{i\mathbf{J}\theta} \right\}$$

$$= \sum_{j=0}^{m-n} \beta_j T_j(x) + \mathcal{R}e \left\{ \sum_{s=1}^{n} \frac{\alpha_s (1-\sigma_s e^{-\mathbf{J}\theta})}{(1-\sigma_s e^{\mathbf{J}\theta})(1-\sigma_s e^{-\mathbf{J}\theta})} \right\}$$

$$= \sum_{j=0}^{m-n} \beta_j T_j(x) + \sum_{s=1}^{n} \frac{\alpha_s(1-\sigma_s x)}{(1+\sigma_s^2-2\sigma_s x)} ,$$

since $x=\cos\theta$. This is the same as (3.3).

2. Clenshaw-Lord Approximants[2]

Given the series $\sum_{r=0}^{\infty}{}' c_r T_r(x)$, Clenshaw and Lord define coefficients $\{b_s\}$ through the equations

$$\sum_{s=0}^{n} b_s c_{|r-s|} = 0 \quad , \quad r=m+1,\ldots,m+n . \tag{3.4}$$

The Clenshaw-Lord denominator is then

$$\sum_{s=0}^{n}{}' q_s T_s(x) = \tfrac{1}{2}\mu \left(\sum_{s=0}^{n} b_s z^{-s} \right) \left(\sum_{r=0}^{n} b_r z^{r} \right) , \tag{3.5}$$

where

$$x=\cos\theta \quad , \quad z=e^{i\theta} \quad \text{and} \quad \mu^{-1}=\tfrac{1}{2} \sum_{i=0}^{n} b_i^2 .$$

Their numerator coefficients are

$$p_r = \tfrac{1}{2} \sum_{s=0}^{n}{}' q_s \,(c_{r+s} + c_{|r-s|}) \quad .$$

Using the theory of difference equations, Clenshaw and Lord prove that their approximants satisfy the formal identity

$$\frac{\sum_{r=0}^{m}{}' p_r T_r(x)}{\mu \sum_{s=0}^{n}{}' q_s T_s(x)} = \sum_{r=0}^{\infty}{}' c_r T_r(x) + O\,[T_{m+n+1}(x)] \quad . \qquad (3.6)$$

We will now show for m≥n-1, these approximants are identical to those given by (2.9). The crucial observation is that for m≥n-1, the equations (3.4) determining the b's, are identical to the standard equations for determining the denominator coefficients of the usual Padé approximant. Therefor the b's defined by (3.4) are the same as those appearing in (2.9); this equation gives

$$f^T_{m/n}(x) = \mathit{IRE} \left\{ \frac{\sum_{r=0}^{m} a_r \mathbf{z}^r}{\sum_{s=0}^{n} b_s \mathbf{z}^s} \right\} \qquad (\mathbf{z} = e^{\mathbf{J}\theta})$$

$$= \tfrac{1}{2}\, \frac{\left[\sum_{r=0}^{m} a_r \mathbf{z}^r \sum_{s=0}^{n} b_s \mathbf{z}^{-s} + \sum_{r=0}^{m} a_r \mathbf{z}^{-r} \sum_{s=0}^{n} b_s \mathbf{z}^s \right]}{\sum_{s=0}^{n} b_s \mathbf{z}^s \sum_{t=0}^{n} b_t \mathbf{z}^{-t}}$$

$$= \frac{\sum_{r=0}^{m}{}' \alpha_r T_r(x)}{\mu^{-1} \sum_{s=0}^{n}{}' q_s T_s(x)} \qquad (3.7)$$

from (3.5) where the α_r's may be expressed in terms of $\{a_r\}$ and $\{b_s\}$. Therefore for $m \geq n-1$, the denominators of the two types of approximant are identical.

Also, for $m \geq n-1$ we have, using (3.6) and (2.18) and remembering that a cosine series is a particular type of Fourier series,

$$\mu \left[\sum_{s=0}^{n}{}' q_s T_s(x) \right] \left[f^T_{m/n}(x) - \sum_{r=0}^{m}{}' p_r T_r(x) \right.$$

$$= \mu \sum_{s=0}^{n}{}' q_s T_s(x) \left\{ \sum_{r=0}^{\infty}{}' c_r T_r(x) - \sum_{r=0}^{\infty}{}' c_r T_r(x) \right.$$

$$\left. + O[T_{m+n+1}(x)] \right\} , \qquad (3.8)$$

Therefore

$$\sum_{r=0}^{m}{}' \alpha_r T_r(x) - \sum_{r=0}^{m}{}' p_r T_r(x) = O[T_{m+1}(x)] \quad , \qquad (3.9)$$

so that

$$\alpha_r = p_r \quad , \quad (r=0,1,\ldots,m) \qquad (3.10)$$

and that the numerators of the two types of approximants are identical for $m \geq n-1$. Hence our approximants given by (2.9) are completely identical to those of Clenshaw and Lord for $m \geq n-1$.

3. Laurent-Padé Approximants

Gragg and Johnson have defined generalisations of Padé approximants for Laurent series[7].

They write

$$f(z) = \tfrac{1}{2} [f^+(z) + f^-(z^{-1})] \quad ,$$

with

$$f^{\pm}(z) = \tfrac{1}{2}c_0 + \sum_{r=1}^{\infty} c_{\pm r} z^r . \qquad (3.11)$$

For $m \geq n$, they define the $[m/n]$ Laurent-Padé approximant to be

$$f^L_{m/n}(z) = \frac{p_{m/n}(z)}{q_{m/n}(z)} \equiv \tfrac{1}{2}[f^+_{m/n}(z) + f^-_{m/n}(z^{-1})] , \qquad (3.12)$$

where $f^{\pm}_{m/n}(z)$ is the $[m/n]$ Padé approximant to $f^{\pm}(z)$. When $c_r = c_{-r}$, (c_r can be complex),

$$f(z) = \sum_{r=0}^{\infty}{}' c_r T_r(x)$$

if $x = \cos\theta$, $z = e^{i\theta}$. In this case,

$$f^+_{m/n}(z) = f^-_{m/n}(z) = \frac{\sum_{r=0}^{m} a_r z^r}{\sum_{s=0}^{n} b_s z^s} \qquad (3.13)$$

where the a_r's and b_r's are identical to those in (2.9) as they are the numerator and denominator coefficients of the $[m/n]$ Padé approximant to the power series $\sum_{r=0}^{\infty}{}' c_r z^r$.

Therefore

$$f^L_{m/n}(z) = \tfrac{1}{2}\left\{\frac{\sum_{r=0}^{m} a_r e^{ir\theta}}{\sum_{s=0}^{n} b_s e^{is\theta}} + \frac{\sum_{r=0}^{m} a_r e^{-ir\theta}}{\sum_{r=0}^{n} b_s e^{-is\theta}}\right\}$$

$$= \frac{\sum_{r=0}^{m}\sum_{s=0}^{n} a_r b_s \cos(r-s)\theta}{\sum_{t=0}^{n} b_t^2 + 2\sum_{t=1}^{n}\sum_{\substack{r=0\\ r<t}}^{n} b_r b_t \cos(t-r)\theta}$$

$$= f^T_{m/n}(x) \qquad (3.14)$$

from (2.10). Therefore when the Laurent series reduced to a Chebychev series, our approximants are identical to the Laurent-Padé approximant for $m \geq n$.

When $z = e^{i\theta}$ the general Laurent-series (3.11) reduced to the general Fourier series

$$s(\theta) = \sum_{r=0}^{\infty} (f_r \cos r\theta + g_r \sin r\theta)$$

$$= \tfrac{1}{2} \sum_{r=0}^{\infty} [(f_r - ig_r)z^r + (f_r + ig_r)z^{-r}]$$

$$= \tfrac{1}{2} [f^+(z) + f^-(1/z)] \tag{3.15}$$

where $f^{\pm}(z)$ are given by (3.11) if

$$c_0 = f_0 \quad , \quad c_{\pm r} = f_r \mp ig_r \quad (r>0) .$$

As a final result of this work we will show that our approximants $s^F_{m/n}(\theta)$ to $s(\theta)$ given by (2.17) are identical for $m \geq n$ to those given by the Laurent-Padé approximant to $\tfrac{1}{2} [f^+(z) + f^-(z^{-1})]$ which are

$$f^L_{m/n}(z) = \tfrac{1}{2} [f^+_{m/n}(z) + f^-_{m/n}(z^{-1})]$$

$$= \tfrac{1}{2} \frac{\begin{vmatrix} f_{m-n+1} - ig_{m-n+1} & \cdots \\ \vdots & \\ \sum_{r=n}^{m} (f_{r-n} - ig_{r-n}) e^{ir\theta} & \cdots \end{vmatrix}}{\begin{vmatrix} f_{m-n+1} - ig_{m-n+1} & \cdots & \\ \vdots & & \vdots \\ e^{in\theta} & \cdots & 1 \end{vmatrix}} + \tfrac{1}{2} \frac{\begin{vmatrix} f_{m-n+1} + ig_{m-n+1} & \cdots \\ \vdots & \\ \sum_{r=0}^{m} (f_{r-n} + ig_{r-n}) e^{-ir\theta} & \cdots \end{vmatrix}}{\begin{vmatrix} f_{m-n+1} + ig_{m-n+1} & \cdots & \\ \vdots & & \vdots \\ e^{-in\theta} & \cdots & 1 \end{vmatrix}}$$

$$= \frac{1}{2}\ \mathit{IRE}\left\{ \frac{\begin{vmatrix} f_{m-n+1} - \mathbf{J}g_{m-n+1} & \cdots \\ \sum_{r=n}^{m} (f_{r-n} - \mathbf{J}g_{r-n})e^{r\mathbf{J}\theta} & \cdots \end{vmatrix}}{\begin{vmatrix} f_{m-n+1} - \mathbf{J}g_{m-n+1} & \cdots \\ \vdots & \\ e^{n\mathbf{J}\theta} & \cdots \end{vmatrix}} + \frac{\begin{vmatrix} f_{m-n+1} + \mathbf{J}g_{m-n+1} & \cdots \\ \sum_{r=n}^{m} (f_{r-n} + \mathbf{J}g_{r-n})e^{-r\mathbf{J}\theta} & \cdots \end{vmatrix}}{\begin{vmatrix} f_{m-n+1} + \mathbf{J}g_{m-n+1} & \cdots \\ \vdots & \\ e^{-\mathbf{J}n\theta} & \cdots \end{vmatrix}} \right\}$$

$$= \mathit{IRE}\left\{ \frac{\sum_{r=0}^{m} \mathbf{A}_r e^{r\mathbf{J}\theta}}{\sum_{s=0}^{n} \mathbf{B}_s e^{s\mathbf{J}\theta}} \right\}$$

$$= s^F_{m/n}(\theta) \tag{3.16}$$

from (2.14), where the $\mathbf{A}_r$'s and $\mathbf{B}_r$'s above are defined by (2.15). Conversely, if we replace $\mathbf{A}_r \to \mathbf{A}_r |z|^r$ and $\mathbf{B}_s \to \mathbf{B}_s |z|^s$, the approximants $s^F_{m/n}(\theta)$ become the Laurent-Padé approximants of Gragg and Johnson.

For the special case m=n-1, the denominator of the Laurent-Padé approximant is identical within a scale factor, to the Clenshaw-Lord denominator (3.5) and hence also to the denominator of $f^T_{m/n}(z)$. Then using the arguments of Section 3(iii), the Laurent-Padé approximant may be shown to be identical with $f^T_{m/n}(x)$ when m=n-1 .

IV. CONCLUSIONS

Using the concept of "JT" numbers we have generalised Padé approximants to Chebychev and general Fourier series in a very simple way: the denominator and numerator coefficients appearing in the approximants (2.9) (2.14) are determined from the coefficients of the corresponding series by the standard set of linear equations (2.15). The coefficients $\{\mathbf{A}_r\}$ and $\{\mathbf{B}_s\}$ in (2.14) are 2×2 matrices.

In the previous section we have shown that the Chebychev series approximants are equivalent to approximants defined by the "generating function" method for all $m,n>0$, and by Clenshaw-Lord and Gragg-Johnson for $m\geq n-1$, and that there is a similar correspondence for general Fourier series.

For $m<n-1$, our approximants do not have the nice form (1.8) but only the less satisfactory form

$$f^T_{m/n}(x) = \frac{\sum\limits_{r=0}^{n}{}' \alpha_r T_r(x)}{\sum\limits_{s=0}^{n}{}' \beta_s T_s(x)} \tag{4.1}$$

Nevertheless the coefficients in (2.14) and (2.17) depend on the minimum number of terms in the series, and the approximants are of maximal order in the sense of (2.18).

In order to define approximants with the minimum numerator order (m) when $m<n$, Clenshaw-Lord and Gragg-Johnson have modified the definition of their approximants for $m<n-1$, to retain the form (1.8). However the straight-forward connection with the ordinary Padé approximants is then lost, which may make the proof of convergence theorems more difficult for these values of m,n.

A C K N O W L E D G E M E N T S

We would like to thank Dr. T. Hopkins for assistance with the numerical computation of our approximants.

R E F E R E N C E S

1 Chisholm J. S. R., and Roberts, D. E. "Rotationally Covariant Approximants derived from Double Power Series Proc.R.Soc.Lond., A351 1976 585-591

2 Clenshaw C. W. and Lord K. Rational Approximations from Chebychev Series - Studies in Numerical Analysis, B.K.P. Sciafe (ed.), Academic Press, London, 1973.

3 Common A. K. Properties of Legendre Expansions Related to Series of Stieltjes and Applicatios to π-π Scattering Nuovo Cimento 63 (1969) 863-891

4 Field D. A. Pade-Chebychev Approximants University of Kent Preprint, 1976

5 Frankel A. P. and Gragg W. B. Algorithm for almost best uniform rational Approximations with Error Bounds (Abstract) SIAM Rev. 15 (1973) 418-419

6 Garibotti C. R. and Grinstein F. F. A Summation Procedure for Expansions in Orthogonal Polynomials Revista Brasileira de Fisica, 7 (1977) 557-567

7 Gragg W. B. and Johnson G. D. The Laurent-Pade Table "Information Processing 74", Proc. IFIP Congress 74, North Holland, Amsterdam (1974), 632-637.

Gragg W. B. Laurent, Fourier and Chebychev-Padé Tables "Pade and Rational Approximation, Theory and Applications", Saff E. B. and Varga, R. S. (Eds.), Academic Press, New York (1977) 61-72

8 Maehly, H. J. Rational Approximations for Transcendental Functions - Proceedings of the International Conference on Information Processing - UNESCO Butterworth, London, 1960

* * * * * *

SUR LA RÉGULARITÉ DES PROCÉDÉS δ^2 D'AITKEN ET W DE LUBKIN

F. CORDELLIER

UNIVERSITE DE LILLE I, UER d'IEEA - INFORMATIQUE
F-59650 VILLENEUVE d'ASCQ (FRANCE)

1 - INTRODUCTION

Nous présentons ici quelques résultats topologiques élémentaires relatifs aux plus simples des transformations non linéaires de suites de nombres complexes, les procédés δ^2 d'Aitken [1] et W de Lubkin [9]. Bien que l'objectif majeur de telles transformations soit presque toujours l'accélération de la convergence des suites auxquelles elles sont appliquées, nous ne nous intéressons pas ici à cette question et nous nous restreignons aux autres propriétés topologiques de ces transformations, à savoir :

- la régularité, c'est-à-dire la conservation de la convergence et de la limite,
- la quasi-régularité, c'est-à-dire la conservation de la limite dès qu'il y a celle de la convergence,
- la compatibilité de deux procédés, c'est-à-dire l'identité des antilimites de la même suite fournies par deux procédés distincts.

Le paragraphe qui suit est consacré à la présentation des notations et à la définition de concepts généraux, tels que le noyau, le domaine de régularité ou d'accélération d'une transformation de suites tandis que le troisième consiste en la description des procédés élémentaires étudiés. Les trois derniers sont respectivement destinés à l'étude des trois propriétés précitées.

Dans le paragraphe 4 relatif à la régularité, nous commençons par montrer qu'une vaste classe de procédés non linéaires ne jouit pas de la classique propriété de régularité. Ceci nous conduit à caractériser les suites convergentes transformées par un procédé donné en suites convergentes de même limite, suites dont l'ensemble sera appelé domaine de régularité du procédé, mais nous ne cherchons pas ici à mettre en évidence des classes de suites incluses dans un tel domaine de régularité comme l'on fait Lubkin [9], Gray, Clark et Adams [3], Tucker [15] ou l'auteur [5]. Dans le paragraphe suivant nous traitons de la quasi-régularité que Lubkin [9] avait introduite dans le cas du procédé δ^2 appliqué à des suites de réels. Tucker [15] a étendu ce résultat aux suites de complexes tandis que Rice [13] l'établissait pour un procédé similaire en lui donnant le nom de "joint convergence". Nous montrerons ici que le procédé W de Lubkin est quasi-régulier pour les suites de nombres réels. Notons en passant que la quasi-régularité de la transformation E_k de Shanks [14] est une conséquence immédiate d'un résultat de Montessus de Ballore [10] relatif à la convergence ponctuelle d'approximants de Padé ; malheureusement la démonstration de ce dernier résultat est sujette à caution et, pour $k \geq 2$, la quasi-régularité de E_k n'est encore qu'une conjecture.

Dans le 6ème et dernier paragraphe nous montrons que si les deux procédés δ^2 et W transforment une suite divergente en une suite convergente, alors les limites des suites transformées sont les mêmes. Cette question de la compatibilité des limites de suites transformées par des procédés non linéaires ne semble pas avoir été abordée jusqu'à présent.

2 - NOTATIONS ET VOCABULAIRE

Nous désignons par K un corps commutatif de caractéristique infinie qui pratiquement sera $\mathbb{R}$ ou $\mathbb{C}$ et nous introduisons les 2 éléments complémentaires suivants:

. l'élément à l'infini (unique) ∞ qui compactifie K pour donner : $\bar{K} = K \cup \{\infty\}$, avec les conventions opératoires classiques :

$$x + \infty = \infty + x = x - \infty = \infty - x = \infty,\ \forall x \in K$$
$$x \times \infty = \infty \times x = \infty,\ \forall x \in K \setminus \{0\}$$
$$x / 0 = \infty \text{ et } x / \infty = 0,\ \forall x \in K \setminus \{0\}$$

. un élément fictif dit indéfini et noté ω, résultat des opérations suivantes :

$$\infty + \infty,\ \infty - \infty,\ 0 \times \infty,\ \infty \times 0,\ 0/0,\ \infty/\infty$$

ainsi que de toute opération dont un opérande est ω :

$$x + \omega,\ \omega + x,\ x - \omega,\ \omega - x,\ \omega \times x,\ x \times \omega,\ \omega/x,\ x/\omega,\ \forall x \in \bar{K}.$$

Nous poserons alors :

$$K' = K \cup \{\omega\}$$
$$\bar{K}' = \bar{K} \cup \{\omega\}$$

$\mathbb{N}$ désignant l'ensemble des entiers naturels et L une partie de $\bar{K}'$, on peut définir

$$S(L) = \{x : \mathbb{N} \to L\}$$

ensemble de toutes les suites à valeurs dans L.

Si $\omega \in L$, certains termes peuvent être indéfinis (un nombre fini, une sous-suite, toute la suite). Pratiquement nous aurons : $K \subset L \subset \bar{K}'$.

On note σ_n la fonctionnelle qui à une suite $x \in S(L)$ associe son $n^{\text{ième}}$ terme : $\sigma_n(x) = x_n \in L$, et σ_∞ la fonction qui lui associe l'ensemble de ses points d'accumulation. Si la suite x converge, sa limite sera notée x^*. On a alors : $\sigma_\infty(x) = \{x^*\}$. Il n'est bien entendu pas exclu que l'on ait $x^* = \infty$. Par contre, l'élément ω étant étranger à la topologie définie sur K (et induite sur $\bar{K}$), on ne peut avoir $x^* = \omega$. Une suite convergente ne peut donc comporter qu'un nombre fini de termes indéfinis. Nous noterons $C(L)$ l'ensemble des suites convergentes dans L :

$$C(L) = \{x \in S(L) \;/\; \sigma_\infty(x) = \{x^*\}\}$$

On notera u et Θ les suites définies par

$$\sigma_n(u) = 1,\ \sigma_n(\Theta) = 0,\ \forall\, n \in \mathbb{N}.$$

. Soit K un corps commutatif de caractéristique infinie et soit L une partie de $\bar{K}'$ contenant K. On notera $T(L)$ (respectivement : $\bar{T}(L)$, $T'(L)$, $\bar{T}'(L)$) l'ensemble des applications qui, à $x \in S(L)$ associent $y = T(x) \in S(K)$ (respectivement : $S(\bar{K})$, $S(K')$, $S(\bar{K}')$).

. Suivant Germain-Bonne [7], nous dirons que $T \in \bar{T}'(\bar{K})$ est quasi-linéaire si :
$\forall\, x \in S(K)$, $\forall$ a, b $\in$ K tels que T(x) et T(ax+bu) $\in S(K)$, on a : $T(ax+bu) = aT(x) + bu$.

. Pour $T \in T(K)$, nous définissons :

$$N_K(T) = \{x \in S(K) \;/\; T(x) = \Theta\}$$

$$R_K(T) = \{x \in C(K) \;/\; T(x) \in C(K) \text{ avec } T(x)^* = x^*\}$$

$$A_K(T) = \{x \in C(K) \;/\; \lim_{n\to\infty} \frac{\sup_{i\ge n} |T(x_i) - x^*|}{\sup_{i\ge n} |x_i - x^*|} = 0\}$$

$$L_K(T) = \{x \in S(K) \;/\; T(x) \in C(K)\}.$$

Par analogie avec les transformations linéaires, $N_K(T)$ sera appelé noyau du procédé T.

Les trois autres sous-ensembles seront appelés respectivement domaines de régularité, d'accélération et de limitabilité du procédé T.

. Notons $H(\bar{K})$ l'ensemble des transformations homographiques régulières qui appliquent $\bar{K}$ sur $\bar{K}$:
$h \in H(\bar{K}) \iff z = h(x) = \dfrac{ax+b}{cx+d}$, où chacune des 4 paires (a,b), (a,c), (b,d), (c,d) comporte un élément non nul.

A toute transformation homographique régulière $h \in H(\bar{K})$, on peut associer une transformation de suite T_h par :

$$x \to y = T_h(x) \text{ où } \sigma_n(y) = h(\sigma_n(x)),\ \forall\, x \in \mathbb{N}.$$

L'ensemble de ces transformations de suites sera noté $H(\bar{K})$.

Nous noterons $H(K)$ le sous-ensemble de celles qui laissent invariant le point à l'infini, c'est-à-dire :

$$T_h \in H(K) \iff x \to y = T_h(x) \text{ où } \sigma_n(y) = a\sigma_n(x) + b, \forall a,\, b \in K \text{ avec } a \neq 0.$$

• Nous noterons (a,b ; c,d) = -1 pour exprimer le fait que les quatre nombres a,b,c,d forment une division harmonique, c'est-à-dire : (c-a)/(c-b) + (d-a)/(d-b) = 0.
On sait qu'une telle relation est invariante dans une transformation homographique, c'est-à-dire :

$$(a,b\,;\,c,d) = -1 \implies (h(a),h(b)\,;\,h(c),\,h(d)) = -1,\ \forall\, h \in H(\bar{K})$$

• Si $t \in T(K)$ et $P \in S(K)$, on posera : $t(P) = \bigcup_{u \in P} t(u)$.

Le lecteur vérifiera sans peine que :

$\forall\, \ell \in H(K)$, $\forall\, T \in T(K)$, on a :

$$\begin{cases} \ell(N_K(T)) \cap C(K) \subset A_K(T) \subset R_K(T) \subset L_K(T) \cap C(T) \\ \ell(N_K(T)) \subset L_K(T). \end{cases}$$

3 - LES PROCÉDÉS UTILISÉS

3.1 - Le procédé γ_α

• Définissons tout d'abord le procédé suivant qui dépend du paramètre $\alpha \in \mathbb{C}$:

$$\gamma_\alpha : x \in S(\mathbb{C}) \to y \in S(\bar{\mathbb{C}}')$$

$$\text{avec } y_n = x_{n+1} - \alpha \frac{\Delta x_n\, \Delta x_{n+1}}{\Delta^2 x_n}$$

où : $\Delta u_n = u_{n+1} - u_n$, $\forall\, n \in \mathbb{N}$.

On vérifie aisément que

$$\begin{cases} y_n = \omega \iff x_n = x_{n+1} = x_{n+2} \\ y_n = \infty \iff \Delta x_n = \Delta x_{n+1} \neq 0 \end{cases}$$

On peut encore l'écrire sous la forme :

$$y_n = x_{n+1} + \alpha\, \Delta(1/\Delta x_n)$$

• Dans le cas où $\alpha = 1$, c'est le procédé δ^2 d'Aitken [1] qu'on peut encore écrire :

$$y_n = \begin{vmatrix} x_n & \Delta x_n \\ x_{n+1} & \Delta x_{n+1} \end{vmatrix} \Big/ \begin{vmatrix} 1 & \Delta x_n \\ 1 & \Delta x_{n+1} \end{vmatrix} = E_1(x_n)$$

Ce procédé a été généralisé par Shanks [14] qui a proposé la transformation $E_k(x_n)$ pour la mise en oeuvre de laquelle Wynn [18] a proposé le maintenant classique ε-algorithme. Rappelons que cette transformation est étroitement connectée à la notion de table de Padé [14].

Le procédé δ^2 s'explicite de nombreuses façons : l'écriture $y_n = x_{n+1} + 1/\Delta(1/\Delta x_n)$ met en lumière le fait qu'il s'agit de la première étape de l'ε-algorithme tandis que l'écriture $y_n = \Delta(x_n/\Delta x_n)/\Delta(1/\Delta x_n)$ montre qu'il s'agit de la première étape de la transformation T de Levin [8] et que l'écriture $y_n = (x_n \, \Delta x_{n+1} - x_{n+1} \, \Delta x_n)/\Delta^2 x_n$ le fait apparaître comme la première étape du schéma d'extrapolation Neville-Aitken pour obtenir la valeur à l'origine du polynôme qui interpole les couples (x_{n+i}, y_{n+i}) selon un algorithme proposé par Germain-Bonne [6]. Cette dernière écriture montre encore que c'est la première étape du procédé d'Overholt [11].

• Pour $\alpha = 2$, le procédé γ_2 apparaît comme la première étape du ρ-algorithme de Wynn [19]. Il correspond à la valeur en $t = \infty$ de la fraction rationnelle de degré (1/1) qui interpole les 3 couples $(n+i, x_{n+i})$ pour i=0, 1 et 2.

3.2 - Le procédé W

Ce procédé s'explicite par :

$$W : x \in S(\mathbb{C}) \to y \in S(\bar{\mathbb{C}}')$$

avec

$$y_n = \frac{x_{n+1} \, \Delta x_{n+2} \, \Delta^2 x_n - x_{n+2} \, \Delta x_n \, \Delta^2 x_{n+1}}{\Delta x_{n+2} \, \Delta^2 x_n - \Delta x_n \, \Delta^2 x_{n+1}}$$

à moins que $y_n = \infty$ ou $y_n = \omega$.

Nous ne préciserons pas ici les cas où l'on a $y_n = \infty$ ou ω. On peut encore écrire : $y_n = x_{n+1} - \Delta x_n \, \Delta x_{n+1} \, \Delta^2 x_{n+1} \, / \, (\Delta x_{n+2} \, \Delta^2 x_n - \Delta x_n \, \Delta^2 x_{n+1})$ ou (avec certaines restrictions) : $y_n = \Delta^2(x_n/\Delta x_n) \, / \, \Delta^2(1/\Delta x_n)$.

Introduite par Lubkin [9] sans autre justification apparente que son efficacité (réelle) sur des exemples accompagnés de quelques résultats théoriques intéressants, ce procédé est retrouvé par Germain-Bonne [6] comme l'extrapolé linéaire à l'origine des deux couples $(1/\Delta(1/\Delta x_{n+i}), x_{n+1+i})$ (i=0 et 1).

Notons en passant que Tucker [16] a étudié le procédé W de Lubkin et proposé un procédé W1 qui n'en diffère que par un décalage d'indices puisque $W1(x_n) = W(x_{n+1})$, $\forall\, n \in \mathbb{N}$.

Le procédé W peut également être considéré comme la première étape de diverses transformations de suites plus élaborées comme le θ-algorithme de Brezinski [2] ou les transformations U et V de Levin [8]. C'est la raison pour laquelle certains auteurs le désignent parfois sous le nom de procédé θ_2 [4].

4 - RÉGULARITÉ

Très étudiée dans le contexte des transformations linéaires de suites où le théorème de Toeplitz est un outil appréciable, cette notion ne l'est guère pour les transformations non linéaires. Notons déjà que, même dans le cas des transformations linéaires, cette notion n'est pas indispensable puisqu'il existe des transformations linéaires non régulières [17] dont l'expérimentation vient confirmer l'efficacité [15]. Nous commencerons par montrer qu'une vaste classe de transformations rationnelles n'admet pas $C(\mathbb{C})$ comme domaine de régularité. La simplicité des procédés γ_α nous permet ensuite de donner une caractérisation de leur domaine de régularité $\mathcal{R}_{\mathbb{C}}(\gamma_\alpha)$ et de fournir quelques informations supplémentaires sur la structure de ce domaine. Sur le procédé W, nous nous contenterons de donner ici une caractérisation de son domaine de régularité.

4.1 - Non régularité des transformations rationnelles

Proposition 1 Si une transformation de suite $y = T(x)$ est définie par :

$$y_n = x_n + \Delta x_n \times R\left(\frac{\Delta x_{n+1}}{\Delta x_n}, \ldots, \frac{\Delta x_{n+k-1}}{\Delta x_{n+k-2}} \right)$$

où R est une fraction rationnelle fixée de k-1 variables vérifiant :

$$(\mathcal{R}) \qquad R(\rho,\ldots,\rho) = \frac{1}{1-\rho}, \ \forall\, \rho \neq 1$$

alors, il existe $x \in C(\mathbb{C})$ telle que $T(x) \notin C(\mathbb{C})$.

Démonstration : Il suffit de construire une telle suite x. Imposons lui de vérifier :

$$\left.\begin{array}{l} \Delta x_{i+kj} = \Delta x_i \times K^j, \ |K| < 1, \ \forall\, j \in \mathbb{N} \\ \Delta x_i = \Delta x_0 \neq 0 \end{array}\right\} \text{ pour } i=0,1,\ldots,k-1$$

La convergence dans $\mathbb{C}$ d'une telle suite est immédiate.

D'autre part, les nombres $\rho_n = \frac{\Delta x_{n+1}}{\Delta x_n}$ vérifient :

$$\rho_{i+kj} = \begin{cases} 1 \text{ pour } i=0,\ldots,k-2 \\ K \text{ pour } i=k-1 \end{cases}, \forall j \in \mathbb{N}$$

d'où

$$R(\rho_{kj}, \ldots, \rho_{kj+k-2}) = \infty$$

implique, compte tenu de $\Delta x_n \neq 0$, $\forall n \in \mathbb{N}$:

$$T(x_{kj}) = \infty.$$

Et le fait qu'une sous-suite de T(x) tende vers ∞ interdit à la suite T(x) de converger dans $\mathbb{C}$.

Remarque 1 Le type de transformation utilisé dans la proposition précédente est dû à Pennacchi [12]. En fait Pennacchi a introduit une classe un peu plus générale et il a montré que, dans cette classe, la condition (R) était nécessaire et suffisante pour garantir l'accélération de la convergence d'une certaine classe de suites convergentes, les suites à convergence linéaire. L'intérêt de la proposition 1 provient du fait que ces suites à convergence linéaire interviennent très souvent en analyse numérique.

Remarque 2 Nombre de procédés classiques entrent dans le cadre des transformations de suites du type précité : outre les procédés élémentaires étudiés dans ce travail, on peut citer la transformation E_k de Shanks, les transformations colonne associées à la mise en oeuvre du θ-algorithme de Brezinski, à celle du ρ-algorithme, à celle du procédé d'Overholt, ou encore à celle du procédé de Germain-Bonne.

4.2 - Caractérisation de $R_{\mathbb{C}}(\gamma_\alpha)$

Grâce à la simplicité des transformations de type γ_α, on établit aisément la :

Proposition 2 $R_{\mathbb{C}}(\gamma_\alpha) = \{x \in C(\mathbb{C}) \mid \lim_{x\to\infty} \Delta(1/\Delta x_n) = \infty\}$.

Puisque cette caractérisation montre que $R_{\mathbb{C}}(\gamma_\alpha)$ ne dépend pas de α, il s'ensuit que :

Corollaire 1 Le domaine de régularité de toutes les transformations γ_α est le même.

D'autre part, on peut donner de la transformation γ_2 une interprétation géométrique très simple. En effet, la relation :

$$2(\gamma_2(x_n) - x_{n-1})^{-1} = (x_{n+2} - x_{n+1})^{-1} + (x_n - x_{n+1})^{-1}$$

traduit le fait que $\gamma_2(x_n)$ est le conjugué harmonique de x_{n+1} par rapport à x_n et x_{n+2}. D'où :

$$R_{\mathbb{C}}(\gamma_2) = \{x \in C(\mathbb{C}) \mid u \in C(\mathbb{C}),\ (x_n, x_{n+2};\ x_{n+1}, u_n) = -1,\ \forall\, n \in \mathbb{N},\ u^* = x^*\}$$

Grâce à la quasi-régularité du procédé γ_2 (qu'on établira au paragraphe suivant), la dernière restriction est inutile. Compte tenu du corollaire précédent, on peut alors énoncer la

Proposition 3

$$R_{\mathbb{C}}(\gamma_\alpha) = \{x \in C(\mathbb{C}) \mid u \in C(\mathbb{C}),\ (x_n, x_{n+2};\ x_{n+1}, u_n) = -1,\ \forall\, n \in \mathbb{N}\}$$

4.3 - Structure de $R_{\mathbb{C}}(\gamma_\alpha)$

On a souvent tendance à vouloir caractériser une partie d'un espace vectoriel en termes de sous-espace ou de partie convexe. Ce type de démarche est inutilisable ici puisque l'on a la :

Proposition 4 $R_{\mathbb{C}}(\gamma_\alpha)$ n'est ni un sous-espace vectoriel, ni une partie convexe de $C(\mathbb{C})$.

Pour établir ce résultat il suffit de mettre en évidence une paire de suites x et y telles que :

$$\begin{cases} x, y \in R_C(\gamma_\alpha) \\ x + y \notin R_C(\gamma_\alpha) \end{cases}$$

Le lecteur vérifiera que ceci est réalisé pour les deux suites :

$x_n = \sum_{i=1}^{n} u_i$ et $y_n = \sum_{i=1}^{n} v_i$ avec :

$$u_n = 1/n^2 \text{ et } v_n = \pi^2/6 - 1 + \sum_{j=1}^{n} (-1)^j (1/j^2 - 1/(j+1)^2)$$

La somme z de ces deux suites convergentes est transformée par γ_α en une suite ayant 2 points d'accumulation : $x^* + y^*$ et ∞.

On a toutefois un résultat plus constructif avec la :

Proposition 5 $R_{\bar{K}}(\gamma_\alpha)$ est globalement invariant dans toute transformation homographique régulière, c'est-à-dire :

$$x \in R_{\bar{K}}(\gamma_\alpha) \cap h \in H(\bar{K}) \Longrightarrow y = h(x) \in R_{\bar{K}}(\gamma_\alpha).$$

Cette propriété est une conséquence de l'invariance du birapport de 4 nombres dans une transformation homographique régulière.

4.4 - Régularité de W

En raison de sa plus grande complexité, l'étude du noyau de W est moins aisée que celle de γ_α. On a toutefois la :

Proposition 6

$$R_C(W) = \{x \in C(\mathbb{C}) \mid \lim_{n\to\infty} \left| \left(\frac{\Delta x_{n+2}\, \Delta^2 x_n}{\Delta x_n\, \Delta^2 x_{n+1}} - 1\right) / \Delta x_{n+1} \right| = \infty\}$$

dont la démonstration résulte d'un calcul élémentaire. On peut d'ailleurs expliciter de nombreuses caractérisations équivalentes de l'appartenance du noyau de W, comme par exemple :

$$\lim_{n\to\infty} [\Delta x_{n+1}\, \Delta(1/\Delta x_{n+1}) / \Delta^2(1/\Delta x_n)] = 0.$$

Divers auteurs ont cherché à donner des conditions suffisantes d'appartenance au domaine de régularité des procédés γ_α ou W. C'est le cas des travaux de Lubkin [9], Gray, Clarke et Adams [3], Tucker [16] ou Cordellier [5]. Cet aspect ne sera pas développé ici.

5 - QUASI-RÉGULARITÉ

L'analyse qui précède a montré que le domaine de régularité d'un procédé ne représente le plus souvent qu'une partie plus ou moins grande de l'ensemble des suites convergentes. L'efficacité des procédés n'est qu'une conséquence du fait que les suites auxquelles on les applique sont des suites particulières qui appartiennent presque toujours au domaine de régularité et parfois au domaine d'accélération.

Observons que, si une suite convergente est transformée par un procédé donné en une suite qui ne converge plus, ce n'est généralement pas très grave dans la mesure où l'on dispose de critères pratiques permettant de s'en rendre compte. Beaucoup plus grave car moins aisément décelable est la situation où une suite convergeant lentement vers une limite x^* est transformée en une suite convergeant vers une autre limite y^*.

D'où la notion de quasi-régularité introduite par Lubkin [9] et que Rice [13] utilise sous le nom de "joint convergence" :

Définition Un procédé $T \in \mathcal{T}(K)$ sera dit quasi-régulier sur K si et seulement si :

$$x \in \mathcal{C}(K) \cap T(x) \in \mathcal{C}(K) \Longrightarrow T(x^*) = x^*.$$

Les résultats que nous allons établir reposent sur le

Lemme 1 Si la suite de nombres complexes x vérifie :

$$(L) \qquad \lim_{n\to\infty} \Delta(1/\Delta x_n) = \ell \in \mathbb{C} \setminus \{0\}$$

alors : $x \notin \mathcal{C}(\mathbb{C})$.

Ce résultat a été établi dans $\mathbb{R}$ par Lubkin [9], et Tucker [16] l'a démontré pour $\mathbb{C}$. En voici une preuve plus élémentaire :

L'hypothèse (L) se traduit par :

$\exists\ \varepsilon \in \mathcal{C}(\mathbb{C})$ telle que $\varepsilon^* = 0$ et $\Delta(1/\Delta x_n) = \ell + \varepsilon_n$, $\forall\ n \geq n_0 \in \mathbb{N}$.

On peut choisir $n_1 \in \mathbb{N}$ pour que $n \geq n_1 \Longrightarrow |\varepsilon_n| < |\ell| / 4$.

Pourvu que $n > m \geq n_1$, on a : $(\Delta x_n)^{-1} - (\Delta x_m)^{-1} = (n-m)\ell + \sum_{i=m}^{n-1} \varepsilon_i$,

d'où : $\ell\Delta x_n = (n-m)^{-1} [1 + ((\Delta x_m)^{-1} + \sum_{i=m}^{n-1} \varepsilon_i)(n-m)^{-1}\ell^{-1}]^{-1}$. L'entier $m \geq \max(n_0,n_1)$ étant fixé, choisissons n pour que $n-m > 4 / |\ell\Delta x_m|$.

Alors $|((\Delta x_m)^{-1} + \sum_{i=m}^{n-1} \varepsilon_i)\ \ell^{-1}(n-m)^{-1}| \leq 1/4 + 1/4 = 1/2$.

Dans le plan complexe, l'image du nombre $z = 1 + \ell^{-1}(n-m)^{-1} [(\Delta x_m)^{-1} + \sum_{i=m}^{n-1} \varepsilon_i]$ appartient donc au cercle de centre 1 et de rayon 1/2. L'image de z^{-1} appartient donc au cercle de centre 4/3 et de rayon 2/3, de sorte que sa partie réelle est minorée par 2/3, d'où :

$$\mathrm{Re}(\ell\Delta x_n) = \mathrm{Re}(z(n-m)^{-1}) \geq 2/3(n-m)^{-1}$$

$\forall\ k \geq n$, nous avons alors :

$$\mathrm{Re}(\ell x_k) = \mathrm{Re}(\ell x_n) + 2/3 \sum_{j=n}^{k-1} (j-m)^{-1}$$

Le second membre tendant vers ∞ avec k, nous avons :

$$\lim_{k\to\infty} \mathrm{Re}(\ell x_k) = \infty$$

et la suite x ne peut converger dans $\mathbb{C}$.

Ce lemme permet d'établir trivialement :

Théorème 1 Pour tout $\alpha \in \mathbb{C} \setminus \{0\}$, le procédé γ_α est quasi-régulier sur $\mathbb{C}$.

Pour $\alpha = 1$, c'est le résultat de Tucker [16].

La quasi-régularité du procédé W est également une conséquence du lemme 1, mais la preuve en est moins aisée et nous n'avons su l'établir que dans le cas réel. Des raisonnements très élémentaires que nous ne détaillons pas ici permettent de montrer que :

Lemme 2 S'il existe $n_1 \in \mathbb{N}$ au-delà duquel $\Delta(1/\Delta x_n)$ reste positif, alors il existe $n_2 \geq n_1$ au-delà duquel la suite (Δx_n) est monotone décroissante et conserve le même signe.

Corollaire 2 S'il existe $n_1 \in \mathbb{N}$ au-delà duquel $\Delta(1/\Delta x_n)$ conserve le même signe, alors il existe $n_2 \geq n_1$ au-delà duquel la suite (Δx_n) est monotone et conserve le même signe.

Corollaire 3 Si la suite convergente (x_n) est telle qu'il existe $n_1 \in \mathbb{N}$ au-delà duquel $\Delta(1/\Delta x_n)$ conserve un signe constant alors la série de terme général Δx_n converge absolument.

Nous sommes maintenant en mesure de prouver le :

Théorème 2 Le procédé W est quasi-régulier sur $\mathbb{R}$.

Démonstration : Supposons le théorème faux :
$\exists\, x \in C(\mathbb{R})$ telle que $y = W(x) \in C(\mathbb{R})$ avec $x^* \neq y^*$.

Puisque $y_n = x_{n+2} + \Delta x_{n+1}\, \Delta(1/\Delta x_n) / \Delta^2(1/\Delta x_n)$, nous avons :

$$\lim_{n\to\infty} (y_n - x_{n+2}) = \lim_{n\to\infty} (\Delta x_{n+1}\, \Delta(1/\Delta x_n) / \Delta^2(1/\Delta x_n)) = y^* - x^* \neq 0.$$

D'où : $\exists\, \varepsilon \in C(\mathbb{R})$ telle que $\varepsilon^* = 0$ et $\Delta x_{n+1}\, \Delta(1/\Delta x_n) / \Delta^2(1/\Delta x_n) = \ell + \varepsilon_n$ où

$\ell = y^* - x^* \neq 0$.

Puisque $\ell \neq 0$ et $\varepsilon_n \to 0$, $\exists\, n_0 \in \mathbb{N}$ tel que $n \geq n_0 \Longrightarrow \ell(\ell+\varepsilon_n) > 0$. Alors les quantités $\Delta(1/\Delta x_n)$ sont liées par la récurrence :

$$\Delta(1/\Delta x_{n+1}) = \Delta(1/\Delta x_n)(1 + \Delta x_{n+1} / (\ell + \varepsilon_n)),\ \forall\, n \geq n_0.$$

Comme ε_n et Δx_n tendent vers 0, il existe $n_1 \geq n_0$ tel que :

$$n \geq n_1 \Longrightarrow 1 + \Delta x_{n+1} / (\ell + \varepsilon_n) > 1/2.$$

Alors, pour $n \geq n_1$, on a $\Delta(1/\Delta x_{n+1})\ \Delta(1/\Delta x_n) > 0$ et le corollaire 2 garantit la convergence absolue de la série de terme général Δx_n. Il en est de même de la série de terme général $u_n = \Delta x_{n+1} / (\ell + \varepsilon_n)$, ce qui assure la convergence du produit infini : $\Pi_{i=n_1}^{\infty} (1+u_i)$. Alors $\Delta(1/\Delta x_n) = \Delta(1/\Delta x_{n_1})\ \Pi_{i=n_1}^{n-1} (1 + \Delta x_{i+1} / (\ell + \varepsilon_i))$ tend vers une limite finie non nulle et le lemme 1 entraîne la divergence de la suite x, ce qui infirme l'hypothèse : le théorème est établi.

6 - COMPATIBILITÉ DES TRANSFORMATIONS DE SUITES

Si une suite divergente x est transformée en une suite convergente y par le procédé δ^2, on a coutume de dire que la limite y^* de la suite transformée est l'anti-limite de la suite initiale x. Cette appellation introduite par Shanks [14] pour illustrer les propriétés des transformations E_k s'interprète également, dans certains cas au moins [9] comme le prolongement analytique d'une fonction en dehors de son disque de convergence. Les procédés linéaires de sommation de séries divergentes relèvent du même point de vue.

Convenons d'appeler anti-limite de la suite x vis à vis du procédé T la limite (si elle existe) de la transformée y = T(x). Par analogie avec les procédés linéaires, nous dirons encore que la suite x est T-limitable de T-limite y^*, ou que la série de terme général Δx_n est T-sommable, de T-somme y^*. La question fondamentale qui se pose alors est la suivante : peut-on affirmer que les anti-limites d'une suite relativement à deux procédés distincts sont identiques ?

L'objet de ce paragraphe est de montrer que dans le cas où les deux procédés sont le procédé δ^2 et le procédé W, alors la réponse est oui.

Théorème 3 Si une suite $x \in S(\mathbb{R})$ est simultanément γ_1-limitable et W-limitable, alors sa γ_1-limite et sa W-limite sont identiques.

Démonstration : Supposons le théorème faux :

$\exists\, x \in S(\mathbb{R})$ telle que $y = \gamma_1(x) \in C(\mathbb{R})$ et $z = W(x) \in C(\mathbb{R})$ avec $\ell = z^* - y^* \neq 0$.

En raison de la quasi-régularité de γ_1 et W, la suite x diverge.

1) Montrons qu'il existe $n_1 \in \mathbb{N}$ au-delà duquel $\Delta(1/\Delta x_n)$ conserve un signe constant :

D'après la définition des procédés, nous avons :

$$y_n = \Delta(x_n/\Delta x_n) \;/\; \Delta(1/\Delta x_n)$$
$$z_n = \Delta^2(x_n/\Delta x_n) \;/\Delta^2(1/\Delta x_n).$$

Donc : $\exists\, \varepsilon, \eta \in C(\mathbb{R})$ telles que $\varepsilon^* = \eta^* = 0$ avec :

$$\Delta(x_n/\Delta x_n) = (y^* + \varepsilon_n)\, \Delta(1/\Delta x_n)$$
$$\Delta^2(x_n/\Delta x_n) = (z^* + \eta_n)\Delta^2(1/\Delta x_n)$$

Ceci implique : $(z^* + \eta_n)\, \Delta^2(1/\Delta x_n) = \Delta[(y^* + \varepsilon_n)\;\; \Delta(1/\Delta x_n)]$,

soit : $(z^* + \eta_n - y^* - \varepsilon_{n+1})\, \Delta(1/\Delta x_{n+1}) = (z^* + \eta_n - y^* - \varepsilon_n)\, \Delta(1/\Delta x_n)$.

Puisque $\varepsilon^* = \eta^* = 0 \neq \ell$, $\exists\, n_0 \in \mathbb{N}$ tel que $n \geq n_0 \Longrightarrow \ell + \eta_n - \varepsilon_{n+1} \neq 0$.

D'où : $\Delta(1/\Delta x_{n+1}) = (1 + \Delta\varepsilon_n \,/\, (\ell + \eta_n - \varepsilon_{n+1}))\, \Delta(1/\Delta x_n)$, $\forall\, n \geq n_0$.

Comme $\lim_{n\to\infty} \Delta\varepsilon_n = 0$, $\exists\, n_1 \geq n_0$ tel que $n \geq n_1 \Longrightarrow |\Delta\varepsilon_n| < |\ell + \eta_n - \varepsilon_{n+1}|$,
ce qui implique : $\Delta(1/\Delta x_{n+1}) * \Delta(1/\Delta x_n) > 0$, c'est-à-dire : au-delà d'un certain rang n_1, $\Delta(1/\Delta x_n)$ conserve son signe.

2) Montrons que ce signe ne peut être positif :

Et pour cela, supposons qu'il le voit : $\Delta(1/\Delta x_n) > 0$, $\forall\, n \geq n_1$.

D'aprés le lemme 2, la suite (Δx_n) est monotone décroissante.

Si Δx_n reste positif, $\forall\, n \geq n_1$, la divergence de la suite x implique que x_n tende vers $+\infty$. Alors $y_n = x_{n+1} + 1/\Delta(1/\Delta x_n) \geq x_{n+1}$ tend aussi vers $+\infty$, en contradiction avec $y^* \in \mathbb{R}$. Donc au-delà d'un certain rang, $\Delta x_n < 0$, ce qui implique : $\lim_{n\to\infty} x_n = -\infty$.

Puisque $y^* + \varepsilon_n = x_{n+1} + 1/\Delta(1/\Delta x_n)$, nous avons : $\Delta(1/\Delta x_n) = 1/(y^* + \varepsilon_n - x_{n+1})$, et $\lim_{n\to\infty} x_n = -\infty$ implique : $\lim_{n\to\infty} \Delta(1/\Delta x_n) = +0$, d'où $\lim_{n\to\infty} \Delta^2(1/\Delta x_n) = 0$. D'autre part, on a : $\Delta^2(1/\Delta x_n) = (\Delta x_{n+1} - \Delta\varepsilon_n)(y^* + \varepsilon_{n+1} - x_{n+2})^{-1}(y^* + \varepsilon_n - x_{n+1})^{-1}$. Puisque Δx_n est monotone, décroissante et négative à partir d'un certain rang alors que $\Delta\varepsilon_n$ tend vers 0, il existe un rang au-delà duquel le facteur $\Delta x_{n+1} - \Delta\varepsilon_n$ reste négatif alors que chacun des 2 autres facteurs reste positif (car $x_n \to -\infty$). D'où $\lim_{n\to\infty} \Delta^2(1/\Delta x_n) = -0$. Alors $\Delta\varepsilon_n/(\ell + \eta_n - \varepsilon_{n+1}) = \Delta^2(1/\Delta x_n) \;/\; \Delta(1/\Delta x_n)$ tend vers 0 en restant négatif (au-delà d'un certain rang) quand $n \to \infty$, ce qui implique que, au-delà d'un certain rang, $\Delta\varepsilon_n$ conserve un signe constant, donc que la suite ε_n soit monotone.

Ceci assure la convergence absolue de la série de terme général $\Delta\varepsilon_n$ donc celle de terme général $u_n = \Delta\varepsilon_n \;/\; (\ell + \eta_n - \varepsilon_{n+1})$. Il s'ensuit que le produit infini $\prod_{i=n_0}^{\infty} (1+u_i)$ converge, donc que la suite $\Delta(1/\Delta x_n) = \Delta(1/\Delta x_{n_0}) \prod_{i=n_0}^{n-1} (1+u_i)$ tend vers une valeur finie non nulle, ce qui infirme le fait que $\lim_{n\to\infty} \Delta(1/\Delta x_n) = 0$.

3) Le signe de $\Delta(1/\Delta x_n)$ ne peut être négatif :

Sinon, la suite $x' = -x$ qui est telle que $y' = \gamma_1(x') = -y$ et $z' = W(x') = -z$ avec $y'^* \neq z'^*$ vérifie $\Delta(1/\Delta x'_n) > 0$, $\forall\, n \geq n_1$; et le raisonnement précédent montre que c'est impossible.

*

* *

RÉFÉRENCES

[1] A.C. AITKEN.- *On Bernoulli's numerical solution of algebraic equations.* Proc. Roy. Soc. Edinb. 46 (1926) 289-305.

[2] C. BREZINSKI.- *Etudes sur les ε- et ρ-algorithmes.* Numer. Math. 17 (1971) 153-162.

[3] W.D. CLARK, H.L. GRAY, and J.E. ADAMS.- *A note on the T-transformation of Lubkin.* J. Res. NBS 73 B (1969) 25-29.

[4] F. CORDELLIER.- *Caractérisation des suites que le procédé θ² transforme en suites constantes.* CRAS Paris 284 A (1977) 389-392.

[5] F. CORDELLIER.- *Analyse numérique des transformations de suites et de séries.* Thèse (à paraître).

[6] B. GERMAIN-BONNE.- *Estimation de la limite de suites et formalisation de procédés d'accélération de la convergence.* Thèse (Lille) 1978.

[7] B. GERMAIN-BONNE.- *Transformations de suites.* R.A.I.R.O., (1973), 84-90.

[8] D. LEVIN.- *Development of Non-Linear transformation for Improving Convergence of Sequences.* Intern. J. Computer Math. B3 (1973) 371-388.

[9] S. LUBKIN.- *A method for summing infinite series.* J. Res. NBS 48 (1952) 228-254.

[10] R. de MONTESSUS de BALLORE.- *Sur les fractions continues algébriques.* Bull. Soc. Math. 22 (1902) 28-36.

[11] K.J. OVERHOLT.- *Extended Aitken acceleration.* B.I.T. 5 (1965) 122-132.

[12] R. PENNACCHI.- *Le trasformazioni razionali di una successione.* Calcolo 5 (1968) 37-50.

[13] J.R. RICE.- *Sequence transformations based on Tchebycheff approximations.* J. Res. N.B.S. 64B (1960) 227-235.

[14] D. SHANKS.- *Nonlinear transformations of divergent and slowly convergent sequences.* J. Math. Phys. 34 (1955) 1-42.

[15] D.A. SMITH, W.F. FORD.- *Acceleration of linear and logarithmic convergence.* S.I.A.M. J. Numer. Anal., 16 (1979) 223-240.

[16] R.R. TUCKER.- *The δ^2-process and related Topics.*
(part.I) Pacif. J. Math. 22 (1967) 349-359
(part.II) Pacif. J. Math. 28 (1969) 455-463.

[17] J. WIMP.- *Some transformations of monotone sequences.* Math. Comp. 26 (1972) 251-254.

[18] P. WYNN.- *On a device for computing the $e_m(S_n)$ transformation.* MTAC 10 (1956) 91-96.

19] P. WYNN.- *On a procrustean technique for the numerical transformation of slowly convergent sequences and series.*
Proc. Comb. Phil. Soc. 52 (1956) 663-671.

DÉMONSTRATION ALGÉBRIQUE DE L'EXTENSION DE L'IDENTITÉ DE WYNN AUX TABLES DE PADÉ NON NORMALES

F. CORDELLIER

UNIVERSITE DE LILLE I, UER d'IEEA - INFORMATIQUE
F - 59650 VILLENEUVE d'ASCQ (FRANCE)

1 - INTRODUCTION

Le calcul des coefficients des approximants de Padé a connu un grand développement ces dernières années, comme en témoignent les papiers de Claessens [9], Bussonnais [8] et Graves-Morris [18]. Jusqu'à un passé récent, les schémas de calcul récursif comme ceux de Baker [1] ou Brezinski [2] supposaient la normalité de la table de Padé, mais divers travaux (Claessens et Wuytack [10], Cordellier [13], McEliece et Shearer [19], Bultheel [3-7]) ont montré que l'on pouvait presque toujours s'affranchir de cette restriction. Cela est dû à ce que les classiques identités de Frobenius sur lesquelles s'appuient les algorithmes récursifs se généralisent le plus souvent au cas des tables non normales. On montre en [13] par exemple qu'il suffit de considérer la table des formes de Padé (définie en [16]) au lieu de la table des fractions réduites pour étendre les identités de Frobenius sur lesquelles repose l'algorithme de Baker.

Le calcul des valeurs ponctuelles des approximants de Padé repose essentiellement sur la liaison entre la table des valeurs et le tableau associé à la mise en oeuvre de la transformation $E_k(S_n)$ de Shanks [21]. Cette liaison a été établie par Shanks dans le cas de la moitié inférieure de la table. Après avoir montré que l'ε-algorithme [22] assurait une mise en oeuvre efficace de la transformation $E_k(S_n)$ Wynn [23] a montré que cet ε-algorithme permettait de calculer toutes les valeurs de la table de Padé. Cette liaison a par ailleurs permis à Wynn de mettre en évidence une identité [25] qui lie 5 approximants de Padé voisins, identité que Gragg [17] appelle "the missing identity of Frobenius". Bien que les règles particulières [24] permettent de calculer les valeurs ponctuelles de tables de Padé non normales dans certains cas particuliers (blocs de taille n'excédant pas 2), la maintenant classique identité de Wynn n'est valable que pour des éléments normaux d'une table de Padé. En présentant les règles singulières généralisées pour l'ε-algorithme vectoriel [11], l'auteur a signalé que l'identité de Wynn se généralise assez simplement aux tables de Padé non normales. Toutefois la preuve (non publiée) qui en a été donnée repose sur des propriétés topologiques et des propriétés d'invariance dans des transformations anallagmatiques (conservation du birapport), et elle consiste essentiellement en l'application du classique théorème de prolongement par continuité. C'est pour remédier à cette situation désagréable où une propriété typiquement algébrique est établie par une voie détournée faisant appel à des notions topologiques que l'auteur [12] a proposé une démonstration algébrique reposant sur la classique identité de Sylvester [15]. C'est cette démonstration qui va être présentée ici. Signalons que dans le cadre d'une étude de l'extension de l'algorithme QD aux tables de Padé non normales, Claessens et Wuytack [10] proposent une autre méthode pour établir cette généralisation de l'identité de Wynn.

Le paragraphe 2 est consacré au rappel des notions d'approximant et de forme de Padé, notions que nous empruntons à Gilewicz [16] en les modifiant légèrement. On y rappelle en particulier la structure en blocs de la table de Padé. Dans le troisième paragraphe, on s'intèresse à la table des valeurs ponctuelles des formes de Padé dont le lien avec la transformation de Shanks est rappelé : on montre en particulier que les valeurs non définies sont localisées dans des blocs carrés et que la table des transformés de Shanks a une structure identique à celle de la table des formes de Padé. Le paragraphe suivant consiste en l'introduction d'un ε-tableau associé à la mise en oeuvre de la transformation de Shanks, tableau dont la structure sera elle aussi liée à celle de la table de Padé. En 5, le développement de certaines identités de déterminants et le fait que l'ε-tableau que nous venons d'introduire a lui aussi une structure en blocs carrés permettent d'établir des identités remarquables au voisinage d'un bloc. La transcription de ces identités dans la table des formes de Padé nous conduit presque naturellement à la généralisation de l'identité de Wynn dans un sixième et dernier paragraphe.

2 - LA TABLE DES FORMES DE PADÉ

Il ne semble pas que la terminologie relative aux approximants de Padé soit suffisamment établie pour qu'on puisse envisager la rédaction d'un papier traitant des tables de Padé non normales sans rappeler certaines définitions. Puisque Gilewicz [16] a proposé un formalisme cohérent permettant de rendre compte sans ambiguïté des propriétés de ces approximants, nous reprenons ici les grandes lignes de ce formalisme.

Dans tout le papier K sera $\mathbb{R}$ ou $\mathbb{C}$, c'est-à-dire un corps commutatif de caractéristique infinie compactitiable par l'adjonction d'un élément unique noté ∞ (pour plus de détails voir [14]). On note $\mathcal{P}_n(K)$ l'ensemble des polynômes à coefficients dans K de degré au plus égal à $n \geq 0$, et $\mathcal{P}_{-1}(K) = \{0\}$, et on posera $\mathcal{P}_n^*(K) = \mathcal{P}_n(K) \setminus \mathcal{P}_{n-1}(K)$. Le couple $B = (P,Q)$ où $P \in \mathcal{P}_p$ et $Q \in \mathcal{P}_q$ sera appelé *bi-polynôme* de degré (p,q), $p,q \geq -1$. Son *degré effectif* est (p,q) si $P \in \mathcal{P}_p^*$ et $Q \in \mathcal{P}_q^*$. Sur l'ensemble $\mathcal{B}(K)$ de tous les bi-polynômes, on peut définir une relation d'équivalence $\sim$ par :

$$(P,Q) \sim (P',Q') \Longleftrightarrow \exists\, a \in P_0^* \text{ tel que } P' = aP \cap Q' = aQ.$$

Cette relation induit sur $\mathcal{B}^*(K) = \mathcal{B}(K) \setminus \{(0,0)\}$ un espace quotient noté $\mathcal{F}(K)$ dont tout élément sera appelé *forme rationnelle*. Soit ϕ une application de $\mathcal{B}(K) \setminus \{(0,0)\}$ dans $\mathbb{R}_+^*$ telle que $\phi((aP,aQ)) = |a|\ \phi\ (P,Q)$, $\forall\ a \in K \setminus \{0\}$. Un bi-polynôme (P,Q) est *ϕ-unitaire* si $\phi((P,Q)) = 1$. Puisque chaque classe de $\mathcal{F}(K)$ contient un élément ϕ-unitaire et un seul, il y a isomorphisme entre $\mathcal{F}(K)$ et l'ensemble des éléments ϕ-unitaires de $\mathcal{B}^*(K)$. Cet élément ϕ-unitaire est appelé *ϕ-représentant* de la forme rationnelle.

En raison de sa compatibilité avec la 1ère relation d'équivalence $\sim$, la seconde relation d'équivalence $\approx$ définie sur $\mathcal{B}^*(K)$ par :

$$(P,Q) \approx (P',Q') \Longleftrightarrow PQ' = QP'$$

induit une relation d'équivalence quotient sur les formes rationnelles et permet de définir un espace quotient noté $\mathcal{R}(K)$ dont tout élément sera appelé *fraction rationnelle*. Une forme rationnelle est dite *réduite* si son ϕ-représentant (P,Q) est tel que P et Q soient premiers entre eux. Chaque classe de $\mathcal{R}(K)$ contient une forme rationnelle réduite unique et par suite, $\mathcal{R}(K)$ peut être mis en correspondance biunivoque avec l'ensemble des bi-polynômes ϕ-unitaires (P,Q) tels que P et Q soient premiers entre eux. Ces éléments sont les ϕ-représentants des fractions rationnelles. Le *degré effectif* d'une forme ou d'une fraction rationnelle est le degré effectif de son ϕ-représentant.

Si la série formelle à coefficients dans K : $f(z) = \sum_{i=0}^{\infty} c_i\ z^i$ vérifie

$$c_i = 0,\ i = 0,\ \ldots, n$$

nous écrirons :

$$f(z) = 0(z^{n+1}).$$

Nous appelons *forme de Padé de degré* (p,q) de la série formelle $f(z) = \sum_{i=0}^{\infty} c_i\ z^i$ toute forme rationnelle de degré (p,q) dont le ϕ-représentant $R = (N,D)$ vérifie : $N = fD + 0(z^{p+q+1})$. Notons que si $f(z) = 0(z^{n_0})$ où $n_0 > 0$, alors la série f s'écrit $f(z) = z^{n_0}\ g(z)$ avec $g(z) = \sum_{i=0}^{\infty} c_{i+n_0}\ z^i$. Toute forme de Padé de degré (p,q) de g dont le ϕ-représentant est (N,D) est alors associée de façon biunivoque à une forme de Padé de degré $(p+n_0,q)$ de f dont le ϕ-représentant est $(z^{n_0}N,D)$: on ne restreint donc nullement la généralité en supposant $c_0 \neq 0$, ce que nous ferons désormais.

Rappelons le *théorème de Frobenius* : $\forall$ p, q $\geq$ 0, la série f admet au moins une forme de Padé de degré (p,q).

Notons que la fraction rationnelle de ϕ-représentant $(z^p,0)/\phi((z^p,0))$ (respectivement : $(0,z^q) / \phi((0,z^q)))$ peut être considérée comme une forme de Padé d'ordre (p,-1) (respectivement : (-1,q)) de toute série formelle f, ce qui permet d'étendre le théorème de Frobénius au cas où l'un des deux polynômes (et un seul) est nul, et d'en déduire des algorithmes de calcul récursif des formes de Padé dans le cadre de tables non normales [13]. C'est pour permettre la description de tels algorithmes que nous avons étendu la définition des formes de Padé de Gilewicz au cas où le numérateur ou le dénominateur sont nuls et que nous avons évité de nous limiter à une ϕ-représentation particulière des formes et des fractions. Toutes les autres notions de ce paragraphe sont dans [16].

Pour p,q fixés, la série f peut avoir plusieurs formes de Padé de degré (p,q) qui correspondent toutes à la même fraction rationnelle [16]. Si le ϕ-représentant de cette fraction rationnelle est le ϕ-représentant d'une ϕ-forme de Padé de degré (p,q), cette fraction rationnelle est l'approximant de Padé (unique) de degré (p,q).

Par abus de langage, nous ne distinguerons plus une forme ou une fraction rationnelle, ni une forme ou un approximant de Padé de leur ϕ-représentant.

On appelle *table de Padé* de la série formelle f le tableau dépendant de 2 indices p,q $\geq$ -1 (p+q $\geq$ -1) dont l'élément d'indice (p,q) est la fraction rationnelle $R_{p,q}$ correspondant à la forme de Padé de degré (p,q) de la série f. L'indice p répétera les lignes et l'indice q les colonnes, les indices progressant de la droite vers la gauche et de haut en bas.

$$
\begin{array}{ccccc}
 & R_{-1,0} & R_{-1,1} & R_{-1,2} & \cdots \\
R_{0,-1} & R_{0,0} & R_{0,1} & R_{0,2} & \cdots \\
R_{1,-1} & R_{1,0} & R_{1,1} & R_{1,2} & \cdots \\
R_{2,-1} & R_{2,0} & R_{2,1} & R_{2,2} & \cdots \\
\vdots & \vdots & \vdots & \vdots &
\end{array}
$$

<u>fig.1</u> - Table de Padé

L'attention du lecteur est attirée sur le fait que beaucoup d'auteurs utilisent la convention contraire.

Une table de Padé est dite *normale* si tous ses éléments sont des approximants de Padé, *non normale* dans le cas contraire. Une table de Padé non normale peut être partitionnée en blocs carrés tels que :

. toutes les fractions rationnelles dérivées des formes de Padé de ce bloc sont égales

. aucune autre forme de Padé de la table n'est égale à cette fraction rationnelle

. les formes de Padé dont les indices correspondent au bord de ce carré sont uniques (Gilewicz [16]).

Pour préciser ces propriétés, appelons *indice* d'un bloc (carré) le degré effectif de la fraction rationnelle à laquelle se réduisent toutes les formes de Padé du bloc. On a un bloc carré de *côté* n et d'indice (p,q) si et seulement si la forme de Padé de degré (p,q) a un ϕ-représentant (P,Q) qui vérifie :

$$\begin{cases} P = Qf + O(z^{p+q+n}),\ P \in \mathcal{P}_p^*,\ Q \in \mathcal{P}_q^* \\ P \text{ et } Q \text{ premiers entre eux.} \end{cases}$$

Proposition 1 Dans un bloc carré de côté n et d'indice (p,q) les formes de Padé dont les indices correspondent au bord du bloc sont uniques et telles que les ϕ-représentant $R_{r,s} = N_{r,s} / D_{r,s}$ vérifient les relations suivantes :

$$\left.\begin{array}{l} d^\circ(N_{p,q+\ell}) = d^\circ(N_{p+\ell,q}) = p \\ d^\circ(D_{p,q+\ell}) = d^\circ(D_{p+\ell,q}) = q \\ d^\circ(N_{p+n-1,q+\ell}) = d^\circ(N_{p+\ell,q+n-1}) = p + \ell \\ d^\circ(D_{p+n-1,q+\ell}) = d^\circ(D_{p+\ell,q+n-1}) = q + \ell \end{array}\right\} \text{ pour } \ell = 0,\dots,n-1$$

Démonstration : L'unicité de ces formes de Padé a été établie par Gilewicz [16 (théorème 5.6, prop. vii)]. Le reste de la preuve est immédiat.

Nous appellerons *table des formes de Padé* un tableau indicé comme une table de Padé dont tous les éléments sont des formes de Padé. D'après la proposition 1, tous les éléments qui appartiennent au bord d'un bloc carré sont parfaitement définis par leur unicité. Par contre les éléments intérieurs à un bloc sont indéfinis. A un bloc carré d'indice (p,q) et de côté n correspond la situation décrite par la figure 2.

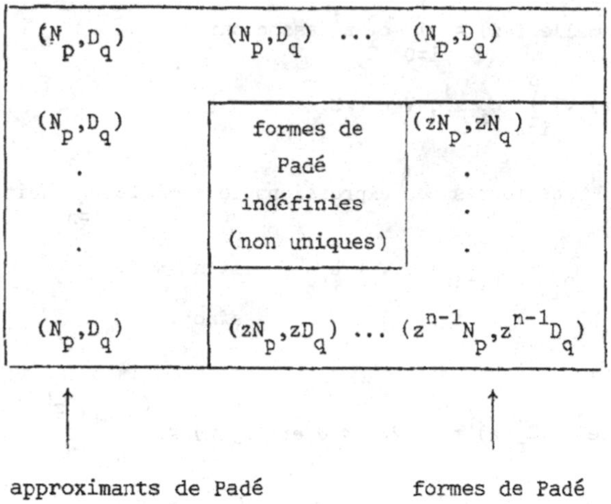

fig.2 - Bloc dans la table des formes de Padé

3 - VALEUR PONCTUELLE D'UNE TABLE DE PADÉ

On sait (voir par exemple [17]) qu'une forme de Padé de degré (p,q) de la série formelle $f(z) = \sum_{i=0}^{\infty} c_i z^i$ peut être représentée par le bi-polynôme $(P_{p,q}, Q_{p,q})$ suivant (Jacobi) :

$$P_{p,q}(z) = \text{dét}\left(\begin{bmatrix} \sum_{i=0}^{p-q} c_i z^{i+q} & \sum_{i=0}^{p-q+1} c_i z^{i+q-1} & \cdots & \sum_{i=0}^{p} c_i z^i \\ c_{p-q+1} & c_{p-q+2} & \cdots & c_{p+1} \\ \vdots & \vdots & & \vdots \\ c_p & c_{p+1} & & c_{p+q} \end{bmatrix}\right)$$

$$Q_{p,q}(z) = \text{dét}\left(\begin{bmatrix} z^q & z^{q-1} & \cdots & 1 \\ c_{p-q+1} & c_{p-q+2} & \cdots & c_{p+1} \\ \vdots & \vdots & & \vdots \\ c_p & c_{p+1} & \cdots & c_{p+q} \end{bmatrix}\right)$$

A la série formelle $f(z) = \sum_{i=0}^{\infty} c_i z^i$ associons :

$$S_n(z) = \sum_{i=0}^{n} c_i z^i, \forall n \in Z.$$

Remarque : Dans toutes ces expressions le symbole $\sum_{i=n_1}^{n_2}$ doit être interprété comme

$$\begin{cases} u_{n_1} + u_{n_1+1} + \ldots + u_{n_2} & \text{si } n_1 \le n_2 \\ 0 & \text{sinon.} \end{cases}$$

Il s'ensuit que $S_n(z) = 0, \forall n < 0$ et $\Delta S_n(z) = \begin{cases} c_{n+1} z^{n+1} & \text{si } n \ge 0 \\ 0 & \text{si } n < 0 \end{cases}$

Pour simplifier l'exposé de ce qui suit nous appellerons "suite" une application de $\mathbb{Z}$ dans K.

A tout "suite" S_n associons les déterminants de Hankel :

$$H_k(S_n) = \det \left(\begin{bmatrix} S_n & S_{n+1} & \ldots & S_{n+k-1} \\ \vdots & & & \\ S_{n+k-1} & S_{n+k} & \ldots & S_{n+2k-2} \end{bmatrix} \right) \quad \begin{matrix} \forall n \in \mathbb{Z} \\ \forall k \ge 1 \end{matrix}$$

avec $H_0(S_n) = 1$

On vérifie immédiatement que :

$$H_{q+1}(S_{p-q}(z)) = \det \left(\begin{bmatrix} \sum_{i=0}^{p-q} c_i z^i & \sum_{i=0}^{p-q+1} c_i z^i & \ldots & \sum_{i=0}^{p} c_i z^i \\ c_{p-q+1} z^{p-q+1} & c_{p-q} z^{p-q} & \ldots & c_{p+1} z^{p+1} \\ \vdots & \vdots & & \vdots \\ c_p z^p & c_{p+1} z^{p+1} & \ldots & c_{p+q} z^{p+q} \end{bmatrix} \right)$$

$$H_q(\Delta^2 S_{p-q}(z)) = \det \left(\begin{bmatrix} 1 & 1 & \cdots & 1 \\ c_{p-q+1} z^{p-q+1} & c_{p-q} z^{p-q} & \cdots & c_{p+1} z^{p-1} \\ \vdots & \vdots & & \vdots \\ c_p z^p & c_{p+1} z^{p+1} & \cdots & c_{p+q} z^{p+q} \end{bmatrix} \right)$$

Par des manipulations de lignes et de colonnes immédiates, il vient

$$H_{q+1}(S_{p-q}(z)) = P_{pq}(z) \times z^{pq}$$

$$H_q(\Delta^2 S_{p-q}(z)) = Q_{pq}(z) \times z^{pq}$$

Rappelons que $H_{q+1}(S_{p-q}(z)) \;/\; H_q(\Delta^2 S_{p-q}(z)) = E_q(S_{p-q}(z))$ n'est autre que la transformée de Shanks [21] de la "suite" $S_n(z)$.

Proposition 2 Une forme de Padé de degré (p,q) est unique si et seulement si $E_q(S_{p-q}(z))$ est défini, $\forall\, z \in K \setminus \{0\}$.

Démonstration : Si la forme de Padé est unique, alors $P_{p,q}$ et $Q_{p,q}$ ne peuvent s'annuler simultanément pour une valeur $\bar{z} \neq 0$: sinon les quotients respectifs $P'_{p,q}$ et $Q'_{p,q}$ de $P_{p,q}$ et $Q_{p,q}$ par $z-\bar{z}$ définiraient une autre forme de Padé. Donc $H_{q+1}(S_{p-q}(z))$ et $H_q(\Delta^2 S_{p-q}(z))$ ne peuvent s'annuler simultanément et $E_q(S_{p-q}(z))$ est défini. La réciproque est immédiate.

Nous dirons que deux éléments d'une table de formes de Padé d'indices respectifs (p,q) et (p',q') sont voisins si $|p-p'| + |q-q'| = 1$.

Nous pouvons alors énoncer le résultat suivant dû à Wynn [26] :

Proposition 3 Si deux éléments R et R' d'une table de formes de Padé sont voisins sans appartenir au même bloc carré, alors

(i) les 2 formes de Padé sont définies

(ii) leur valeur est différente, $\forall\, z \neq 0$.

Démonstration : Le point (i) provient du fait que 2 éléments indéfinis appartenant à des blocs différents ne peuvent être voisins.

Etablissons (ii) : puisque R d'indice (p,q) et R' d'indice (p',q') sont voisins, on a : $|p+q-p'-q'| = 1$ et (p=p' ou q=q'). Il est loisible de supposer $p'+q'-p-q = 1$. Alors R = N/D et R' = N'/D' vérifient :

$$N = fD + O(z^{p+q+1}) \text{ et } N' = fD' + O(z^{p+q+2})$$

$$\text{avec } d^o(N) = p,\ d^o(D) = q,\ d^o(N') = p',\ d^o(D') = q'.$$

Alors $P = ND' - DN' \in P_{p+q+1}$ car max (p+q', p'+q) = p+q+1 et on a :
$P = (fD + O(z^{p+q+1}))\, D' - (fD' + O(z^{p+q+2}))\, D = O(z^{p+q+1})$. C'est donc un monôme de degré p+q+1 qui ne peut s'annuler que pour z = 0. Donc $z \neq 0 \Longrightarrow R(z) \neq R'(z)$.

La proposition 3 montre que la structure de la table des valeurs ponctuelles est la même que celle de la table des formes de Padé. Tandis que la proposition 2 établit que les éléments indéfinis de la table des formes correspond à des éléments indéfinis du tableau $E_q(S_{p-q}(z))$. L'identité des structures est donc parfaite.

4 - L'ε-TABLEAU ASSOCIÉ À LA TABLE DES VALEURS PONCTUELLES

Après que Shanks [21] ait montré que la mise en oeuvre de la transformation de Shanks s'identifiait avec le calcul des valeurs ponctuelles de la moitié inférieure de la table de Padé, Wynn [23] a montré qu'on pouvait utiliser l'ε-algorithme pour calculer les valeurs ponctuelles des deux moitiés de la table de Padé dans le cas où cette table est normale (ou admet des blocs singuliers de taille inférieure à 2). Notre intention est de rebâtir une table complète analogue et de montrer qu'on peut calculer tous ses éléments définis par des relations de récurrence qui généralisent celles de Wynn.

4.1 - Définition formelle de l'ε-tableau associé à la "suite" (S_n) $n \in \mathbb{Z}$

(S_n) étant une suite d'éléments de K indicée par $\mathbb{Z}$, posons :

$$\left.\begin{aligned} \varepsilon_{2k}^{(n)} &= E_k(S_n) = H_{k+1}(S_n) \,/\, H_k(\Delta^2 S_n) \\ \varepsilon_{2k+1}^{(n)} &= 1/E_k(\Delta S_n) = H_k(\Delta^3 S_n)/H_{k+1}(\Delta S_n) \end{aligned}\right\} \quad \forall\, k \in \mathbb{N},\ \forall\, n \in \mathbb{Z}$$

avec $\varepsilon_{-1}^{(n)} = 0$ et $\varepsilon_{-2}^{(n)} = \infty$, $\forall\, n \in \mathbb{Z}$.

Seuls seront indéfinis les éléments qui proviennent du quotient de deux déterminants nuls. Si le seul dénominateur est nul, l'élément est infini.

Ces éléments sont placés dans un tableau à 2 indices (fig.3), l'ε-tableau associé à la "suite" (S_n).

$$
\begin{array}{ccccccc}
\cdot & \cdot & \cdot & & \cdot & \cdot & \cdot \\
\cdot & \cdot & \cdot & & \cdot & \cdot & \cdot \\
\varepsilon_{-2}^{(n+1)} = \infty & & \varepsilon_0^{(n)} = S_n & & \varepsilon_2^{(n-1)} & & \cdots \\
& \varepsilon_{-1}^{(n+1)} = 0 & & \varepsilon_1^{(n)} = 1/\Delta S_n & & \varepsilon_3^{(n-1)} & \cdots \\
\varepsilon_{-2}^{(n+2)} = \infty & & \varepsilon_0^{(n+1)} = S_{n+1} & & \varepsilon_2^{(n)} & & \cdots \\
& \varepsilon_{-1}^{(n+2)} = 0 & & \varepsilon_1^{(n+1)} = 1/\Delta S_{n+1} & & \varepsilon_3^{(n)} & \cdots \\
\varepsilon_{-2}^{(n+3)} = \infty & & \varepsilon_0^{(n+2)} = S_{n+2} & & \varepsilon_2^{(n+1)} & & \cdots \\
\cdot & \varepsilon_{-1}^{(n+3)} = 0 & \cdot & \varepsilon_1^{(n+1)} = 1/\Delta S_{n+2} & \cdot & \varepsilon_3^{(n+1)} & \cdots \\
\cdot & & \cdot & & \cdot & & \\
\cdot & \cdot & \cdot & & \cdot & \cdot & \cdot
\end{array}
$$

fig.3 - ε-tableau associé à la "suite" (S_n)

4.2 - Liaison avec les formes de Padé

D'après le paragraphe précédent on a :

$$\varepsilon_{2k}^{(n)} = E_k(S_n) = R_{n+k,k} \quad (1)$$

$$\varepsilon_{2k+1}^{(n)} = 1/E_k(S_n) = 1/R'_{n+k,k} \quad (1)$$

où $R_{n+k,k}$ et $R'_{n+k,k}$ sont respectivement les formes de Padé des séries :

$$\sum_{i=0}^{\infty} \Delta S_{i-1} z^i \text{ et } \Delta S_0 + \sum_{i=1}^{\infty} \Delta^2 S_{i-1} z^i$$

Il s'ensuit que la structure du sous-tableau $\varepsilon_{2k}^{(n)}$ ou celle de $\varepsilon_{2k+1}^{(n)}$ sera celle d'une table des valeurs des formes de Padé. Chacune de ces tables comportera des blocs carrés avec des valeurs égales sur le bord et des valeurs indéfinies à l'intérieur. Il reste à préciser la liaison de ces 2 sous-tableaux.

4.3 - *Proposition 4* Si les 4 éléments $\varepsilon_k^{(n)}$, $\varepsilon_k^{(n+1)}$, $\varepsilon_{k-1}^{(n+1)}$ et $\varepsilon_{k+1}^{(n)}$ sont définis dans $\bar{K}$, alors on a :

$$(J) \qquad (\varepsilon_{k+1}^{(n)} - \varepsilon_{k-1}^{(n+1)})(\varepsilon_k^{(n+1)} - \varepsilon_k^{(n)}) = 1.$$

Ce résultat a été établi par Wynn [21] en utilisant deux identités déterminantales (Schweins et Sylvester) et il est à la base du célèbre ε-algorithme. On a le

Corollaire : Les 2 termes de l'un des 2 facteurs de J sont égaux si et seulement si l'autre facteur comporte un terme infini et un seul.

4.3 - Structure de l'ε-tableau

Le corollaire précédent permet d'établir la proposition qui suit au moyen de raisonnements élémentaires que nous omettrons.

Proposition 5 : Dans l'ε-tableau $\varepsilon_k^{(n)}$, considérons les 2 blocs carrés emboîtés de côtés respectifs p+1 et p

B =

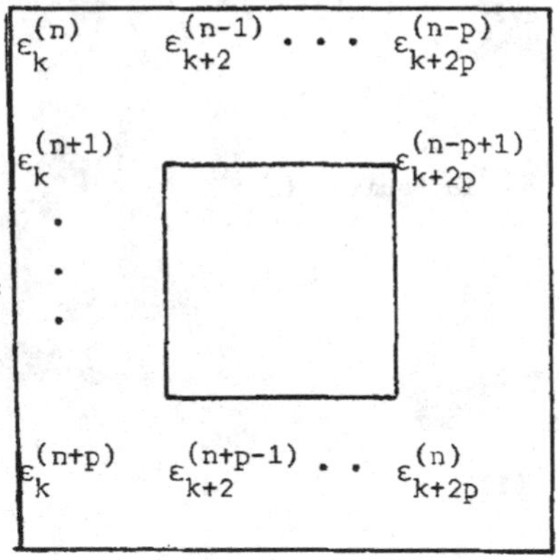

B' =

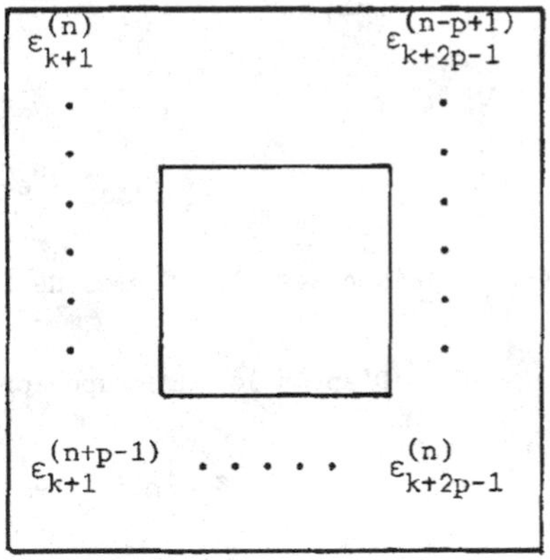

(i) Les 4p valeurs de B sont égales et distinctes de l'infini si et seulement si les 4(p-1) valeurs de B' sont égales à ∞.

(ii) Les 4p valeurs de B sont infinies ou indéfinies si et seulement si les 4'(p-1) valeurs de B' sont indéfinies.

Ce résultat montre qu'au voisinage d'un bloc singulier, on a la situation suivante :

$$\varepsilon_{k-1}^{(n)}=N_0 \qquad \varepsilon_{k+1}^{(n-1)}=N_1 \quad \cdot \cdot \quad \varepsilon_{k+2p-1}^{(n-p)}=N_p \qquad \varepsilon_{k+2p+1}^{(n-p-1)}=N_{p+1}$$

$$\varepsilon_k^{(n)}=C \qquad \varepsilon_{k+2}^{(n-1)}=C \quad \cdot \cdot \cdot \cdot \quad \varepsilon_{k+2p}^{(n-p)}=C$$

$$\varepsilon_{k-1}^{(n+1)}=W_1 \qquad \varepsilon_{k+1}^{(n)}=\infty \quad \cdot \cdot \cdot \quad \varepsilon_{k+2p-1}^{(n-p+1)}=\infty \qquad \varepsilon_{k+2p+1}^{(n-p)}=E_1$$

$$\varepsilon_k^{(n+1)}=C \qquad \varepsilon_{k+2p}^{(n-p+1)}=C$$

Valeurs indéfinies

$$\varepsilon_{k-1}^{(n+p)}=W_p \qquad \varepsilon_{k+1}^{(n+p-1)}=\infty \quad \cdot \cdot \cdot \quad \varepsilon_{k+2p-1}^{(n)}=\infty \qquad \varepsilon_{k+2p+1}^{(n-1)}=E_p$$

$$\varepsilon_k^{(n+p)}=C \qquad \varepsilon_{k+1}^{(n+p-1)}=C \quad \cdot \cdot \cdot \cdot \quad \varepsilon_{k+2p}^{(n)}=C$$

$$\varepsilon_{k-1}^{(n+p+1)}=S_0 \qquad \varepsilon_{k+1}^{(n-p)}=S_1 \quad \cdot \cdot \cdot \quad \varepsilon_{k+2p-1}^{(n+1)}=S_p \qquad \varepsilon_{k+2p+1}^{(n)}=S_{p+1}$$

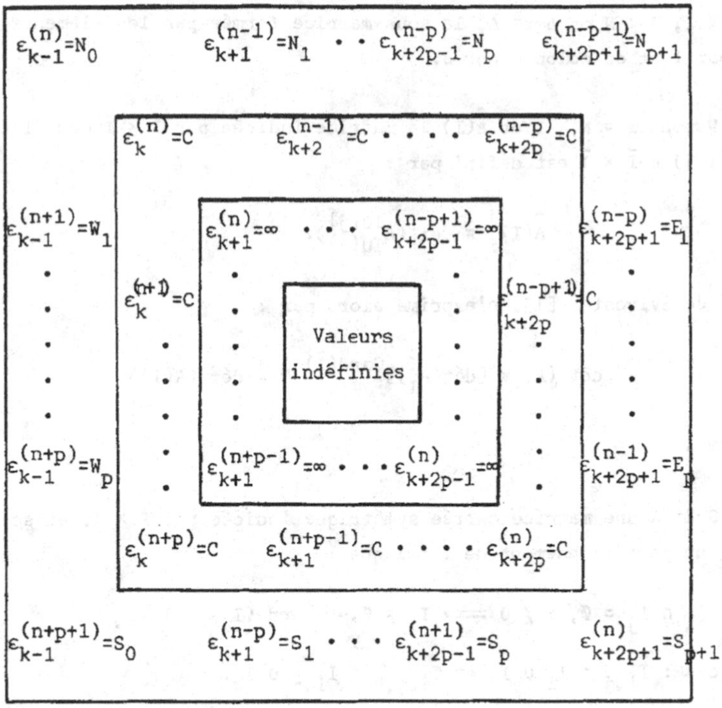

<u>fig.4</u> - Bloc singulier dans un ε-tableau

où C et chacune des valeurs N_i, S_i (i=0,...,p+1) W_i, E_i (i=1,...,p) sont distinctes de ∞.

5 - IDENTITÉS DÉTERMINANTALES

Nous avons regroupé dans ce paragraphe les questions techniques liées à l'utilisation des déterminants. Après un bref rappel de notations matricielles que nous utilisons et de l'identité de Sylvester, nous établissons deux lemmes qui sont utilisés dans les 2 propositions qui suivent pour établir la règle particulière que nous venons d'évoquer.

5.1 - Notations matricielles et identité de Sylvester

Soit A une matrice indicée en ligne et en colonne par un ensemble fini d'indices que nous notons N.

$\forall$ I, J $\subset$ N on note A_I^J la sous-matrice formée par les éléments de A indicés en ligne par I et en colonne par J.

Notons $\bar{I} = N \setminus I$ et $\hat{A}(I)$ la matrice indicée par $\bar{I} \times \bar{I}$ dont l'élément d'indice $(i,j) \in \bar{I} \times \bar{I}$ est défini par :

$$\hat{A}(I)_i^j = \det\,(A_{I\cup\{i\}}^{I\cup\{j\}}).$$

L'identité de Sylvester [15] s'exprime alors par :

$$\text{dét}\,(A) \times [\text{dét}(A_I^I)]^{\text{Card}(\bar{I})-1} = \text{dét}\,(\hat{A}(I))$$

5.2 - **Lemme 1**

Soit A une matrice carrée symétrique indicée par I × I, et soit $I = \bigcup_{k=0}^{3} I_k$ un partitionnement de I tel que :

$i \neq j \Longrightarrow I_i \cap I_j = \emptyset$, $i \neq 0 \Longrightarrow I_i \neq \emptyset$, et card $(I_1) = 1$.

Nous poserons : $I_{i,j} = I_i \cup I_j$ et $I_{i,j,k} = I_{i,j} \cup I_k$.

Si la matrice A vérifie : $\text{rg}(A_{I_{03}}^{I_{02}}) \leq \text{rg}(A_{I_0}^{I_0}) = \text{card}\,(I_0)$ alors on a :

$$(E)\begin{cases} [\text{dét}(A_I^I)\,\text{dét}(A_{I_0}^{I_0}) + \text{dét}(A_{I_{01}}^{I_{01}})\,\text{dét}(A_{I_{023}}^{I_{023}})]\,\text{dét}(A_{I_{02}}^{I_{02}})\,\text{dét}(A_{I_{03}}^{I_{03}}) \\ -\,[\text{dét}(A_{I_{012}}^{I_{012}})\,\text{dét}(A_{I_{03}}^{I_{03}}) + \text{dét}(A_{I_{013}}^{I_{013}})\,\text{dét}\,(A_{I_{02}}^{I_{02}})]\,\text{dét}(A_{I_{023}}^{I_{023}}\,\text{dét}(A_{I_0}^{I_0}) = 0 \end{cases}$$

<u>Démonstration</u> : Notons Δ l'expression qui figure dans le premier membre de (E) et dont nous voulons établir la nullité. Posons γ_i = card (I_i) (i=0,1,2,3) et Δ_0 = dét $(A_{I_0}^{I_0})$ (non nul par hypothèse) et formons

$\Delta' = \Delta \times \Delta_0^{\gamma_1+2\gamma_2+2\gamma_3-4}$.

Il vient :

$$\Delta' = [\text{dét}(A_I^I)\,\Delta_0^{\gamma_1+\gamma_2+\gamma_3-1} + \text{dét}(A_{I_{01}}^{I_{01}})\,\Delta_0^{\gamma_1-1} \times \text{dét}(A_{I_{023}}^{I_{023}})\,\Delta_0^{\gamma_2+\gamma_3-1}]$$

$$\times\ \text{dét}(A_{I_{02}}^{I_{02}})\,\Delta_0^{\gamma_2-1} \times \text{dét}(A_{I_{03}}^{I_{03}})\,\Delta_0^{\gamma_3-1}$$

$$-\ [\text{dét}(A_{I_{012}}^{I_{012}})\,\Delta_0^{\gamma_1+\gamma_2-1} \times \text{dét}(A_{I_{03}}^{I_{03}})\,\Delta_0^{\gamma_3-1} + \text{dét}(A_{I_{013}}^{I_{013}})\,\Delta_0^{\gamma_1+\gamma_3-1} \times \text{dét}(A_{I_{02}}^{I_{02}})\Delta_0^{\gamma_2-1}]$$

$$\times\ \text{dét}(A_{I_{023}}^{I_{023}})\ \Delta_0^{\gamma_2+\gamma_3-1}.$$

Appliquons alors l'identité de Sylvester. Si nous notons $B = \hat{A}(I_0)$ la matrice carrée symétrique indicée par I_{123}, nous obtenons :

$$\Delta' = [\text{dét}(B_{I_{123}}^{I_{123}}) + \text{dét}(B_{I_1}^{I_1}) \times \text{dét}(B_{I_{23}}^{I_{23}})]\ \text{dét}(B_{I_2}^{I_2})\ \text{dét}(B_{I_3}^{I_3})$$

$$-\ [\text{dét}(B_{I_{12}}^{I_{12}})\ \text{dét}(B_{I_3}^{I_3}) + \text{dét}(B_{I_{13}}^{I_{13}})\ \text{dét}(B_{I_2}^{I_2})]\ \text{dét}(B_{I_{23}}^{I_{23}})$$

L'hypothèse $\text{rg}(A_{I_{03}}^{I_{02}}) \le \text{rg}(A_{I_0}^{I_0})$ assure la nullité de $B_{I_2}^{I_3}$, d'où :

$$\text{dét}(B_{I_{23}}^{I_{23}}) = \text{dét}(B_{I_2}^{I_2})\ \text{dét}(B_{I_3}^{I_3})$$

et :

$$\Delta' = [\text{dét}(B_{I_{123}}^{I_{123}}) + \text{dét}(B_{I_1}^{I_1}) \times \text{dét}(B_{I_{23}}^{I_{23}}) - \text{dét}(B_{I_{12}}^{I_{12}})\ \text{dét}(B_{I_3}^{I_3}) - \text{dét}(B_{I_{13}}^{I_{13}})\ \text{dét}(B_{I_2}^{I_2})]$$

$$\times\ \text{dét}(B_{I_{23}}^{I_{23}}).$$

Compte tenu de la nullité de $B_{I_2}^{I_3} = B_{I_3}^{I_2}$ et du fait que $\gamma_1 = 1$, nous avons :

$$\text{dét}(B_{I_{123}}^{I_{123}}) = \text{dét}\left(\begin{bmatrix} B_{I_1}^{I_1} & B_{I_1}^{I_2} & B_{I_1}^{I_3} \\ B_{I_2}^{I_1} & B_{I_2}^{I_2} & 0 \\ B_{I_3}^{I_1} & 0 & B_{I_3}^{I_3} \end{bmatrix}\right)$$

$$= \det \begin{pmatrix} B_{I_1}^{I_1} & 0 & 0 \\ B_{I_2}^{I_1} & B_{I_2}^{I_2} & 0 \\ B_{I_3}^{I_1} & 0 & B_{I_3}^{I_3} \end{pmatrix} + \text{dét} \begin{pmatrix} 0 & B_{I_1}^{I_2} & 0 \\ B_{I_2}^{I_1} & B_{I_2}^{I_2} & 0 \\ B_{I_3}^{I_1} & 0 & B_{I_3}^{I_3} \end{pmatrix} + \text{dét} \begin{pmatrix} 0 & 0 & B_{I_1}^{I_3} \\ B_{I_2}^{I_1} & B_{I_2}^{I_2} & 0 \\ B_{I_3}^{I_1} & 0 & B_{I_3}^{I_3} \end{pmatrix}$$

$$\det(B_{I_1}^{I_1})\det(B_{I_{23}}^{I_{23}}) = \det \begin{pmatrix} B_{I_1}^{I_1} & 0 & 0 \\ 0 & B_{I_2}^{I_2} & 0 \\ 0 & 0 & B_{I_2}^{I_2} \end{pmatrix}$$

$$\text{dét}(B_{I_{12}}^{I_{12}})\ \text{dét}(B_{I_3}^{I_3}) = \det \begin{pmatrix} B_{I_1}^{I_1} & B_{I_1}^{I_2} & 0 \\ B_{I_2}^{I_1} & B_{I_2}^{I_2} & 0 \\ 0 & 0 & B_{I_3}^{I_3} \end{pmatrix} =$$

$$= \det \begin{pmatrix} B_{I_1}^{I_1} & 0 & 0 \\ B_{I_2}^{I_1} & B_{I_2}^{I_2} & 0 \\ 0 & 0 & B_{I_3}^{I_3} \end{pmatrix} + \det \begin{pmatrix} 0 & B_{I_1}^{I_2} & 0 \\ B_{I_2}^{I_1} & B_{I_2}^{I_2} & 0 \\ 0 & 0 & B_{I_3}^{I_3} \end{pmatrix}$$

$$\text{dét}(B_{I_{13}}^{I_{13}})\ \text{dét}(B_{I_2}^{I_2}) = \text{dét} \begin{pmatrix} B_{I_1}^{I_1} & 0 & B_{I_1}^{I_3} \\ 0 & B_{I_2}^{I_2} & 0 \\ B_{I_3}^{I_1} & 0 & B_{I_3}^{I_3} \end{pmatrix}$$

$$= \det \begin{pmatrix} B_{I_1}^{I_1} & 0 & 0 \\ 0 & B_{I_2}^{I_2} & 0 \\ B_{I_3}^{I_1} & 0 & B_{I_3}^{I_3} \end{pmatrix} + \text{dét} \begin{pmatrix} 0 & 0 & B_{I_1}^{I_3} \\ 0 & B_{I_2}^{I_2} & 0 \\ B_{I_3}^{I_1} & 0 & B_{I_3}^{I_3} \end{pmatrix}$$

Après calcul, nous obtenons donc :

$$\Delta' = \text{dét}(B_{I_{23}}^{I_{23}}) \times \left[\text{dét} \begin{pmatrix} 0 & B_{I_1}^{I_2} & 0 \\ 0 & B_{I_2}^{I_2} & 0 \\ B_{I_3}^{I_1} & 0 & B_{I_3}^{I_3} \end{pmatrix} + \text{dét} \begin{pmatrix} 0 & 0 & B_{I_1}^{I_3} \\ B_{I_2}^{I_1} & B_{I_2}^{I_2} & 0 \\ 0 & 0 & B_{I_3}^{I_3} \end{pmatrix} \right]$$

$$= 2\,\text{dét}(B_{I_{23}}^{I_{23}}) \times \text{dét} \begin{pmatrix} 0 & B_{I_1}^{I_2} & 0 \\ 0 & B_{I_2}^{I_1} & 0 \\ B_{I_3}^{I_1} & 0 & B_{I_3}^{I_3} \end{pmatrix}$$

Il suffit alors de noter que la matrice carrée d'ordre $\gamma_1+\gamma_2+\gamma_3$:

$$\begin{pmatrix} 0 & B_{I_1}^{I_2} & 0 \\ 0 & B_{I_2}^{I_2} & 0 \\ B_{I_3}^{I_1} & 0 & B_{I_3}^{I_3} \end{pmatrix}$$

est la somme de 2 matrices de rang maximal respectif γ_2 et γ_3. D'où la nullité de son déterminant, donc de Δ'. Alors $\Delta_0 \neq 0$ entraîne $\quad = 0$, cqfd.

Notons $H_{p,q}(u_n)$ la matrice à p lignes et q colonnes dont le terme général h_{ij} est égal à :

$$h_{i,j} = u_{n+i+j-2} \qquad (i=1,\ldots,p \; ; \; j=1,\ldots,q)$$

<u>*Lemme 2*</u> : Si la suite (u_n) est telle que :

$$H_k(u_{n+\ell}) \neq 0 \text{ et } H_{k+1}(u_{n+\ell-1}) = 0 \qquad \text{pour } \ell=1,\dots,p,$$

alors $k' = \text{rang }(H_{k+p-\ell+1,k+\ell}(u_n)) = k$ pour $\ell=1,\dots,p$.

<u>Démonstration</u> : On procèdera en deux temps : on établit tout d'abord que les termes de la suite qui sont impliqués dans ce lemme vérifient une relation de récurrence linéaire homogène d'ordre k, et on en déduit que le rang de la matrice ne peut excéder k.

1. Notons U_n le vecteur de K^{k+1} dont la j^e composante est u_{n+j} $(j=0,\dots,k)$. L'hypothèse $H_{k+1}(u_{n+\ell-1}) = 0$ se traduit par l'existence d'un vecteur non nul $V_\ell \in K^{k+1}$ orthogonal à $U_{n+\ell+i-1}$ pour $i=0,\dots,k$. Pour $\ell=1,\dots,p-1$, les vecteurs V_ℓ et $V_{\ell+1}$ sont donc orthogonaux aux k vecteurs $U_{n+\ell+i}$ $(i=0,\dots,k-1)$. Puisque l'hypothèse $H_k(u_{n+\ell}) \neq 0$ implique l'indépendance linéaire des k vecteurs $U_{n+\ell+i}$ $(i=0,\dots,k-1)$, on en déduit que V_ℓ et $V_{\ell+1}$ appartiennent à la même variété linéaire de dimension 1 : ils sont donc proportionnels. Puisqu'ils sont définis à une constante multiplicative près, on peut les choisir égaux. Il existe donc un vecteur non nul $V \in K^{k+1}$ orthogonal à tous les vecteurs U_{n+i}, $i=0,1,\dots,p+k-1$, d'où en notant v_i $(i=0,\dots,k)$ les composantes de ce vecteur :

$$\sum_{j=0}^{k} v_j \, u_{n+i+j} = 0, \quad i=0,\dots,p+k-1.$$

2. On vérifie aisément que v_0 et v_k sont non nuls (sinon $H_k(u_{n+\ell}) = 0$, $\forall \ell = 1,\dots,p$) ce qui permet d'imposer $v_k = 1$, d'où :

$$u_{n+i+k} = - \sum_{j=0}^{k-1} v_j \, u_{n+i+j}$$

Si nous notons U'_n le vecteur de $K^{k+\ell}$ dont la j^e composante est u_{n+j} $(j=0,\dots,k+\ell-1)$, nous avons :

$$U'_{n+k+i} = - \sum_{j=0}^{k-1} v_j \, U'_{n+i+j} \qquad (i=0,\dots,p-\ell).$$

Ceci implique que la variété engendrée par les U'_{n+i} $(i=0,\dots,k+p-\ell)$ est de dimension inférieure ou égale à k, d'où $k' \leq k$. Enfin on a $k' \geq k$ car la matrice contient des sous déterminants $H_k\,(u_{n+\ell})$ non nuls.

<u>*Proposition 6*</u> Si pour $n \in \mathbb{Z}$, $k \geq 0$ et $p \geq 1$, on a :

$$\begin{cases} H_k\,(\Delta^2 x_{n+i}) \neq 0, \; H_{k+1}(\Delta^2 x_{n+i-1}) = 0 & i=1,\dots,p \\ H_{k+1}(\Delta^2 x_{n-1}) \neq 0, \; H_{k+1}(\Delta^2 x_{n+p}) \neq 0 & \end{cases}$$

alors

$$\frac{H_{k+p+2}(x_{n-\ell})}{H_{k+p+1}(\Delta^2 x_{n-\ell})} + \frac{H_{k+1}(x_{n-\ell})}{H_k(\Delta^2 x_{n-\ell})} = \frac{H_{k+\ \ +1}(x_{n-\ell})}{H_{k-\ell}(\Delta^2 x_{n-\ell})} + \frac{H_{k+p-\ \ +1}(x_{n+\ell})}{H_{k+p-\ell+1}(\Delta^2 x_{n+\ell})}$$

pour $\ell=1,\ldots,p$.

<u>Démonstration</u> : Considérons la matrice A suivante :

$$\begin{array}{c|c|ccc|ccc|ccc|}
 & I_1 & & I_2 & & & I_0 & & & I_3 & \\
\hline
I_1 & x_{n+\ell} & \Delta x_n & \cdots & \Delta x_{n+\ell-1} & \Delta x_{n+\ell} & \cdots & \Delta x_{n+k+\ell-1} & \Delta x_{n+k+\ell} & \cdots & \Delta x_{n+k+p} \\
\hline
 & \Delta x_n & \Delta^2 x_{n-\ell} & \cdots & \Delta^2 x_{n-1} & \Delta^2 x_n & \cdots & \Delta^2 x_{n+k-1} & \Delta^2 x_{n+k} & \cdots & \Delta^2 x_{n+k+p-\ell} \\
I_2 & \vdots & \vdots & & \vdots & \vdots & & \vdots & \vdots & & \vdots \\
 & \Delta x_{n+\ell-1} & \Delta^2 x_{n-1} & \cdots & \Delta^2 x_{n+\ell-2} & \Delta^2 x_{n+\ell-1} & \cdots & \Delta^2 x_{n+k+\ell-2} & \Delta^2 x_{n+k+\ell-1} & \cdots & \Delta^2 x_{n+k+p-1} \\
\hline
 & \Delta x_{n-\ell} & \Delta^2 x_n & \cdots & \Delta^2 x_{n+\ell-1} & \Delta^2 x_{n+\ell} & \cdots & \Delta^2 x_{n+k+\ell-1} & \Delta^2 x_{n+k+\ell} & \cdots & \Delta^2 x_{n+k+p} \\
I_0 & \vdots & \vdots & & \vdots & \vdots & & \vdots & \vdots & & \vdots \\
 & \Delta x_{n+k+\ell-1} & \Delta^2 x_{n+k-1} & \cdots & \Delta^2 x_{n+k+\ell-2} & \Delta^2 x_{n+k+\ell-1} & \cdots & \Delta^2 x_{n+2k+\ell-3} & \Delta^2 x_{n+2k+\ell-1} & \cdots & \Delta^2 x_{n+2k+p-1} \\
\hline
 & \Delta x_{n+k-\ell} & \Delta^2 x_{n+k} & \cdots & \Delta^2 x_{n+k+\ell-1} & \Delta^2 x_{n+k+\ell} & \cdots & \Delta^2 x_{n+2k+\ell-1} & \Delta^2 x_{n+2k+\ell} & \cdots & \Delta^2 x_{n+2k+p} \\
I_3 & \vdots & \vdots & & \vdots & \vdots & & \vdots & \vdots & & \vdots \\
 & \Delta x_{n+k+p} & \Delta^2 x_{n+k+p-\ell} & \cdots & \Delta^2 x_{n+k+p-1} & \Delta^2 x_{n+k+p} & \cdots & \Delta^2 x_{n+2k+p-1} & \Delta^2 x_{n+2k+p} & \cdots & \Delta^2 x_{n+2k+2p-\ell} \\
\hline
\end{array}$$

Compte tenu des hypothèses, le lemme 2 montre que l'on a :

$$\text{rang}\,(A_{I_{02}}^{I_{03}}) = \text{rang}\,(A_{I_0}^{I_0}) = k$$

Nous sommes alors en mesure d'appliquer le lemme 1. Il est alors aisé de voir que :

$$\det(A_{I_0}^{I_0}) = H_k(\Delta^2 x_{n+\ell}),\ \det(A_{I_{02}}^{I_{02}}) = H_{k+\ell}(\Delta^2 x_{n-\ell})$$

$$\det(A_{I_{03}}^{I_{03}}) = H_{k+p-\ell+1}(\Delta^2 x_{n+\ell}),\ \det(A_{I_{023}}^{I_{023}}) = H_{k+p+1}(\Delta^2 x_{n-\ell})$$

$$\det(A_{I_{01}}^{I_{01}}) = H_{k+1}(x_{n+\ell}),\ \det(A_{I_{012}}^{I_{012}}) = H_{k+\ell+1}(x_{n-\ell})$$

$$\det(A_{I_{013}}^{I_{013}}) = H_{k+p-\ell+2}(x_{n+\ell}),\ \det(A_{I_{0123}}^{I_{0123}}) = H_{k+p+2}(x_{n-\ell})$$

d'où :

$$[H_{k+p+2}(x_{n-\ell})\, H_k(\Delta^2 x_{n+\ell}) + H_{k+1}(x_{n+\ell})\, H_{k+p+1}(\Delta^2 x_{n-\ell})]\, H_{k+}\ (\Delta^2 x_{n-\ell})\, H_{k+p-\ell+1}(\Delta^2 x_{n+\ell})$$

$$= [H_{k+\ell+1}(x_{n-\ell})\, H_{k+p-\ell+1}(\Delta^2 x_{n+\ell}) + H_{k+p-\ell+2}(x_{n+\ell})\, H_{k+}\ (\Delta^2 x_{n-\ell})]$$

$$H_{k+p+1}(\Delta^2 x_{n-\ell})\, H_k(\Delta^2 x_{n+\ell})$$

et il suffit alors de montrer que les 4 nombres $H_k(\Delta^2 x_{n+\ell})$, $H_{k+\ell}(\Delta^2 x_{n-\ell})$ $H_{k+p+1}(\Delta^2 x_{n-\ell})$ et $H_{k+p-\ell+1}(\Delta^2 x_{n+\ell})$ sont non nuls pour obtenir le résultat annoncé.

Bien qu'on puisse établir ce résultat directement par des manipulations de déterminants, il est plus simple de constater que les hypothèses faites impliquent que les nombres en question sont des dénominateurs des approximants de Padé qui jouxtent un carré singulier de valeur infinie : ils ne peuvent donc être nuls (proposition 5,i)

Proposition 7 Si pour $n \in \mathbb{Z}$, $k \geq 0$, $p \geq 1$ on a

$$H_k(\Delta x_{n+i}) \neq 0,\ H_{k+1}(\Delta\, x_{n+i}) = 0 \qquad i = 1,\ldots,$$

$$H_{k+1}(\Delta^2 x_{n-i}) \neq 0,\ H_{k+1}(\Delta^2 x_{n+p}) \neq 0$$

alors $$\frac{H_{k+p}(\Delta^3 x_{n-\ell})}{H_{k+p+1}(\Delta x_{n-\ell})} + \frac{H_{k-1}(\Delta^3 x_{n+\ell})}{H_k(\Delta x_{n+\ell})} = \frac{H_{k+\ell-1}(\Delta^3 x_{n-\ell})}{H_{k+\ell}(\Delta x_{n-\ell})} + \frac{H_{k+p-\ell}(\Delta^3 x_{n+\ell})}{H_{k+p-\ell+1}(\Delta x_{n+\ell})}$$

pour $\ell = 1,\ldots,p$

<u>Démonstration</u> : Elle est très semblable à la précédente. Ainsi que l'a noté Paquet [20], il suffit de reprendre le raisonnement précédent avec la matrice A suivante :

$$
\begin{array}{c|c|ccc|ccc|ccc|}
 & I_1 & & I_2 & & & I_0 & & & I_3 & \\
\hline
I_1 & 0 & 1 & \cdots & 1 & 1 & \cdots & 1 & 1 & \cdots & 1 \\
\hline
 & 1 & \Delta x_{n-\ell} & \cdots & \Delta x_{n-1} & \Delta x_n & \cdots & \Delta x_{n+k-1} & \Delta x_{n+k} & \cdots & \Delta x_{n+k+p-\ell} \\
I_2 & \vdots & \vdots & & \vdots & \vdots & & \vdots & \vdots & & \vdots \\
 & 1 & \Delta x_{n-1} & \cdots & \Delta x_{n+\ell-2} & \Delta x_{n+\ell-1} & \cdots & \Delta x_{n+k+\ell-2} & \Delta x_{n+k+\ell-1} & \cdots & \Delta x_{n+k+p-1} \\
\hline
 & 1 & \Delta x_n & \cdots & \Delta x_{n+\ell-1} & \Delta x_{n+\ell} & \cdots & \Delta x_{n+k+\ell-1} & \Delta x_{n+k+\ell} & \cdots & \Delta x_{n+k+p} \\
I_0 & \vdots & \vdots & & \vdots & \vdots & & \vdots & \vdots & & \vdots \\
 & 1 & \Delta x_{n+k-1} & \cdots & \Delta x_{n+k+\ell-2} & \Delta x_{n+k+\ell-1} & \cdots & \Delta x_{n+2k+\ell-2} & \Delta x_{n+2k+\ell-1} & \cdots & \Delta x_{n+2k+p-1} \\
\hline
 & 1 & \Delta x_{n+k} & \cdots & \Delta x_{n+k+\ell-1} & \Delta x_{n+k+\ell} & \cdots & \Delta x_{n+2k+\ell-1} & \Delta x_{n+2k+\ell} & \cdots & \Delta x_{n+2k+p} \\
I_3 & \vdots & \vdots & & \vdots & \vdots & & \vdots & \vdots & & \vdots \\
 & 1 & \Delta x_{n+k+p-\ell} & \cdots & \Delta x_{n+k+p-1} & \Delta x_{n+k+p} & \cdots & \Delta x_{n+2k+p-1} & \Delta x_{n+2k+p} & \cdots & \Delta x_{n+2k+2p-\ell} \\
\hline
\end{array}
$$

et de noter que
$$\left\{\begin{array}{lcl}
\text{dét } (A) = & = & -H_{k+p}(\Delta^3 x_{n-\ell}) \\
\text{dét } (A_{I_{10}}^{I_{10}}) & = & -H_{k-1}(\Delta^3 x_{n+\ell}) \\
\text{dét } (A_{I_{012}}^{I_{012}}) & = & -H_{k+\ell-1}(\Delta^3 x_{n-\ell}) \\
\text{dét } (A_{I_{013}}^{I_{013}}) & = & -H_{k+p-\ell}(\Delta^3 x_{n+\ell})
\end{array}\right.$$

On peut regrouper les 2 résultats précédents en la :

Proposition 8 Si dans un ε-tableau, il existe k, n et p tels que :

$$\begin{cases} \varepsilon_k^{(n+i)} \neq \infty \text{ et } \varepsilon_{k+2}^{(n+i-1)} \neq \infty & \text{pour } i=1,\ldots,p \\ \varepsilon_{k+2}^{(n-i)} \neq \infty \ , \ \varepsilon_{k+2}^{(n+p)} \neq \infty & \end{cases}$$

alors

$$\text{(S)} \qquad \varepsilon_{k+2(p+1)}^{(n-\ell)} + \varepsilon_k^{(n+\ell)} = \varepsilon_{k+2\ell}^{(n-\ell)} + \varepsilon_{k+2(p-\ell+1)}^{(n+\ell)} \quad \text{pour } \ell=1,\ldots,p$$

6 - L'IDENTITÉ DE WYNN

La relation précédente permet la mise en oeuvre de la transformation de Shanks pour calculer les valeurs ponctuelles d'une table de Padé non normale. D'un point de vue purement algorithmique, elle est suffisante. Il est toutefois possible de montrer que l'identité de Wynn s'étend aisément aux tables de Padé non normales, bien que son utilité numérique effective soit moindre que celle de la relation (S).

Proposition 9 Si dans un ε-tableau, il existe k, n et p entiers et $x \in K$ (non compactifié) tels que :

$$\begin{cases} \varepsilon_k^{(n+i)} \neq x, \ \varepsilon_{k+2}^{(n+i-1)} = x & \text{pour } i=1,\ldots, p \geq 2 \\ \varepsilon_{k+2}^{(n-1)} \neq x, \ \varepsilon_{k+2}^{(n+p)} \neq x & \end{cases}$$

alors

$$(\varepsilon_{k+2p+2}^{(n-\ell)} - x)^{-1} + (\varepsilon_k^{(n+\ell)} - x)^{-1} = (\varepsilon_{k+2\ell}^{(n-\ell)} - x)^{-1} + (\varepsilon_{k+2(p-\ell)+2}^{(n+\ell)} - x)^{-1}$$

pour $\ell = 1,\ldots,p$

Démonstration : Les hypothèses impliquent l'existence d'un bloc carré de côté p et compte tenu de la proposition 5, on a :

$$\begin{cases} \varepsilon_{k+1}^{(n+i)} \neq \infty, \ \varepsilon_{k+3}^{(n+i-1)} = \infty & \text{pour } i = 1,\ldots,p-1 \\ \varepsilon_{k+3}^{(n-1)} \neq \infty, \ \varepsilon_{k+3}^{(n+p-1)} \neq \infty & \end{cases}$$

D'après la proposition 8, on a donc :

$$\varepsilon_{k+2p+1}^{(n-\ell)} + \varepsilon_{k+1}^{(n+\ell)} = \varepsilon_{k+2\ell+1}^{(n-\ell)} + \varepsilon_{k+2(p-\ell)+1}^{(n+\ell)} \quad \text{pour } \ell=1,\ldots,p-1$$

Cette relation est encore véfifiée pour $\ell = 0$ et $\ell = p$.

Compte tenu des 4 relations :

$$\left.\begin{array}{l}
\varepsilon_{k+1}^{(n+\ell)} - \varepsilon_{k+1}^{(n+\ell-1)} = (\varepsilon_{k+2}^{(n+\ell-1)} - \varepsilon_{k}^{(n+\ell)})^{-1} \\
\varepsilon_{k+2p+1}^{(n-\ell)} - \varepsilon_{k+2p+1}^{(n-\ell+1)} = (\varepsilon_{k+2p}^{(n-\ell+1)} - \varepsilon_{k+2p+2}^{(n-\ell)})^{-1} \\
\varepsilon_{k+2\ell+1}^{(n-\ell)} - \varepsilon_{k+2\ell-1}^{(n-\ell+1)} = (\varepsilon_{k+2}^{(n-\ell+1)} - \varepsilon_{k+2}^{(n-\ell)})^{-1} \\
\varepsilon_{k+2(p-\ell)+1}^{(n+\ell)} - \varepsilon_{k+2(p-\ell)+3}^{(n+\ell-1)} = (\varepsilon_{k+2(p-\ell)+2}^{(n+\ell-1} - \varepsilon_{k+2(p-\ell)+2}^{(n+\ell)})^{-1}
\end{array}\right\} \text{pour } \ell=1,\dots,p$$

les relations :

$$\varepsilon_{k+2p+1}^{(n-\ell)} - \varepsilon_{k+2p+1}^{(n-\ell+1)} + \varepsilon_{k+1}^{(n+\ell)} - \varepsilon_{k+1}^{(n+\ell-1)} = \varepsilon_{k+2\ell+1}^{(n-\ell)} - \varepsilon_{k+2\ell-1}^{(n-\ell+1)} + \varepsilon_{k+2(p-\ell)+1}^{(n+\ell)} - \varepsilon_{k+2(p-\ell)+3}^{(n+\ell-1)}$$

$$(\ell=1,\dots,p)$$

conduisent au résultat annoncé.

Théorème Si une table des formes de Padé comporte un bloc d'indice (p,q) et de côté n, alors les formes $R_{i,j}$ de degré (i,j) qui jouxtent ce carré sont liées par les identités suivantes :

$$(\mathcal{W}) \qquad [(R_{p-1,q-1+\ell}-C)^{-1}+(R_{p+n,q+n-\ell}-C)^{-1}]-[(R_{p-1+\ell,q-1}-C)^{-1}+(R_{p+n-\ell,q+n}-C)^{-1}]=0$$

$$\text{pour } \ell = 1,\dots,n,$$

avec $C = R_{p,q}$.

Démonstration : Les formes rationnelles qui interviennent dans cette relation sont bien définies (unicité). Compte tenu de la liaison entre la table des formes de Padé et l'ε-tableau associé à chaque valeur ponctuelle, la proposition 9 implique la nullité de la fraction rationnelle définie par le premier membre de $(\mathcal{W})$ pour toute valeur z qui n'est pas un pôle d'un de ses quatre composants. Cette fraction rationnelle est donc identiquement nulle.

<u>Remarque</u> : Il est clair que la transformation de Shanks d'une suite peut conduire à des quotients indéfinis alors que l'approximant de Padé classique associé est parfaitement défini. Comme on l'a vu, cette anomalie apparente disparaît si on considère les seules formes de Padé uniquement définies. Il est intéressant de constater que c'est cette même notion de forme de Padé qui avait permis d'étendre l'algorithme d'Euclide [19] et l'algorithme de Baker aux tables de Padé non normales [13]. Ainsi, à trois reprises, la notion de forme de Padé a permis de résoudre des problèmes posés par la non normalité alors que la notion classique d'approximant paraît en défaut. Puisque la notion de forme de Padé s'identifie à celle d'approximant dans le cas d'une table normale, et que la première paraît plus fructueuse dans les autres cas, il paraît souhaitable de ne plus s'intéresser désormais qu'aux seules formes de Padé.

++++++

RÉFÉRENCES

1 G.A. BAKER Jr.- *Recursive calculation of Padé approximants* dans "Padé Approximants and their applications" Graves-Morris P.R. ed., Academic press, London (1973) 83-91.

2 C. BREZINSKI.- *Computation of Padé approximants and continued fractions.* Journ. of Comput. Appl. Math. 2 (1976) 113-123.

3 A. BULTHEEL.- *Recursive algorithms for non normal Padé tables.* Appl. Math. Prog. Div., Kath. Univ. Leuven, Report TW40, 1978.

4 A. BULTHEEL.- *Division algorithms for continued fractions and the Padé Table.* Appl. Math. Prog. Div., Kath. Univ. Leuven, Report TW41, 1978.

5 A. BULTHEEL.- *Fast algorithms for the factorisation of Hankel and Toeplitz matrices and the Padé approximation problem.* Appl. Math. Prog. Div., Kath. Univ. Leuven, Report TW42, 1978.

6 A. BULTHEEL.- *Remark on " A new look at the Padé Table and different methods for computing its elements".* Journal of Comput. Appl. Math. 5 (1979) 67.

7 A. BULTHEEL.- *Recursive algorithms for the Padé Table : two approaches.* (These procedings).

8 D. BUSSONNAIS.- *Tous les algorithmes de calcul par récurrence des approximants de Padé d'une série. Construction des fractions continues correspondantes.* Exposé au "Colloque sur les approximants de Padé", Lille (1978).

9 G. CLAESSENS.- *A new look at the Padé table and different methods for computing its elements.* Journal of Comput. Appl. Math. 1 (1975) 141-152.

10 G. CLAESSENS, L. WUYTACK.- *On the computation of non normal Padé approximants.* Depart. Wiskunde, Univ. Inst. Antwerpen, Report 77-41 (1977).

11 F. CORDELLIER.- *L'ε-algorithme vectoriel : interprétation géométrique et règles singulières.* Exposé au "Colloque d'Analyse Numérique de Gourette", France (1974).

12 F. CORDELLIER.- *Détermination algébrique des règles singulières pour l'ε-algorithme scalaire et la table de Padé.* Exposé au "Colloque sur les approximants de Padé", Lille (1977).

13 F. CORDELLIER.- *Deux algorithmes de calcul récursif des éléments d'une table de Padé non normale.* Exposé au "Colloque sur les approximants de Padé", Lille (1978).

14 F. CORDELLIER.- *Sur la régularité des procédés δ^2 d'Aitken et W de Lubkin.* (Ces comptes rendus).

15 F.R. GANTMACHER.- *The theory of matrices.* Vol.1. Chelsea, New York, 1960.

16 J. GILEWICZ.- *Approximants de Padé.* Lecture Notes in Math. 667, Springer Verlag (1978).

17 W.B. GRAGG.- *The Padé Table and its relation to certain algorithms of numerical analysis.* SIAM Review 14 (1972) 1-62.

18 P.R. GRAVES-MORRIS.- *Numerical calculation of Padé approximants.* (Ces comptes rendus).

19 R.J. McELIECE, J.B. SHEARER.- *A property of Euclid's algorithm and an application to Padé approximation.* SIAM Journ. Appl. Math. 34 (1978) 611-615.

20 J.B. PAQUET.- Communication privée.

21 D. SHANKS.- *Non-linear transformation of divergent and slowly convergent sequences.* J. Math. Phys. 34 (1955) 1-42.

22 P. WYNN.- *On a device for computing the $e_m(S_n)$ transformation.* M.T.A.C. 10 (1956) 90-96.

23 P. WYNN.- *L'ε-algorithmo e la tavola di Padé.* Rend. di Mat. Roma 20 (1961) 403-408.

24 P. WYNN.- *Singular rules for certain non-linear algorithms.* Nord Tid. for Inf. Beh. 3 (1963) 175-195.

25 P. WYNN.- *Upon systems of recursions which obtain among the quotients of the Padé table.* Numer. Math. 8 (1966) 264-269.

26 P. WYNN.- Communication privée.

To my father Pierre F.M. Cuyt.

ABSTRACT PADÉ-APPROXIMANTS IN OPERATOR THEORY**

by

ANNIE A.M. CUYT

DEPARTMENT OF MATHEMATICS
UNIVERSITY OF ANTWERP
UNIVERSITEITSPLEIN 1
B - 2610 WILRIJK (BELGIUM)

The use of Padé-approximants for the solution of mathematical problems in science has great development. Padé-approximants have proved to be very useful in numerical analysis too : the solution of a nonlinear equation, acceleration of convergence, numerical integration by using nonlinear techniques, the solution of ordinary and partial differential equations. Especially in the presence of singularities the use of Padé-approximants has been very interesting.
Yet we have tried to generalize the concept of Padé-approximant to operator theory, departing from "power-series-expansions" as is done in the classical theory*.
A lot of interesting properties of classical Padé-approximants remain valid and the classical Padé-approximant is now a special case of the theory. The notion of abstract Padé-table is introduced; it also consists of squares of equal elements as in the classical theory.

* Roman figures between brackets refer to a work in the reference-list.

** This work is supported by I.W.O.N.L. (Belgium)

0. NOTATIONS

R_0^+	{positive real numbers}
X,Y	always normed vectorspaces or Banach-spaces or Banach-algebras with unit
L(X,Y)	{linear bounded operators $L : X \to Y$}
$L(X^k,Y)$	{k-linear bounded operators $L : X \to L(X^{k-1},Y)$}
Λ	field R or C
$\lambda,\mu,\ldots$	elements of Λ
0	unit for addition in a Banach-space, or multilinear operator $L \in L(X^k,Y)$ such that $Lx_1\ldots x_k=0 \quad \forall (x_1,\ldots,x_k) \in X^k$
I	unit for multiplication in a Banach-algebra
1	unit for multiplication in Λ
F,G,...	non-linear operators $: X \to Y$
$B(x_0,r)$	open ball with centre $x_0 \in X$ and radius $r > o$
$\overline{B}(x_0,r)$	closed ball with centre $x_0 \in X$ and radius $r > o$
P,Q,R,S,T,...	non-linear operators $: X \to Y$, usually abstract polynomials
$\partial P, \partial Q,\ldots$	exact degree of the abstract polynomial P,Q,...
$F^{(k)}(x_0)$	k^{th} Fréchet-derivative of the operator $F : X \to Y$ in x_0
D(G)	$\{x \in X \mid G(x)$ is regular in Y$\}$ for the operator $G : X \to Y$ (=Banach-algebra)
A_i,B_j,C_k,D_s	i-linear, j-linear, k-linear, s-linear operators

1. INTRODUCTION

A lot of attempts have been made to generalize in some way classical Padé-approximants. We refer e.g. to quadratic Padé-approximants (X,XV), Chebyshev-Padé or

Legendre-Padé (VII), operator Padé-approximants for formal power series in a parameter with non-commuting elements of a certain algebra as coefficients (VI), N-variable rational approximants (VIII, IX, XI, XII, XIII, XIV).

Another genralisation now is the following one.

Let X and Y be Banach-spaces (same field Λ). We always work in the norm-topology.

We define $L(X^k,Y) = \{L \mid L \text{ is a k-linear bounded operator, } L : X \to L(X^{k-1},Y)\}$ and $L(X^0,Y) = Y$. So $Lx_1 \ldots x_k = (Lx_1)(x_2 \ldots x_k) \in Y$ with $x_1,\ldots,x_k \in X$ and $Lx_1 \in L(X^{k-1},Y)$ (V pp. 100). $L \in L(X^k,Y)$ is called symmetric if $Lx_1 \ldots x_k = Lx_{i_1} \ldots x_{i_k}$, $\forall (x_1,\ldots,x_k) \in X^k$ and $\forall$ permutations $(i_1,\ldots,i_k)$ of $(1,\ldots,k)$ (V pp. 103).

We remark that the operator $\bar{L} \in L(X^k,Y)$ defined by $\bar{L}x_1 \ldots x_k = \frac{1}{k!} \sum_{(i_1,\ldots,i_k)} Lx_{i_1} \ldots x_{i_k}$ for a given $L \in L(X^k,Y)$ is symmetric.

Let us identify $y \in Y$ with the constant operator $X \to Y : x \to y$ and call it o-linear.

Definition 1.1. : An abstract polynomial is a non-linear operator $P : X \to Y$ such that

$$P(x) = A_n x^n + \ldots + A_0 \in Y \text{ with } \begin{cases} A_i \in L(X^i,Y) \\ A_i \text{ symmetric} \end{cases}$$

The degree of $P(x)$ is n.

The notation for the exact degree of $P(x)$ is ∂P.

Definition 1.2. : Let X be a Banach-space, Y a Banach-algebra; let $F : X \to Y$ and $G : X \to Y$ be operators.

The <u>product F.G</u> is defined by : $(F.G)(x) = F(x).G(x)$ in Y.

Definition 1.3. : Let $X_1,\ldots,X_p$, $Z_1,\ldots,Z_q$ be vector spaces and Y an algebra (same field Λ). Let $F : X_1 \times \ldots \times X_p \to Y$ be bounded and p-linear, and $G : Z_1 \times \ldots \times Z_q \to Y$ be bounded and q-linear.

The <u>tensorproduct</u> $F \otimes G : X_1 \times \ldots \times X_p \times Z_1 \times \ldots \times Z_q \to Y$ is bounded and (p+q)-linear when defined by $(F \otimes G)x_1 \ldots x_p z_1 \ldots z_q = Fx_1 \ldots x_p . Gz_1 \ldots z_q$ (IIpp.318).

One can easily prove that in a Banach-algebra Y :

$$(F.G)'(x_0) = F'(x_0) \otimes G(x_0) + F(x_0) \otimes G'(x_0) ,$$

where the accent stands for Fréchet-differentiation.

We call $y \in Y$ regular if there exists $y^{-1} \in Y$ such that : $y.y^{-1} = I = y^{-1}.y$;

we call $y \in Y$ singular if it is not regular.

Definition 1.4. : Let $G : X \to Y$ with X a Banach-space and Y a Banach-algebra;

$D(G) = \{x \in X | G(x)$ is regular in $Y\}$ is an open set in X (III pp.31).

The operator $\frac{1}{G}$ is defined by $\frac{1}{G} : D(G) \subset X \to Y : x \to [G(x)]^{-1}$.

One can easily prove that in a commutative Banach-algebra Y :

$$(\frac{1}{G})'(x_0) = -G'(x_0) \otimes (\frac{1}{G}(x_0))^2 .$$

Let again X and Y both be Banach-spaces.

We note the fact that $F^{(k)}(x_0)$, the kth derivative of an operator $F : X \to Y$ in x_0, is a symmetric k-linear operator (V pp. 110).

Abstract polynomials are differentiated as in elementary calculus :

if $P(x) = A_n x^n + ... + A_0$ with $A_i \in L(X^i,Y)$ and A_i symmetric, then

$P'(x_0) = n.A_n x_0^{n-1} + ... + A_1 \in L(X,Y)$

$P^{(2)}(x_0) = n.(n-1).A_n x_0^{n-2} + ... + 2A_2 \in L(X^2,Y)$

$\vdots$

$P^{(n)}(x_0) = n!\, A_n \in L(X^n,Y)$

We now can easily prove the fact that if for an abstract polynomial

$P(x) = \sum_{i=0}^{n} C_i x^i$ with $C_i \in L(X^i,Y)$ and C_i symmetric: $P(x) = 0 \ \forall x \in X$, then $C_i \equiv 0$ $\forall i \in \{0,...,n\}$.

Let $B(x_0,r) = \{x \in X | \|x_0 - x\| < r\}$ for $r \in R_0^+$ and $x_0 \in X$.

Definition 1.5. : The operator $F : X \to Y$ possesses an <u>abstract Taylor-series</u> in x_0 if $\exists B(x_0,r)$ with $r > 0$:

$$F(x_0 + h) = \sum_{k=0}^{\infty} \frac{1}{k!} \cdot F^{(k)}(x_0)\, h^k \text{ for } x_0 + h \in B(x_0,r).$$

We then call F <u>abstract analytic</u> in x_0 (V pp. 113).

2. DEFINITION OF ABSTRACT PADE-APPROXIMANT

To generalize the notion of Padé-approximant we start from analyticity, as in elementary calculus.

Let $F : X \to Y$ be a non-linear operator, X a Banach-space and Y a Banach-algebra. Let F be analytic in $B(x_0,r)$ with $r > 0$.

So F has the following abstract Taylor-series :

$$F(x_0 + x) = \sum_{k=0}^{\infty} \frac{1}{k!} F^{(k)}(x_0)\, x^k \qquad (1)$$

with $\frac{1}{0!} F^{(0)}(x_0) x^0 = F(x_0)$

and $F^{(k)}(x_0) \in L(X^k, Y)$

We give some examples of such series :

a) $C([0,1])$ with the supremum-norm and $(f.g)(x) = f(x).g(x)$ for $f,g \in C([0,1])$, is a commutative Banach-algebra. Consider the Nemyckii-operator $G : C([0,1]) \to C([0,1]) : x \to g(s,x(s))$ with $g \in C^{(\infty)}([0,1] \times C([0,1]))$ (V pp. 95).
Let $I_x : C([0,1]) \to C([0,1]) : x \to x$.
Then clearly $G^{(n)}(x_0) = \frac{\partial^n g}{\partial x^n}(s,x_0(s)).\ \underbrace{I_x \otimes \dots \otimes I_x}_{n \text{ times}}$, n-linear and bounded.

b) Consider the Urysohn integral operator $U : C([0,1]) \to C([0,1])$:
$x \to \int_0^1 f(s,t,x(t))dt$ with $f \in C^{(\infty)}([0,1] \times [0,1] \times C([0,1]))$ (V pp. 97).
Let [ ] indicate a place-holder for $x(t) \in C([0,1])$ (V pp. 90).
Then we write $U^{(n)}(x_0) = \int_0^1 \frac{\partial^n f}{\partial x^n}(s,t,x_0(t))\underbrace{[\,]\dots[\,]}_{n \text{ times}}\, dt$

c) Consider the operator $P : C'([0,T]) \to C([0,T]) : y \to \frac{dy}{dt} - f(t,y)$ in the initial value problem $P(y) = 0$ with $y(o) = a \in R$.
Let $f \in C^{(\infty)}([0,T] \times C'([0,T]))$ and $I_y : C'([0,T]) \to C([0,T]) : y \to y$.
We remark that $C^{(i)}([0,T])$ with the supremum-norm is a Banach space.
We see that $P'(y_0) = \frac{d}{dt} - \frac{\partial f(t,y)}{\partial y}(t,y_0).I_y$ and

$$P^{(n)}(y_0) = \frac{-\partial^n f(t,y)}{\partial y^n}(t,y_0) . \underbrace{I_y \otimes ... \otimes I_y}_{n \text{ times}} \text{ for } n \geq 2.$$

d) Finally let this nonlinear system of 2 real variables $F\binom{\xi}{\eta} = \binom{\xi + \sin(\xi\eta) + 1}{\xi^2 + \eta^2 - 4\xi\eta}$ be given; let $x_0 = \binom{0}{0}$. $\mathbb{R}^2$ with component-wise multiplication is a Banach-algebra with unit $\binom{1}{1}$.

Then $F(x) = \binom{1}{0} + \binom{\xi}{0} + \binom{\xi\eta}{\xi^2 + \eta^2 - 4\xi\eta} + \sum_{k=1}^{\infty} \left(\begin{array}{c} (-1)^k . \frac{(\xi\eta)^{2k+1}}{(2k+1)!} \\ 0 \end{array} \right)$

Definition 2.1. : Let $F : X \to Y$ be an operator with X and Y Banach-spaces.
We say that <u>$F(x) = O(x^j)$</u> if $\exists J \in R_0^+$, $\exists B(0,r)$ with $0<r<1$: $\forall x \in B(0,r) : \|F(x)\| \leq J.\|x\|^j$ $(j \in N)$

Now let $x_0 = 0$ without loss of generality, and let Y be a commutative Banach-algebra.
In Y we can use the fact that for $y,z \in Y$: $y.z$ regular $\Rightarrow$ y regular and z regular.

Definition 2.2. : In <u>Padé-approximation</u> we try to find a couple of abstract polynomials $(P(x),Q(x)) = (A_{n.m+n} x^{n.m+n} + ... + A_{n.m} x^{n.m}, B_{n.m+m} x^{n.m+m} + ... + B_{n.m} x^{n.m})$
such that the abstract power series

$$F(x).(B_{n.m+m} x^{n.m+m} + ... + B_{n.m} x^{n.m}) - (A_{n.m+n} x^{n.m+n} + ... + A_{n.m} x^{n.m}) = O(x^{n.m+n+m+1}).$$

(In 5.f) we justify the choice of (P(x), Q(x)) made here).

Write $\frac{1}{k!}\cdot F^{(k)}(0) = C_k \in L(X^k, Y)$.

The condition in definition 2.2 is equivalent with (1a) and (1b) :

$$(1a)\begin{cases} C_0 \cdot B_{n.m}\, x^{n.m} = A_{n.m}\, x^{n.m} \quad \forall x \in X \\ C_1 x . B_{n.m}\, x^{n.m} + C_0 \cdot B_{n.m+1}\, x^{n.m+1} = A_{n.m+1}\, x^{n.m+1} \quad \forall x \in X \\ \vdots \\ C_n\, x^n . B_{n.m}\, x^{n.m} + C_{n-1}\, x^{n-1} . B_{n.m+1}\, x^{n.m+1} + \dots + C_0 \cdot B_{n.m+n}\, x^{n.m+n} = \\ \qquad A_{n.m+n}\, x^{n.m+n} \quad \forall x \in X \end{cases}$$

with $B_j \equiv 0 \in L(X^j, Y)$ if $j > n.m+m$

$$(1b)\begin{cases} C_{n+1}\, x^{n+1} . B_{n.m}\, x^{n.m} + \dots + C_{n+1-m}\, x^{n+1-m} . B_{n.m+m}\, x^{n.m+m} = 0 \quad \forall x \in X \\ \vdots \\ C_{n+m}\, x^{n+m} . B_{n.m}\, x^{n.m} + \dots + C_n\, x^n . B_{n.m+m}\, x^{n.m+m} = 0 \quad \forall x \in X \end{cases}$$

with $C_k \equiv 0 \in L(X^k, Y)$ if $k < 0$.

For every solution $\{B_{n.m+j}\, x^{n.m+j} \mid j = 0, \dots, m\}$ of (1b), a solution $\{A_{n,m+i}\, x^{n.m+i} \mid i = 0, \dots, n\}$ of (1a) can be computed.

3. EXISTENCE OF A SOLUTION

a) case : $m = 0$

Choose $B_{n.m} = B_0 = I$, unit for the multiplication in Y.

Then $A_i = C_i$ for $i = 0, \dots, n$ are a solution of (1a).

The partial sums of (1) are the sought abstract polynomials.

b) case : $m \neq 0$

Compute $D_{n.m} = \sum_{i_1=1}^{m} \dots \sum_{i_m=1}^{m} [\varepsilon_{i_1 \dots i_m} \bigotimes_{j=1}^{m} C_{n-(j-1)+(i_j-1)}]$

with $i_1,\dots,i_m \in \{1,\dots,m\}$, and $\varepsilon_{i_1\dots i_m} = +1$ when $i_1\dots i_m$ is an even permutation of $1\dots m$, and $\varepsilon_{i_1\dots i_m} = -1$ when $i_1\dots i_m$ is an odd permutation of $1\dots m$, and $\varepsilon_{i_1\dots i_m} = 0$ elsewhere.

Compute for $h = 1,\dots,m$: $D_{n.m+h}$ by replacing in $D_{n.m}$ the operator $C_{n-(h-1)+(i_h-1)}$ by the operator $-C_{n+1+(i_h-1)}$.

Clearly $D_{n.m+h} \in L(X^{n.m+h}, Y)$ for $h = 0,\dots,m$.

Now $D_{n.m+h}\, x^{n.m+h}$ is a solution of system (1b); and $D_{n.m+h}\, x^{n.m+h} = \overline{D}_{n.m+h}\, x^{n.m+h}$.

We thus can consider a symmetric solution, also for (1a).

This is a correct procedure to calculate a solution. But in some cases it can be more practical to solve the system otherwise, e.g. to get the most general form of the solution.

4. UNICITY OF A SOLUTION

From now on $F : X \to Y$ is a nonlinear operator with X a Banach-space and Y a commutative Banach-algebra such that for each polynomial $T : X \to Y$ with $D(T) \neq \phi$, the set $D(T)$ is dense in X (or any other equivalent condition).

This is the case e.g. for $F : R^p \to R^q$; if $T(x) = \left(\sum_{j_1+\dots+j_p = 0}^{m} \alpha_{ij_1\dots j_p} x_1^{j_1} \dots x_p^{j_p},\ i = 1,\dots,q \right)$, $D(T) \neq \phi$, the set $X \setminus \bigcup_{i=1}^{q} \{(x_1,\dots,x_p) \in R^p \mid \sum_{j_1+\dots+j_p = 0}^{m} \alpha_{ij_1\dots j_p} x_1^{j_1} \dots x_p^{j_p} = 0\}$ is dense in X with the norm-topology. We then have the following important lemma.

Lemma 4.1. :

Let U,T be abstract polynomials : $X \to Y$
$U(x).T(x) = 0 \quad \forall x \in X$ ⎫
$\{x \in X \mid T(x)$ regular$\}$ is dense in X ⎭ $\Rightarrow U \equiv 0$

After calculating the solution of (1a) and (1b) we are going to look for an irreducible rational approximant.

Definition 4.1. : Let P and Q be 2 abstract polynomials. We call $\frac{1}{Q}.P$ <u>reducible</u> if there exist abstract polynomials T,R,S such that $P=T.R=R.T$ and $Q=T.S=S.T$ and $\partial T\geq 1$, $\partial R\geq 0$, $\partial S\geq 0$.

For reducible $\frac{1}{Q}.P$ we know that $\forall x\in D(Q) : (\frac{1}{Q}.P)(x)=(\frac{1}{S}.R)(x)$.
It is possible that $\frac{1}{S}$ is defined on a greater domain than $\frac{1}{Q}$.

Lemma 4.2. :

> Let P,Q,R be abstract polynomials : $X\rightarrow Y$
>
> For $R=P.Q$: $\begin{cases} D(R)=D(P)\cap D(Q) \\ D(R)=\phi \Leftrightarrow D(P)=\phi \text{ or } D(Q)=\phi \end{cases}$

Proof : $R(x)$ regular $\Leftrightarrow P(x)$ regular and $Q(x)$ regular

so $D(R)=D(P)\cap D(Q)$

We know that $D(P)$ is open (and so is $D(Q)$)

$D(Q)$ is dense in X if $D(Q)\neq\phi$ (and so is $D(P)$)

If $D(P)=\phi$ or $D(Q)=\phi$ then evidently $D(R)=\phi$.

The second implication is proved by contraposition.

If $D(R)=\phi$ and $\exists x\in D(P)$ then $\exists r_0>0 : B(x,r_0)\subset D(P)$.

Now $\forall x\in X$, $\forall\ r>0 : B(x,r)\cap D(Q)\neq\phi$.

And so $\phi\neq B(x,r_0)\cap D(Q)\subseteq D(P)\cap D(Q)$.

This implies a contradiction.

Definition 4.2. : Let (P,Q) be a couple of abstract polynomials satisfying definition 2.2 and suppose $D(Q)\neq\phi$ or $D(P)\neq\phi$. Possibly $\frac{1}{Q}.P$ is reducible. Let $\frac{1}{Q_\star}.P_\star$ be the irreducible form of $\frac{1}{Q}.P$ such that $0\in D(Q_\star)$ and and $Q_\star(0)=I$, if it exists. We then call $\frac{1}{Q_\star}.P_\star$ an <u>abstract Padé-approximant</u> of order (n,m) for F.

That irreducible form $\frac{1}{Q_\star}.P_\star$ with $Q_\star(0)=I$ is unique because if $P=P_{\star 1}.T_1=P_{\star 2}.T_2$

and $Q = Q_{*1}.T_1 = Q_{*2}.T_2$ with $\frac{1}{Q_{*1}}.P_{*1}$ and $\frac{1}{Q_{*2}}.P_{*2}$ irreducible, $Q_{*1}(0) = I = Q_{*2}(0)$, $D(T_1) \neq \phi$ and $D(T_2) \neq \phi$, then $P_{*1}.Q_{*2} = P_{*2}.Q_{*1}$ because of lemma 4.1 and so we can prove that $\exists$polynomial R $\ni \begin{cases} P_{*1} = R.P_{*2}, \\ Q_{*1} = R.Q_{*2} \\ R(0) = I \end{cases}$ what contradicts the irreducible character of $\frac{1}{Q_{*1}}.P_{*1}$ unless $\partial R = o$.

Call n' the exact degree of P_* and m' the exact degree of Q_*.
When $(P(x) = P_*(x).T(x),\ Q(x) = Q_*(x).T(x))$ is a solution of (1a) and (1b) and $\frac{1}{Q_*}.P_*$ is an abstract Padé-approximant of order (n,m) for F, then $\partial T \geq n.m$ and $n' \leq n$ and $m' \leq m$.
We have the following theorem concerning the solutions of (1a) and (1b).

Theorem 4.1 : If the couples (P,Q) and (R,S) of abstract polynomials both satisfy (1a) and (1b), then $P.S = R.Q$; in other words : $\forall x \in X : P(x).S(x) = R(x).Q(x)$.

Proof : Regard $P(x).S(x)-R(x).Q(x) =$

$$[F(x).S(x)-R(x)].Q(x)-[F(x).Q(x)-P(x)].S(x)$$

Now $(F.Q-P)(x) = 0(x^{n.m+n+m+1})$

$(F.S-R)(x) = 0(x^{n.m+n+m+1})$

But $(P.S-R.Q)(x)$ is an abstract polynomial of degree at most $2n.m+n+m$, while $[(F.S-R).Q-(F.Q-P).S](x) = 0(x^{2n.m+n+m+1})$

So $(P.S-R.Q)(x) = 0 \quad \forall x \in X$.

This theorem implies that $(\frac{1}{Q}.P)(x) = (\frac{1}{S}.R)(x) \quad \forall x \in D(Q) \cap D(S)$.
If $D(Q.S) \neq \phi$ then $D(Q.S)$ is dense in X.
Possibly $\frac{1}{Q}.P$ and $\frac{1}{S}.R$ are reducible. If $P = P_*.T$, $Q=Q_*.T$, $R=R_*.U$, $S=S_*.U$ with

$D(T) \neq \phi$ and $D(U) \neq \phi$, then :

$$P.S = R.Q \Rightarrow P_\star.S_\star = R_\star.Q_\star \quad \text{because of lemma 4.1.}$$

We then know that $(\frac{1}{Q_\star}.P_\star)(x) = (\frac{1}{S_\star}.R_\star)(x) \quad \forall x \in D(Q_\star) \cap D(S_\star)$; if $D(Q_\star.S_\star) \neq \phi$ then $D(Q_\star.S_\star)$ is dense in X.

We can define an equivalence relation ... ~ ... in

$A = \{(P,Q) \mid (P,Q)$ satisfies definition 2.2 and $(D(P) \neq \phi$ or $D(Q) \neq \phi)\} \cup$

$\{(P_\star,Q_\star) \mid (P = P_\star.T,\ Q = Q_\star.T)$ satisfies definition 2.2 and $(D(P) \neq \phi$ or $D(Q) \neq \phi)$

and $\frac{1}{Q_\star}.P_\star$ is irreducible$\}$ where $P_\star, Q_\star, T, P, Q$ are abstract polynomials, by

$(P,Q) \sim (R,S) \Leftrightarrow P(x).S(x) = R(x).Q(x) \ \forall x \in X.$

If there exists a solution $(P,Q) \in A$ such that $Q_\star(0) = I$, then for all equivalent solutions $(R,S) \in A : 0 \in D(S_\star)$ because $P_\star S_\star = R_\star Q_\star$ implies : $\exists$ polynomial $V \ni \begin{cases} R_\star = VP_\star, \\ S_\star = VQ_\star \\ V(0) = S(0) \end{cases}$

what contradicts the irreducible character of $\frac{1}{S_\star}.R_\star$ unless $\partial V = o$ and so $\begin{cases} R_\star = S(0).P_\star; \\ S_\star = S(0).Q_\star \end{cases}$

if now S(0) were not regular then (R,S) were no element of A.

If $S_\star(0) = I = Q_\star(0)$ then $P_\star.S_\star = R_\star.Q_\star$ implies that $\exists$ polynomial $V \ni \begin{cases} P_\star = V.R_\star \\ Q_\star = V.S_\star \\ V(0) = I \end{cases}$

In other words : for $\frac{1}{S_\star}.R_\star$ irreducible we have $\partial V = o$ and so $\frac{1}{Q}.P$ and $\frac{1}{S}.R$ supply the same abstract Padé-approximant of order (n,m) for F when (P,Q) and (R,S) both satisfy (1a) and (1b).

We call $\frac{1}{Q_\star}.P_\star$ satisfying definition 4.2 <u>the abstract Padé-approximant (APA) of order (n,m) for F</u>.

Definition 4.3. : If for all the solutions (P,Q) of (1a) and (1b) with $D(P) \neq \phi$ or $D(Q) \neq \phi$ the irreducible form $\frac{1}{Q_*} . P_*$ (representant of the equivalence relation-class) is such that $D(Q_*) \not\ni 0$, then we call $\frac{1}{Q_*} . P_*$ <u>the abstract rational approximant (ARA) of order (n,m) for F.</u>

(We do come back on abstract rational approximants in 5.f).

We remark that, although $F(0) = C_0$ is defined, $(\frac{1}{Q} . P)(0) = \frac{0}{0}$ is always undefined for (P,Q) satisfying definition 2.2 with $n > o$ and $m > o$, since 0 is always singular in Y.

If for all the solutions (P,Q) of (1a) and (1b) : $0 \notin D(Q_*)$ or $D(Q) = \phi = D(P)$, we shall call the abstract Padé-approximant <u>undefined</u>.

If for the ARA $D(Q_*) = \phi$ then for all solutions (R,S) of (1a) and (1b) : $D(S_*) = \phi$ because $D(P_*) \cap D(S_*) = D(R_*) \cap D(Q_*) = \phi$ and $D(P) \neq \phi$; the ARA is in fact useless then.

An example will prove that it is very well possible that for an operator $F : X \to Y$, the (n,m) Padé-approximant is defined, while the (l,k) Padé-approximant is undefined for $l \neq n$ or $k \neq m$.

Consider the operator $F\begin{pmatrix}\xi\\ \eta\end{pmatrix} = \begin{pmatrix}\xi + \sin(\xi n) + 1\\ \xi^2 + \eta^2 - 4\xi\eta\end{pmatrix} = \begin{pmatrix}1\\ 0\end{pmatrix} + \begin{pmatrix}\xi\\ 0\end{pmatrix} + \begin{pmatrix}\xi\eta\\ \xi^2 + \eta^2 - 4\xi\eta\end{pmatrix} + \ldots$

Then : (1,1)-APA is $\begin{pmatrix}\frac{1+\xi-\eta}{1-\eta}\\ 0\end{pmatrix}$, $P_*(x) = P_*\begin{pmatrix}\xi\\ \eta\end{pmatrix} = \begin{pmatrix}1\\ 0\end{pmatrix} + \begin{pmatrix}1 & -1\\ 0 & 0\end{pmatrix}\begin{pmatrix}\xi\\ \eta\end{pmatrix}$

$$Q_*(x) = Q_*\begin{pmatrix}\xi\\ \eta\end{pmatrix} = \begin{pmatrix}1\\ 1\end{pmatrix} + \begin{pmatrix}0 & -1\\ 0 & 0\end{pmatrix}\begin{pmatrix}\xi\\ \eta\end{pmatrix}$$

$$D(Q_*) = R^2 \setminus \{(\xi, 1) \mid \xi \in R\}$$

(2,1)-APA is $\begin{pmatrix}1+\xi+\xi\eta\\ \xi^2+\eta^2-4\xi\eta\end{pmatrix}$ $P_*(x) = C_0 + C_1 x + C_2 x^2$

$$Q_*(x) = I$$

$$D(Q_*) = R^2$$

(1,2)-APA is undefined.

The next theorem is a summary of the previous results.

Theorem 4.2. :

> For every non-negative value of n and m, the systems (1a) and (1b) are solvable; if the abstract Padé-approximant of order (n,m) for $F : X \to Y$ is defined, it is unique.
> For the (n,m)-APA $\frac{1}{Q_\star} . P_\star$ we know that $P_\star$ and $Q_\star$ are abstract polynomials, respectively of degree at most n and at most m.

Proof : Evident.

From now on, when mentioning abstract Padé-approximants, we consider only the abstract Padé-approximants that are not undefined. Let (P,Q) be a solution of (1a) and (1b). Because of definition 4.2 it is very well possible that $(P_\star, Q_\star)$ itself does not satisfy definition 2.2.

Theorem 4.3. :

> Let $\frac{1}{Q_\star} . P_\star$ be the abstract Padé-approximant of order (n,m) for F.
> Then $\exists s : o \leq s \leq \min(n-n', m-m')$, $\exists$ an abstract polynomial
> $T(x) = \sum_{k=n.m}^{n.m+s} T_k x^k$, $T_{n.m+s} \neq 0$, $D(T) \neq \phi \ni (P_\star . T, Q_\star . T)$ satisfies definition 2.2 ; $\partial(P_\star . T) = n.m+n'+s$ and $\partial(Q_\star . T) = n.m+m'+s$.
> If then $T(x) = T_{n.m+r} x^{n.m+r} + T_{n.m+r+1} x^{n.m+r+1} + \ldots + T_{n.m+s} x^{n.m+s}$ with $D(T_{n.m+r}) \neq \phi$, also $(P_\star . T_{n.m+r}, Q_\star . T_{n.m+r})$ satisfies definition 2.2 and $o \leq r \leq s \leq \min(n-n', m-m')$.

Proof : Because of theorem 4.2 we may consider abstract polynomials P and Q that satisfy (1a) and (1b) and supply $P_\star$ and $Q_\star$. Because of definition 4.2, there exists an abstract polynomial T such that : $P = P_\star . T$ and $Q = Q_\star . T$ and $\partial T \geq n.m$. Because of lemma 4.2 $D(T) \neq \phi$ (otherwise $D(P) = \phi = D(Q)$).
Let $n' = \partial P_\star$, $m' = \partial Q_\star$, $P = \sum_{i=n.m}^{n.m+n} A_i x^i$, $Q = \sum_{j=n.m}^{n.m+m} B_j x^j$.

Consequently $T(x)=\sum_{k=n.m}^{n.m+s} T_k x^k$ with $\begin{cases} \partial T = n.m+s \\ n.m+n'+s \leq n.m+n \\ n.m+m'+s \leq n.m+m \\ s \geq 0 \end{cases}$

and so $0 \leq s \leq \min(n-n', m-m')$.

$F(x).Q(x)-P(x) = T(x).[F(x).Q_\star(x)-P_\star(x)] = O(x^{n.m+n+m+1})$

Because $T(x) = T_{n.m+r}\, x^{n.m+r} + \dots$ with $T_{n.m+r} \in L(X^{n.m+r}, Y)$, $D(T_{n.m+r}) \neq \phi$,

we have that $T_{n.m+r}\, x^{n.m+r}.[F(x).Q_\star(x)-P_\star(x)] = O(x^{n.m+n+m+1})$.

5. REMARKS AND SPECIAL CASES

a) When $X = R = Y$ ($\Lambda = R$), then the definition of abstract Padé-approximant is precisely the classical definition. F is now a real-valued function f of 1 real variable, with a Taylor-series development $\sum_{k=0}^{\infty} c_k.x^k$ with $c_k = \frac{1}{k!} f^{(k)}(o)$.

The k-linear operators $C_k \in L(X^k, Y)$ are :

$$C_k x^k = c_k.\underbrace{x\dots x}_{k} \in R \text{ with } c_k \in R \quad .$$

The j-linear functions $B_j x^j = b_j.\underbrace{x\dots x}_{j} \in R$, $b_j \in R$, $j = n,m,\dots,n.m+m$ and such that :

$$\begin{cases} c_{n+1}.b_{n.m} + \dots + c_{n+1-m}.b_{n.m+m} = 0 \\ \vdots \\ c_{n+m}.b_{n.m} + \dots + c_n.b_{n.m+m} = 0 \end{cases}$$

are a solution of (1b).

The i-linear functions $A_i x^i = a_i.\underbrace{x\dots x}_{i} \in R$, $a_i \in R$, $i = n.m,\dots,n.m+n$ such that :

$$\begin{cases} c_0 \cdot b_{n.m} = a_{n.m} \\ c_1 \cdot b_{n.m} + c_0 \cdot b_{n.m+1} = a_{n.m+1} \\ \vdots \\ c_n \cdot b_{n.m} + \dots + c_0 \cdot b_{n.m+n} = a_{n.m+n} \end{cases}$$

are a solution of (1a).

The irreducible form $\frac{1}{Q_\star} \cdot P_\star$ of $(\frac{1}{Q} \cdot P)(x) = \frac{1}{(\sum_{j=n.m}^{n.m+m} b_j x^j)} \cdot (\sum_{i=n.m}^{n.m+n} a_i x^i)$, such that $Q_\star(o) = 1$, is the irreducible form $\frac{1}{Q_\star} \cdot P_\star$ of $(\sum_{i=0}^{n} a_{i+n.m}\, x^i)/(\sum_{j=0}^{m} b_{j+n.m}\, x^j)$, such that $Q_\star(o) = 1$.

b) When we calculate the abstract Padé-approximant of order (n,o) we find the nth partial sum of the abstract Taylor series.
For if $B_{n.m} = I$ then $A_i x^i = C_i x^i$, $i = 0, \dots, n$ is a solution of system (1a).
This result has also been found in the classical theory.

c) To find equivalent formulations of the problem of Padé-approximating, we consider a couple of abstract polynomials (P,Q) satisfying definition 2.2. We then know that $(F.Q-P)(x) = 0(x^{n.m+n+m+1})$.
The systems (1a) and (1b) are completely equivalent with :

$$(F.Q-P)^{(i)}(0)\ x^i = 0 \quad \forall x \in X \text{ and } i = 0, \dots, n.m+n+m,$$

because clearly $(F.Q-P)^{(i)}(0) \equiv 0 \in L(X^i, Y)$ for $i = 0, \dots, n.m-1$ and $(F.Q-P)^{(i)}(0)\ x^i = 0$ $\forall x \in X$, $i = n.m, \dots, n.m+n$ is system (1a) and $(F.Q-P)^{(i)}(0)\ x^i = 0$ $\forall x \in X$, $i = n.m+n+1, \dots, n.m+n+m$ is precisely system (1b).

d) If $X = R^p$ and $Y = R$ ($\Lambda = R$), then F is a real-valued function of p real variables.
Now $L(X^i, Y)$ is isomorphic with R^{p^i}. Consequently for (P(x),Q(x)) satisfying

definition 2.2 the operator $(\frac{1}{Q}\cdot P)(x)$ has the following form :

$$\frac{\sum_{j_1+\dots+j_p=n.m}^{n.m+n} \alpha_{j_1\dots j_p} x_1^{j_1}\dots x_p^{j_p}}{\sum_{j_1+\dots+j_p=n.m}^{n.m+m} \beta_{j_1\dots j_p} x_1^{j_1}\dots x_p^{j_p}}$$

This form agrees with the form proposed by J. Karlsson and H. Wallin :

$$\frac{\sum_{j_1+\dots+j_p=o}^{n} \alpha_{j_1\dots j_p} x_1^{j_1}\dots x_p^{j_p}}{\sum_{j_1+\dots+j_p=o}^{m} \beta_{j_1\dots j_p} x_1^{j_1}\dots x_p^{j_p}}$$

if $n=o$ or $m=o$ (III).

Let $p=2$.

To calculate the abstract Padé-approximant we have to calculate the $(n.m+1)+\dots+(n.m+n+1)+(n.m+1)+\dots+(n.m+m+1)$ real coefficients $\alpha_{j_1\dots j_p}$ and $\beta_{j_1\dots j_p}$.

Now $(n.m+1)+\dots+(n.m+n+1)+(n.m+1)+\dots+(n.m+m+1) = n.m.(n+m+2) + \frac{(n+1)(n+2)}{2} + \frac{(m+1)(m+2)}{2}$.

The formulation in c) supplies us an amount of conditions on the derivatives of $(F.Q-P)$:

in all $\sum_{i=n.m}^{n.m+n+m} \binom{p+i-1}{i}$ conditions.

For $p=2$ these are $(n.m+1)+\dots+(n.m+n+m+1)$ conditions.

If we use the extra condition of definition 4.2, we have in all $n.m.(n+m+1) + \frac{(n+m+1)(n+m+2)}{2} + 1$ conditions, just enough to calculate the $\alpha_{j_1j_2}$ and $\beta_{j_1j_2}$.

The extra condition is : o-linear term in $Q_*(x)$ is I.

e) If $X = R^p$ and $Y = R^q$ ($\Lambda = R$), then F is a system of q real-valued functions in p real variables.

Now $L(X,Y)$ is isomorphic with $R^{q\times p}$ and $L(X^k,Y)$ isomorphic with $R^{q\times p^k}$ while an element of $R^{q\times p^k}$ is represented by a row of p^{k-1} matrices (blocks), each containing q rows and p columns;

$L = (c_{i_1 \dots i_{k+1}}) \in L(X^k,Y) \Rightarrow$ i_1 is the row-index in the block

$i_2 \dots i_k$ is the number of the block (the most right index grows the fastest)

i_{k+1} is the column-index in the block.

So $L = (c_{i_1 1 \dots 1 1 i_{k+1}} | c_{i_1 1 \dots 1 2 i_{k+1}} | \dots | c_{i_1 1 \dots 1 p i_{k+1}} | c_{i_1 1 \dots 1 2 1 i_{k+1}} | \dots | c_{i_1 p \dots p i_{k+1}})$

The abstract polynomials $(P(x),Q(x))$ satisfying definition 2.2 now have for each of the q components the form of the abstract polynomials of p real variables mentioned in d).

f) When we would try, in order to calculate the (n,m)-APA, to find a couple of abstract polynomials $(A_n x^n + \dots + A_0, B_m x^m + \dots + B_0)$ such that :

$$F(x).(B_m x^m + \dots + B_0) - (A_n x^n + \dots + A_0) = O(x^{n+m+1}) \tag{2}$$

instead of $(A_{n.m+n} x^{n.m+n} + \dots + A_{n.m} x^{n.m}, B_{n.m+m} x^{n.m+m} + \dots + B_{n.m} x^{n.m})$ such that:

$$F(x).(B_{n.m+m} x^{n.m+m} + \dots + B_{n.m} x^{n.m}) - (A_{n.m+n} x^{n.m+n} + \dots + A_{n.m} x^{n.m}) = O(x^{n.m+n+m+1}) \tag{3}$$

we would remark that this problem is not always solvable (except with $Q \equiv 0 \equiv P$).

Consider again the example $F\binom{\xi}{\eta} = \binom{\xi + \sin(\xi\eta) + 1}{\xi^2 + \eta^2 - 4\xi\eta} = \binom{1}{0} + \binom{\xi}{0} + \binom{\xi\eta}{\xi^2 + \eta^2 - 4\xi\eta} + \dots$

and take n=1 and m=2.

The system
$$\begin{cases} C_0 . B_0 = A_0 \\ C_1 x . B_0 + C_0 . B_1 x = A_1 x \\ C_2 x^2 . B_0 + C_1 x . B_1 x + C_0 . B_2 x^2 = 0 \\ C_3 x^3 . B_0 + C_2 x^2 . B_1 x + C_1 x_0 B_2 x^2 = 0 \end{cases}$$
$\forall x \in X$, has only the solution $Q \equiv 0 \equiv P$, and thus is not solvable such that $\frac{1}{Q} . P$ is somewhere defined (for $n=1$, $m=3$ this is the case for the first component of the solution),

while (3) is very well solvable, but the solution (P,Q) is such that the irreducible form of $(\frac{1}{Q}.P)(x)$ is undefined in $\binom{0}{0}$.

So via (3) we find an abstract rational operator $(\frac{1}{Q} . P)(x) = \begin{pmatrix} \frac{\xi - \eta + \xi^2 - 2\xi\eta}{\xi - \eta - \xi\eta + \xi\eta^2} \\ 0 \end{pmatrix}$ that is useful in points in the vicinity of $\binom{0}{0}$.

In other words : (2) does not provide us any solution at all (except $Q \equiv 0 \equiv P$)

(3) does provide an ARA but no APA .

What's more : the situation cannot occur where (2) supplies us the (n,m)-APA while (3) does nct, because for every solution (P,Q) of the systems resulting from (2) such that $Q_\star(0) = I$ and for every $L \in L(X^{n.m}, Y)$:

$$\begin{cases} (L.P, L.Q) \text{ is a solution of (1a) and (1b)} \\ \frac{1}{Q_\star} . P_\star \text{ is the (n,m)-APA} \end{cases}$$

And we have to look for an irreducible form anyhow.

6. COVARIANCE-PROPERTIES OF ABSTRACT PADE-APPROXIMANTS

The first property we are going to prove is the reciprocal covariance of abstract Padé-approximants.

Theorem 6.1. :

> Suppose $F(0)$ is regular in Y and F is continuous in 0 and $\frac{1}{Q}.P$ is the abstract Padé-approximant of order (n,m) for F, then $\frac{1}{P}.Q$ is the abstract Padé-approximant of order (m,n) for $\frac{1}{F}$.

Proof : Since $\{y\in Y|y \text{ is regular}\}$ is an open set in Y, there exists $B(F(0),r_2)$ with $r_2>o$ such that $\forall y\in B(F(0),r_2)$: y is regular. Since F is continuous in 0, there exists $B(0,r_1)$ with $r_1>o$ such that $\forall x\in B(0,r_1)$: $F(x)$ is regular. So $\frac{1}{F}$ is defined in $B(0,r_1)$. We speak about $\frac{1}{Q}.P$ and $\frac{1}{P}.Q$ too only on the set of points on which those operators are defined.

$P(0)=C_o=F(0)$ is regular $\Rightarrow \exists B(0,r):\forall x\in B(0,r):P(x)$ is regular.

So $\frac{1}{P}$ exists in $B(0,r)$.

Let $n'=\partial P$ and $m'=\partial Q$.

$\exists s\in N,\ o\leq s\leq \min(n-n',m-m'), \exists$polynomial $T(x)=\sum_{k=n.m}^{n.m+s} T_kx^k,\ D(T)\neq\phi \ni$

$(P_1(x)=P(x).T(x),\ Q_1(x)=Q(x).T(x))$ satisfies definition 2.2 for F.

$[(F.Q-P).T](x)=(F.Q_1-P_1)(x)=0(x^{n.m+n+m+1})$

$\Rightarrow (\frac{1}{F}.P_1-Q_1)(x)=0(x^{n.m+n+m+1})$ since $\frac{1}{F}(0)=C_o^{-1}\neq 0$ in the abstract Taylor series for $\frac{1}{F}$.

So $\exists s\in N,\ o\leq s\leq\min(n-n',m-m'),\ \exists$ polynomial $T(x)=\sum_{k=n.m}^{n.m+s} T_kx^k,\ D(T)\neq\phi\ni$

$(Q_1(x)=Q(x).T(x),\ P_1(x)=P(x).T(x))$ satisfies definition 2.2 for $\frac{1}{F}$.

The irreducible form of $\frac{1}{P_1}.Q_1$ is $\frac{1}{P}.Q$ $(D(P_1)\neq\phi$ or $D(Q_1)\neq\phi)$.

If we want the o-linear term in the denominator to be I, then

$\frac{1}{(P(x).C_o^{-1})}.(Q(x).C_o^{-1})$ is the abstract Padé-approximant of order (m,n) for $\frac{1}{F}$.

Theorem 6.2. : Suppose $a,b,c,d \in Y$, $c.F(0)+d$ is regular in Y, $a.d-b.c$ is regular in Y, $\frac{1}{Q}.P$ is the (n,n)-APA for F and $D(c.P+d.Q) \neq \phi$ or $D(a.P+b.Q) \neq \phi$, then $\frac{1}{(c.\frac{1}{Q}.P+d)}.(a.\frac{1}{Q}.P+b)$ is the (n,n)-APA for $\frac{1}{(c.F+d)}.(a.F+b)$.

Proof : $c.F(0)+d$ is regular $\Rightarrow c.(\frac{1}{Q}.P)(0)+d$ is regular since $F(0) = (\frac{1}{Q}.P)(0)$.

So $\exists B(0,r)$: $\frac{1}{Q}$ is defined in $B(0,r)$

$$\frac{1}{(c.\frac{1}{Q}.P+d)}.(a.\frac{1}{Q}.P+b) \text{ is defined in } B(0,r)$$

$$\frac{1}{(c.F+d)}.(a.F+b) \text{ is defined in } B(0,r).$$

Let $n' = \partial P$ and $n'' = \partial Q$.

$\exists s \in N : o \leq s \leq \min(n-n',n-n'')$, $\exists$ polynomial $T(x) = \sum_{k=n^2}^{n^2+s} T_k x^k$, $D(T) \neq \phi \ni$

$(P_1(x) = P(x).T(x),\ Q_1(x) = Q(x).T(x))$ satisfies definition 2.2 for F.

In other words : $[(F.Q-P).T](x) = (F.Q_1-P_1)(x) = O(x^{n^2+2n+1})$.

Now where $\frac{1}{(c.\frac{1}{Q}.P+d)}.(a.\frac{1}{Q}.P+b)$ is defined :

$$\frac{1}{(c.\frac{1}{Q}.P+d)}.(a.\frac{1}{Q}.P+b) = \frac{1}{\frac{1}{Q}.(c.P+d.Q)}.(a.P+b.Q).\frac{1}{Q} = \frac{1}{c.P+d.Q}.(a.P+b.Q).$$

Also $(c.P+d.Q)(0) = c.F(0)+d$ is regular in $B(0,r)$.

$$\begin{cases} \partial(a.P+b.Q) \leq \max(\partial P,\partial Q) \text{ and } \partial[(a.P+b.Q).T] \leq n^2+n \\ \partial(c.P+d.Q) \leq \max(\partial P,\partial Q) \text{ and } \partial[(c.P+d.Q).T] \leq n^2+n \end{cases}$$

Since $(F.Q_1-P_1)(x) = O(x^{n^2+2n+1})$ and $c.F(0)+d$ is regular,

$$[(a.d-b.c).\frac{1}{c.F+d}.(F.Q_1-P_1)](x) = O(x^{n^2+2n+1}).$$

Now $\frac{1}{(c.F+d)} \cdot (a.F+b).(c.P+d.Q).T - (a.P+b.Q).T =$

$\frac{1}{(c.F+d)} \cdot T.(F.Q-P).(a.d-b.c) = (a.d-b.c).\frac{1}{c.F+d} \cdot (F.Q_1-P_1)$

and $[(a.d-b.c)\frac{1}{c.F+d} \cdot (F.Q_1-P_1)](x) = O(x^{n^2+2n+1})$.

We now search the irreducible form of $\frac{1}{(c.P+d.Q).T} \cdot (a.P+b.Q).T$.

It is $\frac{1}{c.P+d.Q} \cdot (a.P+b.Q)$, for if $\frac{1}{c.P+d.Q} \cdot (a.P+b.Q)$ were reducible :

$$\begin{cases} a.P+b.Q = U.V & \text{with } U,V,W \text{ abstract polynomials} \\ c.P+d.Q = U.W & \text{and } \partial U \geqslant 1 \end{cases}$$

then : $\begin{cases} (a.d-b.c).P = d.U.V-b.U.W \\ (a.d-b.c).Q = a.U.W-c.U.V \end{cases}$

and so $\frac{1}{Q} \cdot P$ were reducible.

If we want the o-linear term in the denominator to be I,

$\frac{1}{(c.P+d.Q).e} \cdot (a.P+b.Q).e$, with $e = (c.P(0)+d.Q(0))^{-1} = (c.C_0+d)^{-1}$, is the (n,n)-APA for $\frac{1}{(c.F+d)} \cdot (a.F+b)$.

We have to remark that if $\frac{1}{Q} \cdot P$ were the (n,m)-APA for F with $n>m$ for instance, then a.P+b.Q was indeed an abstract polynomial of degree n but c.P+d.Q not necessarily an abstract polynomial of degree m. This clarifies the condition in theorem 6.2 that $\frac{1}{Q} \cdot P$ is the (n,n)-APA for F.

Another property we can prove is the scale-covariance of abstract Padé-approximants.

Theorem 6.3. :

> Let $\lambda \in \Lambda$, $\lambda \neq 0$, $y = \lambda x$ and $\frac{1}{Q} . P$ be the (n,m)-APA for F.
> If $S(x) := Q(\lambda x)$, $R(x) := P(\lambda x)$, $G(x) := F(\lambda x)$, then
> $\frac{1}{S} . R$ is the (n,m)-APA for G.

Proof : We remark that if $L \in L(X^i, Y)$, then $\forall \mu \in \Lambda : \mu L \in L(X^i, Y)$.

Because $\frac{1}{Q} . P$ is the (n,m)-APA for F, $\exists s$, $o \leq s \leq \min(n-n', m-m')$,

$$\exists \text{ polynomial } T(x) = \sum_{k=n.m}^{n.m+s} T_k x^k, D(T) \neq \phi \ni [(F.Q-P).T](x) = O(x^{n.m+n+m+1}).$$

Thus $[(F.Q-P).T]\ (\lambda x) = O(x^{n.m+n+m+1})$.

Now $[(F.Q-P).T]\ (\lambda x) = (G(x).S(x)-R(x)).U(x)$ with $U(x) := T(\lambda x)$ and so $[(G.S-R).U]\ (x) = O(x^{n.m+n+m+1})$.

We can prove that

$$\begin{cases} D(P) = \lambda . D(R) = \{\lambda x \mid R(x) \text{ regular in } Y\} \\ D(Q) = \lambda . D(S) \\ D(T) = \lambda . D(U) \end{cases}$$

So $D(S.U) \neq \phi$ or $D(R.U) \neq \phi$.

The irreducible form of $\frac{1}{S.U} . (R.U)$ is $\frac{1}{S} . R$ and $S(0) = Q(0) = I$, what finally proves the theorem.

7. THE ABSTRACT PADE-TABLE

Let $R_{n,m}$ denote the (n,m)-APA for F if it is not undefined. The $R_{n,m}$ can be ordered for different values of n and m in a table :

$$\begin{matrix} R_{0,0} & R_{0,1} & R_{0,2} & \cdots \\ R_{1,0} & R_{1,1} & R_{1,2} & \cdots \\ R_{2,0} & R_{2,1} & R_{2,2} & \cdots \\ R_{3,0} & \vdots & & \\ \vdots & & & \end{matrix}$$

Gaps can occur in this Padé-table because of undefined elements. An important property of the table is the next one : the abstract Padé-table consists of squares of equal elements (if one element of the square is defined, all the elements are).

We explicitly restrict ourselves now to spaces $X \underset{\neq}{\supset} \{o\}$ (and $Y \supseteq \{0, I\}$ of course). Thus $\exists x \in X : x \neq 0$ and $\forall \lambda \in \Lambda : \lambda . I \in Y$.

Lemma 7.1 :

$$\forall n \in \mathbb{N}, \ \exists D_n \in L(X^n, Y), \ \exists (x_1, \dots, x_n) \in X^n : \quad D_n x_1 \, x_2 \dots x_n = I$$

Proof : The reader must be familiar with the well-known functional analysis theorem of Hahn-Banach (Rudin W., Functional Analysis, Mc Graw-Hill, New York, 1973, pp. 57).

Let $n = 1$.

Take $x_o \in X$, $x_o \neq 0$ and define the linear functional (V pp.34)

$f : M = \{\lambda \, x_o | \lambda \in \Lambda\} \to \Lambda : \lambda x_o \to \lambda$.

Now $|f(\lambda \, x_o)| = |\lambda| = \dfrac{\|\lambda \, x_o\|}{\|x_o\|}$.

Define the norm $p(x) = \dfrac{\|x\|}{\|x_o\|}$ on X. Thus $|f(x)| \leq p(x) \ \forall x \in M$.

This linear functional f can be extended to a linear functional $\tilde{f} : X \to \Lambda$ such that $\tilde{f}(x) = f(x) \ \forall x \in M$ and $|\tilde{f}(x)| \leq p(x) \forall x \in X$.

We now define $D_1 : X \to Y : x \to \tilde{f}(x).I$.

Clearly $D_1 \in L(X,Y)$ and $D_1 \, x_o = I$ since $\tilde{f}(x_o) = f(x_o) = 1$.

If $D_{n-1} \in L(X^{n-1}, Y), (x_1, \dots, x_{n-1}) \in X^{n-1} \succ D_{n-1} \, x_1 \dots x_{n-1} = I$,

then we can define for $x \in X : D_n x = \tilde{f}(x).D_{n-1} \in L(X^{n-1}, Y)$.

Then $D_n \in L(X^n, Y)$ and $D_n \, x_o x_1 \dots x_{n-1} = \tilde{f}(x_o).D_{n-1} \, x_1 \dots x_{n-1} = I$.

This lemma implies that $\forall n \in \mathbb{N}, \exists D_n \in L(X^n, Y) : D(D_n) \neq \phi$. We shall use this result in the proofs of the following theorems.

Theorem 7.1. : Let $\frac{1}{Q} . P = R_{n,m}$ be the abstract Padé-approximant of order (n,m) for F.

Then : a) $(F.Q-P)(x) = \sum_{i=o}^{\infty} J_i x^{n'+m'+t+i+1}$

with $J_i \in L(X^{n'+m'+t+i+1},Y)$,

$t \geq o$ and $J_o \neq 0$

b) $\begin{cases} n \leq n'+t \\ m \leq m'+t \end{cases}$

c) $\forall k,l \supset \begin{cases} n' \leq k \leq n'+t : R_{k,l} = R_{n,m} \\ m' \leq l \leq m'+t \end{cases}$

Proof : a) Suppose $(F.Q-P)(x) = O(x^j)$ with $j < n'+m'+1$.

Then $\forall r \supset o \leq r \leq \min(n-n',m-m') : j+r+n.m < n.m+n+m+1$

This is in contradiction with theorem 4.3.

b) Suppose $n > n'+t$ or $m > m'+t$.

Then $\forall r \in \mathbb{N}$, $o \leq r \leq \min(n-n',m-m')$, $\forall T_{n.m+r} \in L(X^{n.m+r},Y)$, $D(T_{n.m+r}) \neq \phi$, we know that $(F.Q.T_{n.m+r} - P.T_{n.m+r})(x)$ is not $O(x^{n.m+n+m+1})$ since $(F.Q-P)(x) = \sum_{i=o}^{\infty} J_i x^{n'+m'+t+i+1}$ with $J_o \neq 0$ and $n.m+n'+m'+t+r+1 < n.m+n+m+1$.

This is in contradiction with theorem 4.3.

c) Let $\begin{cases} s = \min(k-n',l-m'), \ D_s \in L(X^{k.l+s},Y), \ D(D_s) \neq \phi \\ P_1 = P.D_s \qquad \partial P_1 \leq k.l+k \\ Q_1 = Q.D_s \qquad \partial Q_1 \leq k.l+l \end{cases}$

$(F.Q_1-P_1)(x) = O(x^{n'+m'+1+t+s+k.l})$ because of a).

Now for $k \leq n'+t$ and $l \leq m'+t$: $k.l+k+l+1 \leq k.l+n'+m'+t+s+1$.

So $(F.Q_1-P_1)(x) = O(x^{k.l+k+l+1})$ and $D(P_1) \neq \phi$ or $D(Q_1) \neq \phi$.

Definition 7.1. : The (n,m)-APA for F is called <u>normal</u> if it occurs only once in the abstract Padé-table.

The abstract Padé-table is called normal if each of its elements is normal.

Theorem 7.2. : The (n,m)-APA $R_{n,m} = \frac{1}{Q} . P$ for F is normal if and only if :

a) $\partial P = n$ and $\partial Q = m$

and

b) $(F.Q-P)(x) = \sum_{i=0}^{\infty} J_i x^{n+m+1+i}$

with $J_i \in L(X^{n+m+1+i}, Y)$ and $J_0 \not\equiv 0$

Proof : ⇒

We proof it by contraposition.

Let $n' = \partial P < n$ or $m' = \partial Q < m$.

According to theorem 7.1 a): $(F.Q-P)(x) = O(x^{n'+m'+1})$ at least.

Then $R_{n',m'} = \frac{1}{Q} . P$ (irreducible and satisfying $Q(0)=I$) since for $D \in L(X^{n'.m'}, Y)$, $D(D) \neq \phi$:

$[(F.Q-P).D](x) = O(x^{n'.m'+n'+m'+1})$ and $\partial(P.D) = n'.m'+n'$ and $\partial(Q.D) = n'.m'+m'$.

$R_{n',m'} = R_{n,m}$ contradicts the normality of $R_{n,m}$.

If b) is not valid, then according to theorem 7.1 a) :

$(F.Q-P)(x) = O(x^{n+m+1+t})$ with $t > o$ (for $t = o$ b) would be valid).

This implies that $\forall k,l : n \leqslant k \leqslant n+t$ and $m \leqslant l \leqslant m+t$:

$R_{k,l} = R_{n,m}$ and thus contradicts the normality of $R_{n,m}$.

⇐

The proof goes again by contraposition.

Suppose $R_{k,l} = R_{n,m}$ for k,l such that $k>n$ or $l>m$.

Now b) implies : $(F.Q-P)(x) = O(x^{n+m+1})$.

If $s \in N$, $D_s \in L(X^{k.l+s},Y)$, $D(D_s) \neq \phi \supset (F.Q.D_s - P.D_s)(x) = O(x^{k.l+k+l+1})$,

then $k.l+k+l+1 \leqslant n+m+1+k.l+s$ and thus $s > k-n$ or $s > l-m$.

This is a contradiction with theorem 4.3.

8. INTERPOLATING OPERATORS

Theorem 7.1 a) and 7.2 b) allow us to write down the following conclusions.

If $\frac{1}{Q}.P$ is the (n,m)-APA for F then $(F.Q-P)(x) = O(x^{n'+m'+1+t})$ with $t \geqslant o$.

This implies $(F - \frac{1}{Q}.P)(x) = O(x^{n'+m'+1+t})$ with $t \geqslant o$, since $Q(0) = I$ is regular.

In other words : $(F - \frac{1}{Q}.P)^{(i)}(0) \equiv 0 \in L(X^i,Y)$ for $i = o,\dots,n'+m'+t$.

Thus : for $R_{n,m} = \frac{1}{Q}.P : F^{(i)}(0) = (\frac{1}{Q}.P)^{(i)}(0)$ $i = o,\dots,n'+m'+t$ with $t \geqslant o$.

What is more, if $R_{n,m}$ is normal then $n' = n$, $m' = m$ and $(F.Q-P)(x) = O(x^{n+m+1})$.

Thus : for $R_{n,m} = \frac{1}{Q}.P$ normal : $F^{(i)}(0) = (\frac{1}{Q}.P)^{(i)}(0)$ $i = o,\dots,n+m$.

This also agrees with the classical theory of Padé-approximants.

Acknowledgements

I hereby want to thank Prof. Dr. L. Wuytack who was helpful with his comments, and other future readers whose remarks will be gratefully accepted.

References

(I) de BRUIN, M.G. and van ROSSUM, H.
Formal Padé-approximation. Nieuw Archief voor Wiskunde (3), 23, 1975, pp. 115-130.

(II) GODEMENT, R.
Algebra. Kershaw Publ. Co. Ltd., London, 1969.

(III) KARLSSON, J. and WALLIN, H.
Rational approximation by an interpolation procedure in several variables. In : Saff, E.B. and Varga, R.S. Padé and rational approximation : theory and appl. Academic Press, London, 1977, pp. 83-100.

(IV) LARSEN, R.
Banach-Algebras, an introduction. Marcel Dekker, Inc. New York, 1973.

(V) RALL, L.B.
Computational Solution of Nonlinear Operator Equations. J. Wiley and Sons. New York, 1969.

(VI) BESSIS, J.D. and TALMAN, J.D.
Variational approach to the theory of operator Padé approximants. Rocky Mountain Journ. Math. 4(2), 1974, pp. 151-158.

(VII) CHENEY, E.W.
Introduction to Approximation Theory chapter 5 section 6, McGraw-Hill, New York, 1966.

(VIII) CHISHOLM, J.S.R.
N-variable rational approximants. In : Saff, E.B. and Varga, R.S. Padé and rational approximation : theory and appl. Academic Press, London, 1977, pp. 23-42.

(IX) COMMON, A.K. and GRAVES-MORRIS, P.R.
Some properties of Chisholm Approximants. Journ. Inst. Math. Applics 13, 1974, pp. 229-232.

(X) GAMMEL, J.L.
Review of two recent generalizations of the Padé approximant. In : Graves-Morris, P.R. Padé-approximations and their appl. Academic Press, London, 1973, pp. 3-9.

(XI) GRAVES-MORRIS, P.R. and HUGHES JONES, R. and MAKINSON, G.J.
The calculation of some rational approximants in two variables. Journ. Inst. Math. Applics 13, 1974, pp. 311-320.

(XII) HUGHES JONES, R.
General Rational Approximants in N-Variables. Journal of approximation Theory 16, 1976, pp. 201-233.

(XIII) KARLSSON, J. and WALLIN, H.
Rational approximation by an interpolation procedure in several variables. In : Saff, E.B. and Varga, R.S. Padé and rational approximation : theory and appl. Academic Press, London, 1977, pp. 83-100.

(XIV) LUTTERODT, C.H.
Rational approximants to holomorphic functions in n-dimensions. Journ. Math. Anal. Applic. 53, 1976, pp. 89-98.

(XV) SHAFER, R.E.
On quadratic approximation. SIAM Journ. Num. Anal. 11(2), 1974, pp.447-460.

APPROXIMATION DE PADÉ-HERMITE

J. DELLA DORA
C. DI-CRESCENZO

Laboratoire d'Informatique et de Mathématiques Appliquées de Grenoble
BP 53 - 38041 GRENOBLE CEDEX
FRANCE

On présente dans ce travail les premiers résultats d'une généralisation des approximants de Padé suivant une idée d'Hermite. Le cadre choisi est assez large pour englober les approximants de Padé, de Shafer, et de D-log de Backer.

Après avoir défini ces approximants on indiquera la méthode pour les construire puis on signalera quelques applications et résultats numériques.

Première Partie

1. Définition des approximants P.H.

Dans ce qui suit k désignera un corps commutatif ($\mathbb{R}$ ou $\mathbb{C}$ en général).

Définition : Soit $(f_1, \ldots, f_m)$ un m-uple de séries formelles appartenant à $k[[X]]$

Soit $\bar{n} = (n_1, \ldots, n_m)$ un élément de $\mathbb{N}^m$, dont nous noterons $|\bar{n}| = \sum_1^m n_i$ la longueur.

On appellera forme de Padé-Hermite de type $\bar{n}$ (associée au m-uple (f_i)), tout m-uple de polynômes

$(Y_1^{\bar{n}}, \ldots, Y_m^{\bar{n}})$ vérifiant :

1°) $\partial^0 Y_i^{\bar{n}} \leq n_i$

2°) $\sum_{i=1}^{m} Y_i^{\bar{n}}(X) f_i(X) = X^{|\bar{n}|+m-1} \Gamma_{\bar{n}}(X)$

où $\Gamma_{\bar{n}}$ appartient à $k[[X]]$

Le problème de l'existence d'une telle forme PH, $\bar{n}$ se résoud comme suit : on introduit la matrice d'ordre $(|\bar{n}|+m-1) \times (|\bar{n}|+m)$ $\Delta(\bar{n})$ associée aux séries $f_i(X) = \sum_{j=0}^{+\infty} f_j^i X^j \quad (i = 1, \ldots, m)$

$$\Delta(\bar{n}) = \left| \begin{array}{ccc|c|ccc} f_0^1 & & & & f_0^m & & \\ f_1^1 & \ddots & & & f_1^m & \ddots & \\ f_2^1 & & f_0^1 & \cdots & & & f_0^m \\ \vdots & & \vdots & & \vdots & & \vdots \\ f^1_{|\bar{n}|+m-2} & & f^1_{|\bar{n}|+m-n_1-2} & & f^m_{|\bar{n}|+m-2} & & f^m_{|\bar{n}|+m-n_m-2} \end{array} \right|$$

Puis on ajoute une dernière ligne à ce tableau :

$$(f_1, Xf_1, \ldots, X^{n_1} f_1 \; ; \; f_2, Xf_2, \ldots, X^{n_2} f_2 \; ; \; \ldots \; ; \; f_m, Xf_m, \ldots, X^{n_m} f_m)$$

On obtient alors une matrice carrée $\Delta(n)$ dont on peut considérer le déterminant développé suivant la dernière ligne : on obtient d'une part une somme :

$$\sum_{i=1}^{m} Z_i^{\bar{n}} f_i$$

d'autre part en développant les f_i dans la dernière ligne on obtient :

$$X^{|\bar{n}|+m-2} \left(\sum_{j=0}^{+\infty} \Delta_{\bar{n}}^{j-1} X^j \right).$$

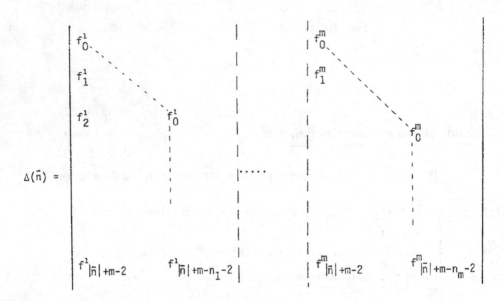

en notant $\Delta_{\bar{n}}^{j}$ le déterminant suivant :

$$
\Delta_{\bar{n}}^{j} = \left| \begin{array}{cc|c|ccc}
f_0^1 & & & f_0 & & \\
 & \ddots & - - - & & \ddots & \\
 & f_0^1 & & & & f_0 \\
 & \vdots & & & & \vdots \\
f^1_{|\bar{n}|+m-2} & f^1_{|\bar{n}|+m-n_1-2} & & f^m_{|\bar{n}|+m-j} & \ldots\ldots & f^m_{|\bar{n}|+m-n_m-2} \\
f^1_{|\bar{n}|+m-j} & f^1_{|\bar{n}|+m-n_1-j} & & f^m_{|\bar{n}|+m-j} & \ldots\ldots & f^m_{|\bar{n}|+m-n_m-j}
\end{array} \right|
$$

$(j = -1, 0, 1, \ldots)$

donc nous avons :

$$\sum_{i=1}^{m} Z_i^{\bar{n}}(X)\, f_i(X) = X^{|\bar{n}|+m-2} \left(\sum_{j=0}^{+\infty} \Delta_{\bar{n}}^{j-1} X^j \right)$$

qui montre :

1°) Qu'il y a toujours existence de la solution

2°) Nous appellerons cette solution forme HP normalisée.

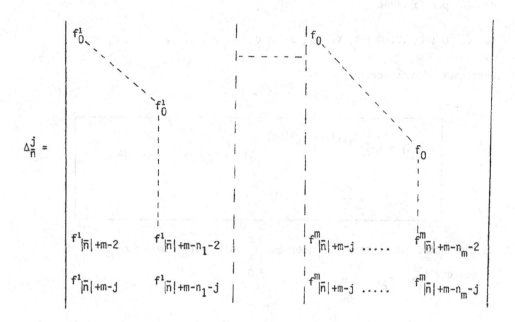

Remarquons enfin que l'on obtient $Z_i^{\bar{n}}$ (X) en remplaçant la dernière ligne de $\hat{\Delta}(\bar{n})$ par la ligne :

$$(0, \ldots 0 \; ; 0, \ldots 0 \; ; 1, X, \ldots, X^{n_i} \; ; 0, \ldots 0 \; ; \ldots \; ; 0, \ldots 0)$$

donc nous pouvons écrire :

$$Z_i^{\bar{n}}(X) = (-1)^{n_{i+1}+\ldots+n_m+(m-i)} \Delta^{-1}_{(n_1, n_2, \ldots, n_i-1, \ldots, n_m)} X^{n_i} + \ldots.$$

Ce qui montre que la condition pour que $Z_i^{\bar{n}}$ soit effectivement de degré n_i est que $\Delta^{-1}_{(n_1, n_2, \ldots, n_i-1, \ldots, n_m)} \neq 0$

Table Δ et Table HP d'ordre m

La table Δ d'ordre m est obtenue en portant en chaque point $\bar{n}$ de $\mathbb{N}^m$ la valeur du déterminant $\Delta_{\bar{n}}^{-1}$, c'est la généralisation naturelle de la C-table de Gragg [1].

La table HP d'ordre m sera la table des formes de Padé normalisées obtenue en portant en chaque point de $\mathbb{N}^m$ le vecteur $(Z_1^{\bar{n}}, \ldots, Z_m^{\bar{n}})$.

On conviendra de plus que : $\bar{n}$ va varier dans $(\mathbb{N} \cup \{-1\})^m$ de manière à considérer aussi les polynômes associés à des indices $\bar{n} = (n_1, \ldots, n_{i-1}, -1, n_{i+1}, \ldots, n_m)$ où $Z_i^{\bar{n}} \equiv 0$ par définition.

De la même façon $\Delta_{\bar{n}}^{-1}$ avec $\bar{n} = (n_1, \ldots, n_{i-1}, -1, n_{i+1}, \ldots, n_m)$

sera le déterminant d'ordre $n_1 + \ldots + n_{i-1} + n_{i+1} + \ldots + n_m + \underline{m - 1}$

obtenu en supprimant f_i.

Exemple : m = 3 si nous considérons les polynômes associés au 3-uple

$(f_1, f_2, 1)$ alors :

$\bar{n} = (-1, q, r) \Longrightarrow \quad Z_2^{-1,q,r} f_2 + Z_3^{-1,q,r} = X^{q+r+1} \Gamma_{-1,q,r}$.

le couple $(Z_2^{-1,q,r} , Z_3^{-1,q,r})$ est une forme de Padé-Frobenius (non normalisée) de f_2.

$\bar{n} = (p, -1, r)$ donnera une forme de P.F. de f_1

On conçoit facilement que la nullité de certains Δ^{-1} posent des problèmes de construction pénibles (inévitables dans les applications !) aussi nous allons nous placer pour l'exposé dans la situation idyllique suivante :

Définition :

Un m-uple $(f_1, \ldots, f_m)$ de m séries formelles sera dit parfait (pour la normalisation choisie) si :

1°) il existe un indice i tel que $f_i(0) \neq 0$

2°) $\forall\, \bar{n} \in (\mathbb{N} \cup \{-1\})^m \quad \Delta_{\bar{n}}^{-1} \neq 0$

2. Relations d'Hermite

Nous allons donner une relation due à Hermite [2] mais en utilisant un lemme qui sera important par ailleurs. Nous donnons d'autre part cette formule pour un triplé (f_1, f_2, f_3) de séries formelles de $k[[X]]$ renvoyant à [2] pour le cas général.

Nous supposerons enfin que (f_1, f_2, f_3) est un triplé parfait.

Soit $\bar{n} = (n_1, n_2, n_3)$ un multi-entier appartenant à $\mathbb{N}^3$, associons lui le tableau :

$$T(\bar{n}) : \begin{pmatrix} n_1 & n_2-1 & n_3-1 \\ n_1-1 & n_2 & n_3-1 \\ n_1-1 & n_2-1 & n_3' \end{pmatrix}$$

qui contient 3 multi-entiers de $(\mathbb{N} \cup \{-1\})^3$ que nous noterons :

$\bar{n}^{(1)} = (n_1, n_2-1, n_3-1)$; $\bar{n}^{(2)} = (n_1-1, n_2, n_3-1)$ et $\bar{n}^{(3)} = (n_1-1, n_2-1, n_3)$

A ce tableau on peut faire correspondre la matrice :

$$M(\bar{n}) = \begin{pmatrix} Z_1^{\bar{n}^{(1)}} & , Z_2^{\bar{n}^{(1)}} & , Z_3^{\bar{n}^{(1)}} \\ Z_1^{\bar{n}^{(2)}} & , Z_2^{\bar{n}^{(2)}} & , Z_3^{\bar{n}^{(2)}} \\ Z_1^{\bar{n}^{(3)}} & , Z_2^{\bar{n}^{(3)}} & , Z_3^{\bar{n}^{(3)}} \end{pmatrix}$$

par définition :

$$M(\bar{n}) \begin{pmatrix} f_1 \\ f_2 \\ f_3 \end{pmatrix} = X^{|\bar{n}|} \begin{pmatrix} \Gamma_{\bar{n}}(1) \\ \Gamma_{\bar{n}}(2) \\ \Gamma_{\bar{n}}(3) \end{pmatrix} \qquad (*)$$

On démontre alors le

lemme :

> Sous l'hypothèse que (f_1, f_2, f_3) est parfait
>
> $$\det(M(\bar{n})) = \varepsilon(\Delta_{\bar{n}}^{-1})^3 \; X^{|\bar{n}|}$$

(ε valant +1 ou -1 mais étant parfaitement déterminé).

d'où l'on déduit que sous l'hypothèse considérée $\det(M(\bar{n}))\,(0) = 0$ (c'est-à-dire que le polynôme $\det(M(\bar{n}))$ est sans terme constant.

3. Relations dans la table PH et dans la table Δ

Afin de construire récursivement la table PH et la table Δ il est nécessaire de généraliser les diverses identités de Frobenius utilisées dans les constructions classiques des tables de Padé.

Nous allons donner dans ce paragraphe un groupe de formules. pour un table à 3 dimensions renvoyant à [2] pour leurs généralisations à m quelconque et à [3] pour d'autres relations permettant une autre approche de la programmation des algorithmes.

L'idée est la suivante : on considère la relation (*) du § 2 et l'on cherche un vecteur $\tau(\bar{n}) \in k^3$ tel que :

$$\langle \tau(\bar{n}), \Delta^j(\bar{n}) \rangle = 0 \qquad j = -1, 0$$

$$\langle \tau(\bar{n}), \Delta^j(\bar{n}) \rangle = \Delta^{j-2}_{\bar{n}} \qquad j = 1, 2$$

où :

$$\Delta^j(\bar{n}) = \begin{pmatrix} \Delta^j_{\bar{n}}(1) \\ \vdots \\ \Delta^j_{\bar{n}}(2) \\ \\ \Delta^j_{\bar{n}}(3) \end{pmatrix} \qquad j \in \mathbb{N} \cup \{-1\}.$$

La solution de ce système d'une infinité d'équations à 3 inconnues existe dans le cas d'un triplé parfait, elle repose sur :

Identités de Frobenius généralisées : on a la relation suivante entre les éléments de la table Δ :

$$\Delta^j_{p,q,r}\,\Delta^{-1}_{p-1,q-1,r} = \Delta^{j+2}_{p-1,q-1,r}\,\Delta^{-1}_{p,q,r-1} - \Delta^{j+2}_{p-1,q,r-1}\,\Delta^{-1}_{p,q-1,r} + \Delta^{j+2}_{p,q-1,r-1}\,\Delta^{-1}_{p-1,q,r}$$

$\forall\, j \geq -1$ entier

(pour la démonstration voir [2] ou d'autres identités sont prouvées aussi).

Si nous prenons alors :

$$\tau_1(p,q,r) = \frac{\Delta^{-1}_{p-1,q,r}}{\Delta^{-1}_{p-1,q-1,r-1}}$$

$$\tau_2(p,q,r) = - \frac{\Delta^{-1}_{p,q,r}}{\Delta^{-1}_{p-1,q-1,r-1}}$$

$$\tau_3(p,q,r) = \frac{\Delta^{-1}_{p,q,r,-1}}{\Delta^{-1}_{p-1,q-1,r-1}}$$

nous résolvons le problème posé dans le cas d'un triplé parfait.

Remarque : Il est intéressant de considérer sur un dessin l'identité de Frobénius généralisée.

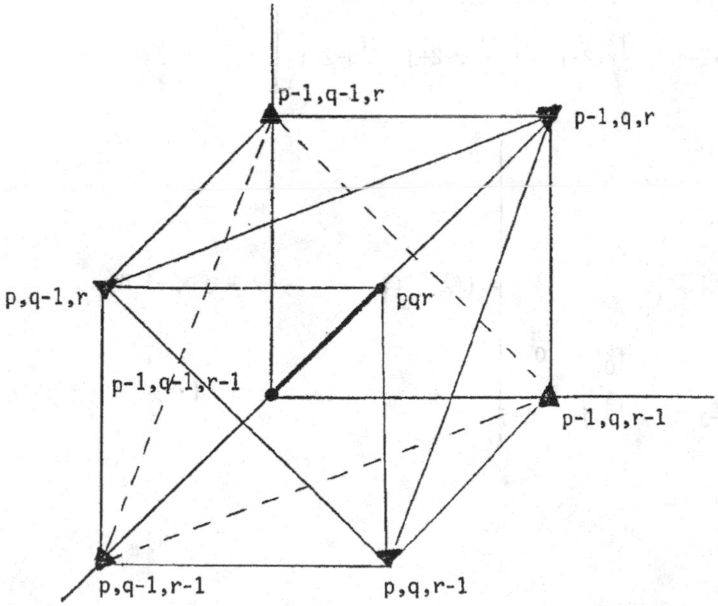

Deuxième Partie

Nous allons illustrer une façon très pratique de construire la Δ-table et la table PH dans le cas d'un triple parfait. Si le triple n'était pas parfait un algorithme de secours peut toujours être mis en place, il est cependant moins économe en taille mémoire.

1. Quelques notations

Nous avons déjà signalé que :

$$\Delta^{j}_{-1,q,r} = \det \left| \begin{array}{cccc:cccc} f_0^2 & & & & f_0^3 & & & \\ \vdots & \ddots & & & & \ddots & & \\ \vdots & & & f_0^2 & & & & f_0^3 \\ \vdots & & & \vdots & & & & \vdots \\ f_{q+r}^2 & & & f_r^2 & f_{q+r}^3 & & & f_q^3 \\ f_{q+r+2+j}^2 & & & f_{r+2+j}^2 & f_{q+r+2+j}^3 & & & f_{q+2+j}^3 \end{array} \right| \quad ; \ \forall\, j = -1,0,\ldots \ ;$$

de même on note

$$\Delta^{j}_{-1,-1,r} = \det \left| \begin{array}{ccc} f_0^3 & & \\ \vdots & \ddots & 0 \\ \vdots & & \\ f_{r-1}^3 & f_0^3 & 0 \\ f_{r+1-j}^3 & f_{2+j}^3 & f_{1+j} \end{array} \right| = (f_0^3)^r\, f_{1+j}^3 \qquad \forall\, j = -1,0,\ldots$$

(**)

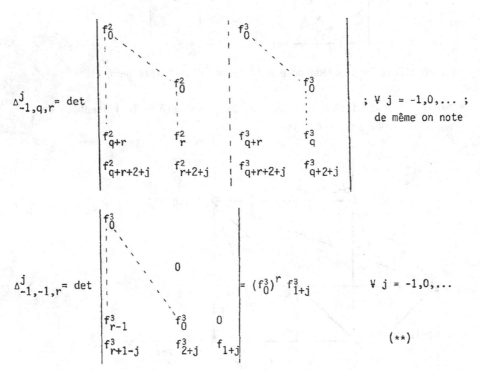

enfin nous conviendrons de noter $\Delta^{j}_{-1,-1,-1} = 1$. Cela nous permet de construire la Δ-table bordée grace à l'utilisation du :

<u>lemme</u>

si $\alpha \in \mathbb{N} \cup \{-1\}$ et si $p,q,r \in \mathbb{N} \cup \{-1\}$ (avec la convention que si p-1 apparaît dans la formule considérée alors p, devra être ≥ 0, de même pour q et r).

On a les 3 formules :

(1) $\Delta^{\alpha}_{pqr}\,\Delta^{-1}_{p,q-1,r-1} = \Delta^{\alpha+1}_{p,q-1,r}\,\Delta^{-1}_{p,q,r-1} - \Delta^{\alpha+1}_{p,q,r-1}\,\Delta^{-1}_{p,q-1,r}$

(2) $\Delta^{\alpha}_{pqr}\,\Delta^{-1}_{p-1,q-1,r} = \Delta^{\alpha+1}_{p-1,q,r}\,\Delta^{-1}_{p,q-1,r} - \Delta^{\alpha+1}_{p,q-1,r}\,\Delta^{-1}_{p-1,q,r}$

(3) $\Delta^{\alpha}_{pqr}\,\Delta^{-1}_{p-1,q,r-1} = \Delta^{\alpha+1}_{p-1,q,r}\,\Delta^{-1}_{p,q,r-1} - \Delta^{\alpha+1}_{p,q,r-1}\,\Delta^{-1}_{p-1,q,r}$

Lemme que l'on peut illustrer :

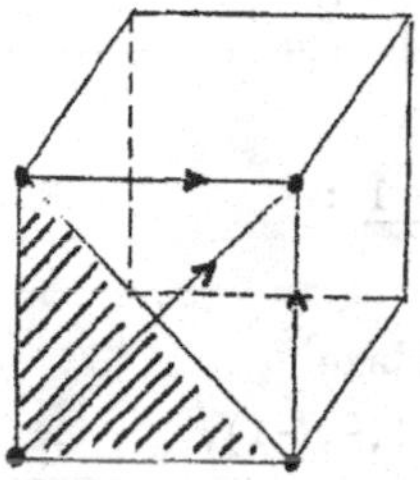

<== (1)

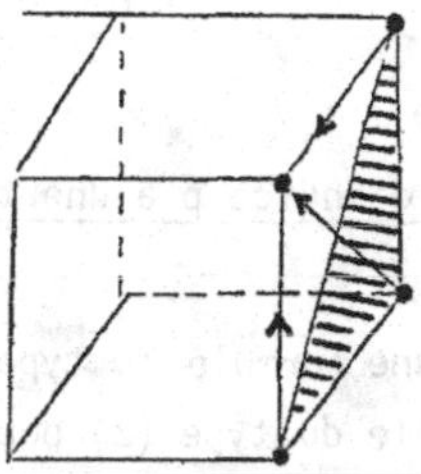

<== (3)

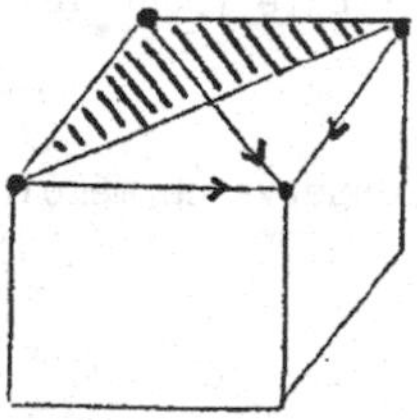

<== (2)

2. CONSTRUCTION

On se donne le même nombre de coefficients dans les 3 séries :

$$f_0^i, \ldots\ldots\ldots\ldots , f_N^i \qquad i = 1, 2, 3$$

Ce qui a pour conséquence de nous limiter au calcul des Δ_{pqr}^{-1} pour : $p + q + r \leq N - 2$ et de façon générale à celui des Δ_{pqr}^{α} pour $1 + q + r + \alpha \leq N - 3$. L'algorithme se conduit suivant le remplissage de plaques triangulaires indexées par l'indice p (par exemple) (on suivra sur les figures page suivante).

<u>On commence par remplir la "plaque" p = -1.</u>

On connaît $\Delta_{-1,-1,r}^{j}$, $\Delta_{-1,q,-1}^{j}$, $\Delta_{r,-1,-1}^{j}$ pour $j \leq N - 1$.

(formule **) ce qui nous permet de remplir les axes centrés en (-1,-1,r), puis l'utilisation de la formule (1) nous permet de remplir complètement cette tranche.

<u>On passe d'une tranche d'indice p à une tranche d'indice p + 1</u> :

En utilisant une formule du type (3) pour remplir l'axe (p+1,-1,-) et une formule du type (2) pour remplir l'axe (p+1,-, -1).

Ayant rempli les deux côtés du triangle on utilise de nouveau des formules du type (1) pour remplir complètement la plaque.

En fait la programmation est numériquement délicate (pour les détails supplémentaires voir [2]).

Il est important de se rappeler que l'on ne conserve en mémoire <u>que</u> les Δ^{-1}.

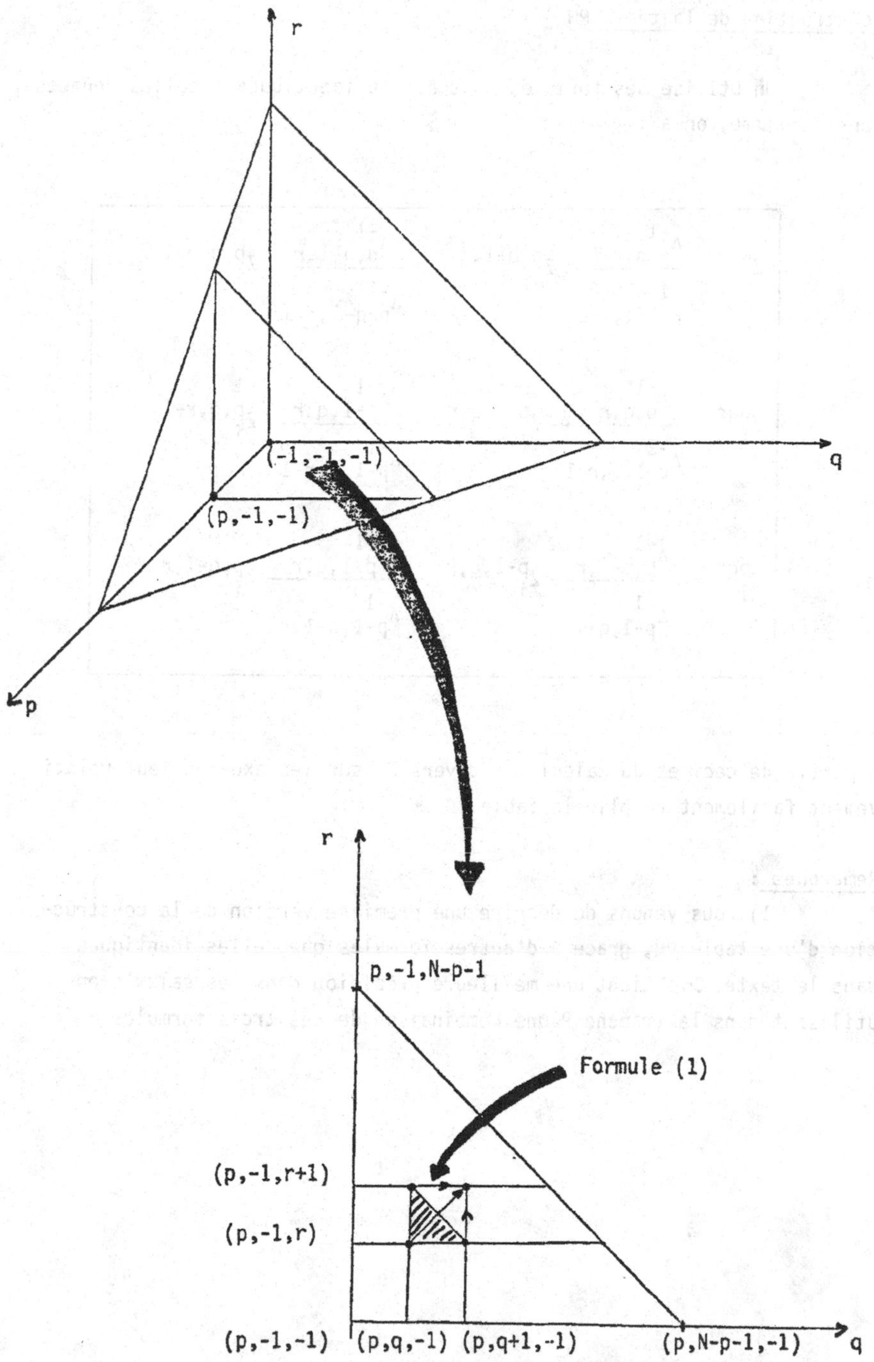
r
q
p
(-1,-1,-1)
(p,-1,-1)
r
p,-1,N-p-1
Formule (1)
(p,-1,r+1)
(p,-1,r)
(p,-1,-1)
(p,q,-1)
(p,q+1,-1)
(p,N-p-1,-1)
q

Construction de la table PH

On utilise des formules tout à fait identiques à celles données dans le lemme, on a :

$$Z_i^{pqr} = \frac{\Delta^{-1}_{p,q,r-1}}{\Delta^{-1}_{p,q-1,r-1}} Z_i^{p,q-1,r} - \frac{\Delta^{-1}_{p,q-1,r}}{\Delta^{-1}_{p,q-1,r-1}} Z_i^{p,q,r-1}$$

$$Z_i^{pqr} = \frac{\Delta^{-1}_{p,q,r-1}}{\Delta^{-1}_{p-1,q,r-1}} Z_i^{p-1,q,r} - \frac{\Delta^{-1}_{p-1,q,r}}{\Delta^{-1}_{p-1,q,r-1}} Z_i^{p,q,r-1}$$

$$Z_i^{pqr} = \frac{\Delta^{-1}_{p,q-1,r}}{\Delta^{-1}_{p-1,q-1,r}} Z_i^{p-1,q,r} - \frac{\Delta^{-1}_{p-1,q,r}}{\Delta^{-1}_{p-1,q-1,r}} Z_i^{p,q-1,r}$$

A partir de ceci et du calcul des divers Z_i sur les axes on peut relativement facilement remplir la table.

Remarques :

1) Nous venons de décrire une première version de la construction d'une table PH, grace à d'autres formules que celles identiques dans le texte. On obtient une meilleure précision dans les calculs en utilisant dans la tranche P une combinaison de ces trois formules :

$$\left|\begin{array}{l} \Delta^{\alpha}_{pqr} = a\,\Delta^{\alpha+1}_{p,q-1,r} + b\,\Delta^{\alpha+1}_{p,q,r-1} \\ \\ Z_i^{pqr} = a\,Z_i^{p,q-1,r} + b\,Z_i^{p,q,r-1} \end{array}\right.$$

avec :

$$\left|\begin{array}{l} 2.a = \dfrac{\Delta^{-1}_{p,q,r-1}}{\Delta^{-1}_{p,q-1,r-1}} + \dfrac{\dfrac{\Delta^{-1}_{p-1,q,r}}{\Delta^{-1}_{p,q-1,r}}}{\dfrac{\Delta^{-1}_{p-1,q,r-1}}{\Delta^{-1}_{p,q,r-1}} - \dfrac{\Delta^{-1}_{p-1,q-1,r}}{\Delta^{-1}_{p,q-1,r}}} \\ \\ 2.b = -\dfrac{\Delta^{-1}_{p,q-1,r}}{\Delta^{-1}_{p,q-1,r-1}} - \dfrac{\dfrac{\Delta^{-1}_{p-1,q,r}}{\Delta^{-1}_{p,q,r-1}}}{\dfrac{\Delta^{-1}_{p-1,q,r-1}}{\Delta^{-1}_{p,q,r-1}} - \dfrac{\Delta^{-1}_{p-1,q-1,r}}{\Delta^{-1}_{p,q-1,r}}} \end{array}\right.$$

2) D'autres algorithmes plus intéressants pour certaines applications sont possibles (Cf. 3).

TROISIÈME PARTIE

Quelques utilisations

Nous allons présenter les premières utilisations possibles des algorithmes PH ce sont les plus évidentes :

1. utilisation en accélération de la convergence
2. utilisation pour la localisation de singularités de fonctions analytiques au voisinage de l'origine.

1. ACCÉLÉRATION DE LA CONVERGENCE

La méthode est de considérer une suite $\{u_n\}$ de nombres complexes et d'essayer de la sommer exactement par le procédé d'Abel c'est-à-dire de former la fonction génératrice de cette suite :

$$\varphi(z) : \sum_{i=0}^{+\infty} u_n z^n$$

puis la fonction :

$f(z) = (1-z)\ \varphi(z)$ et l'on sait que la suite $\{u_n\}$ est sommable au sens d'Abel et a pour somme ℓ si :

$$\lim_{\substack{z \to 1^- \\ z \in S(\alpha)}} (1-z)\ \varphi(z) = \ell$$

($S(\alpha)$ étant le domaine de Stolz d'angle α

$$S(\alpha) : \mathcal{B}\, D(0,1) \cap \{z, \ \mathrm{Arg}(1.z) < \alpha\} \qquad 0 < \alpha < \frac{\pi}{2}$$

Si maintenant la fonction f considérée est une fonction algébrique vérifiant une équation du type

$$P_n(X)f^2 + Q_m(X)f + R_p(X) = 0$$

(P_n, Q_m, R_p étant 3 polynômes de degré respectif n, m, p) alors on voit que f(1) se calculera exactement (à condition de pouvoir choisir la bonne détermination).

L'idée est donc de construire la table HP3 correspondant aux fonctions :

$$f_1 = f^2 \ , \ f_2 = f, \ f_3 = 1$$

et de calculer pour un 3-indice (p,q,r)

$$Z_1^{p,q,r}(1)\, y^2 + Z_2^{p,q,r}(1)\, y + Z_3^{p,q,r}(1) = 0$$

on choisit ensuite la racine convenable.

Le problème est de savoir en fait choisir un cheminement dans la table des y ainsi obtenus, c'est le même problème que celui qui se pose dans la table de Padé, via l'ε-algorithme.

Pour le moment nous nous sommes contenté d'observer les 3 stratégies :

(p,p,p) (diagonale)
(p+1,p,p)
(p,p,p+1)

Les résultats obtenus sont toujours (sur les exemples considérés) au moins aussi bons que les résultats de l'ε-algorithme travaillant avec la même information.

Souvent on améliore de 2 chiffres significatifs les résultats obtenus. Il faut signaler que dans la table complète (mais par forcément sur les chemins choisis) on trouve souvent de bien meilleurs résultats mais si on ne connaît pas le résultat il semble difficile de les retenir.

Remarques : Comme cet exposé sommaire le montre on a encore un très gros travail pratique à réaliser pour que ces algorithmes aient l'efficacité de l'ε-algorithme.

On peut se poser, à ce propos, les questions de savoir quelles suites vont être exactement sommées par cet algorithme (on sait, par exemple, que l'ε-algorithme permet de prendre en compte des suites du type $S_n = \Sigma A_i \alpha_i^n$). Nous pouvons apporter certains éléments de réponse de la manière suivante : nous pensons en effet, que la "bonne" classe de fonctions à considérer pour ces problèmes est la classe des fonctions somme d'une fonction de la classe $\mathcal{S}$ de Nuttal et d'une fonction mé - morphe soit :

$$f(z) = \frac{P(z)}{\sqrt{\pi (z-\alpha_i)^{n_i}}} + \sum_j \frac{Q_i}{(z-\beta_i)^{m_j}} + \varphi$$

A partir de cette simple remarque on peut caractériser certaines suites qui sont exactement sommées. (Cf.3)

2. Localisation des singularités

On peut aborder le problème sous deux aspects :

1. un aspect algébrique
2. un aspect différentiel

Le premier aspect consiste à partir de la fonction $f(z) = \sum_{n=0}^{+\infty} a_n z^n$

connue partiellement au voisinage de l'origine à former la table HP basée sur :

$$f_1 = f^2 \; ; \; f_2 = f \; ; \; f_3 = 1.$$

En un point (p,q,r) de cette table on a donc une équation algébrique

$$Z_1^{(p,q,r)} y^2 + Z_2^{(p,q,r)} y + Z_3^{(p,q,r)} = 0 \qquad (1)$$

On peut alors rechercher les singularités des branches de fonctions algébriques solution de (1), on sait qu'elles sont situées en les zéros de $Z_1^{(p,q,r)}$ et en les zéros du discriminant $(Z_2^{p,q,r})^2 - 4Z_3^{(p,q,r)} Z_1^{(p,q,r)} = \delta(p,q,r)(z)$ et que leur nature est déterminée (de la façon la plus simple) en étudiant

$$\frac{-Z_2^{p,q,r} \pm \sqrt{\delta(p,q,r)}}{2\, Z_1^{p,q,r}}$$

On pourra donc avoir des singularités polaires ou des singularités de branchement (avec des exposants $\frac{1}{2}$ ou $-\frac{1}{2}$ seulement).

On peut alors considérer la table de ces singularités et suivre certaines stratégies comme pour l'accélération de la convergence.

Un deuxième aspect repose sur le désir de pouvoir mettre en évidence des singularités de branchement avec des exposants différents de $\frac{1}{2}$ ou $-\frac{1}{2}$ sans pour autant être obligé de monter en degré dans les approximants algébriques.

Cela nous conduit aux approximants différentiels basés sur les fonctions :

$$f_2 = f', \quad f_1 = f, \quad f_2 = 1$$

l'étude des singularités étant maintenant ramenée à l'étude des zéros de $Z_1^{p,q,r}$.

Des résultats prometteurs ont été obtenus dans ces deux voies.

3. Quelques résultats numériques.

$\frac{1}{1-3x} + \frac{1}{\sqrt{1-x/2}}$ points singuliers en utilisant f^2; f, 1 PADE-HERMITE

(p,p,p)	Racines de Z_1		Racines de Z_2		Racines du résultant	
1	0.33		0.53		0.50	
2	0.333 333 327	2.0136		1.72	{0.479 {0.48	2.014
3	{0.333 333 333 29 {0.333 36	1.999 999 8	0.333 36	2.000 002 7	{0.333 36 ± 0.003i {0.333 37 ± 0.003i	1.999 999 8
4	{0.333 333 332 {0.332 8	2.000 000 19	0.333 09	1.999 994	{0.3245 ; 0.3245 {0.3416 ; 0.3416	2.000 000 19
5	{0.333 333 333 34 {0.35	1.999 993	0.34	2.000 16	{0.3432 ± 0,05i {0,3432 ± 0,05i	1.999 993

$\frac{1}{1-x/2} + \frac{1}{\sqrt{1-x}} + e^x$

Points singuliers en utilisant f^2 , f, 1

	Racines de Z_1 (p+1, p, p)		Racines de Z_2		Racines du résultant	
1	1.73				-2.4	
2	1.619	1.4	1.52		1.87	0.22 ± 0.8i
3	2.89	0.82	0.23 ± 0.88i		1.51	0.95
4	2.268	0.931	0.75		1.37	0.97
5	2.229	1.002 6	2.500	1.006	1.688	1.003

Points singuliers en utilisant f', f, 1

	(p+1, p, p)			
1	2.057			
2	28.1	1.000 003	1.41	
3	1.81	1.001 6	2.08	1.38
4	2.67	0.997 1	2.67	1.10 ± 0.4i
5	2.18	1.000 2	2.29	1.36 ± 0.4i

$\frac{1}{1-x/2} + \frac{1}{\sqrt{1-x}} + e^x$

Points singuliers en utilisant f^2, f, 1

PADE

(p,p)			(p+1, p)			(p, p+1)		
1	1.77		1.	1.9		1	1.86	
2	1.79	-0.53	2	1.5	0.90 ± 1.5i	2	2.20	0.90
3	2.18 ± 1.3i	1.114	3	2.4	1.06	3	2.8	1.07
4	2.26	1.054	4	2.6	1.004	4	1.78	1.04
5	2.95	1.030	5	2.8	1.02	5	2.7	1.02
6	1.45	1.035	6	2.003 9	1.018	6	1.8	1.016
7	1.797	1.016	7	1.7	1.014	7	1.8	1.016
8	2.03	1.011	8	1.94	1.009	8	1.95	1.010

<u>Points singuliers en utilisant f', f, 1</u>

<u>idem</u>

$$\frac{1}{1-x/2} + \frac{1}{\sqrt{1-x}} + e^x$$

Points singuliers en utilisant f^2, f, 1

(p,p,p)	Racines de Z_1		Racines de Z_2		Racines du résultant	
1	1.796		5.89		0.52 ± 0.006i	
2	3.0	0.47	0.896 ± 1.48i		1.75	0.92
3	2.82	0.829	0.06		1.52	0.95
4	2.64	1.016	3.1	1.028	2.18	1.033
5	2.093	1.002 2	2.32	1.005 2	1.94 ± 0.6i	1.0025

Points singuliers en utilisant f', f, 1

(p,p,p)				
1		1.015		1.46
2		0.972		1.29
3	2.11	0.98	2.41	0.919 ± 0.18i
4	2.11	1.000 26	2.21	1.38 ± 0.4i
5	2.18	1.000 23	2.29	1.36 ± 0.4i

$$\frac{1}{1-3x} + \frac{1}{(1-x/2)^{1/3}}$$

Points singuliers en utilisant f^2, f, 1

PADE-HERMITE

(p,p,p)

	Racines de Z_1		Racines de Z_2		Racines du résultant	
1	0.332		0.52		0.60	
2	0.333 333 35	1.80	0.547 ± i		0.47 ± 0.0005i	1.91
3	0.333 333 333 334	1.96	0.63	1.66	0.62 ± 0.00005i	1.981
4	0.333 333 333 333 3	1.97	0.65	1.69	0.60 ± 0.000 004i	1.985
5	0.333 333 333 333 5	1.976		1.818	{0.389 {0.389	1.987

Points singuliers en utilisant f',f,1

1	0.333 01			1.52	0.08	3.3
2	0.333 333 339	1.95		1.801		1.955
3	{0.333 333 333 4 {0.335	2.000 01	0.338 ± 0.09i		0.334 ± 0.001i	2.18 ± 0.5i
4	{0.333 333 333 7 {0.333 7	2.000 01	0.3327 ± 0.04i		0.3335 ± 0.0002i	2.18 ± 0.5i
5	{0.333 333 333 24 {0.329 5	2.000 007	0.43		0.331 ± 0.002i	2.18 ± 0.5i

La série de Leibnitz

$$\pi = 4 - \frac{4}{3} + \frac{4}{5} - \frac{4}{7} + \ldots$$

Cette série est connue pour converger très lentement.

1/

	(p,p,p)		Padé (n,n)
0	4.	0	4.
1	3.141 5	1	3.1
2	3.141 592 6	2	3.14
3	3.141 592 653 58	3	3.141
4	3.141 592 653 58	4	3.141 59
		5	3.141 592 6
		6	3.141 592 65
		7	3.141 592 653

2/

	(p+1,n,p)		Padé (p+1,n)
0	3.14	0	3.
1	3.141 592	1	3.1
2	3.141 592 653 5	2	3.141
3	3.141 592 653 58	3	3.141 5
		4	3.141 592
		5	3.141 592 65
		6	3.141 592 653

3/

	(p,p+1,p)		Padé (p,p+1)
0	3.1	0	2.6
1	3.141 60	1	3.1
2	3.141 592 65	2	3.141
3	3.141 592 653 5	3	3.141 5
4	3.141 592 653 58	4	3.141 592
		5	3.141 592 65
		6	3.141 592 653

4/

	(p,p,p+1)
0	3.06
1	3.141
2	3.141 592 65
3	3.141 592 653 5
4	3.141 592 653 58

BIBLIOGRAPHIE

[1] W.B. GRAGG : The Padé Table and its relation to certain algorithms of numerical analysis.
Siam Review 14. 1-62 (1972).

[2] J. DELLA DORA et C. DI CRESCENZO :
Approximants de Padé-Hermite
Théorie - Programmation - Applications
A paraître.

[3] J. DELLA DORA, C. Di CRESCENZO :
Algorithmes de Padé-Hermite
Construction des tables PH
A paraître comme rapport de recherche du Laboratoire d'Analyse Numérique de Grenoble . [mai 1979]

ENSEMBLES DE SUITES ET DE PROCÉDÉS LIÉS POUR L'ACCÉLÉRATION DE LA CONVERGENCE

B. GERMAIN - BONNE

INTRODUCTION

Dans cet article, qui est une contribution à l'étude des procédés d'accélération de la convergence, nous nous posons la question suivante :

Etant donné une famille de procédés d'accélération de la convergence, existe-t-il un procédé capable d'accélérer toutes les suites susceptibles d'être accélérées ? (Existence d'un algorithme universel).

Pour certaines familles bien précises, la réponse est négative, ce qui permet d'envisager la question : comment caractériser la plus vaste famille de suites dont on peut accélérer la convergence au moyen de certaines familles de procédés.

Dans la première partie nous définissons les notions qui nous sont nécessaires ; dans la deuxième partie nous établissons la non-existence d'algorithme universel pour deux familles de suites.

I - ENSEMBLES LIÉS - DÉFINITIONS - PROPRIÉTÉS

Section 1 - Définitions

Soient $\mathcal{S}$ et $\mathcal{G}$ deux ensembles et $\mathcal{R}(s, g)$ une relation définie pour tout $s \in \mathcal{S}$ et tout $g \in \mathcal{G}$, que nous noterons :

$s \leftrightarrow g$ si la relation est vérifiée (s est lié à g)

$s \not\leftrightarrow g$ dans le cas contraire (s n'est pas lié à g)

* Définition 1 Ensembles liés

Soient $S \subset \mathcal{S}$ et $G \subset \mathcal{G}$ deux sous ensembles. Ils sont **liés** si :

$$\begin{cases} \forall s \in S & \\ & s \longleftrightarrow g \\ \forall g \in G & \end{cases}$$

* <u>Définition 2</u> <u>Ensembles maximaux</u>

Soit $s \in \mathcal{S}$; notons G_s le sous ensemble de $\mathcal{G}$ tel que :

$$\begin{cases} \forall g \in G_s & s \longleftrightarrow g \\ \forall g \notin G_s & s \not\longleftrightarrow g \end{cases}$$

Par définition, G_s est **maximal** pour s.

De même, soit $S \subset \mathcal{S}$; notons :

$$G_S = \bigcap_{s \in S} G_s$$

Par définition G_S est **maximal** pour S.

Cet ensemble est tel que :

$$\begin{cases} \forall g \in G_S & \forall s \in S & s \longleftrightarrow g \\ \forall g \notin G_S & \exists s \in S & s \not\longleftrightarrow g \end{cases}$$

D'une manière analogue, g et G étant respectivement un élément et une partie de $\mathcal{G}$, notons :

S_g et S_G les sous ensembles de $\mathcal{S}$ maximaux pour g et G.

$S_G = \bigcap_{g \in G} S_g$ vérifie :

$$\begin{cases} \forall s \in S_G & \forall g \in G & s \longleftrightarrow g \\ \forall s \notin S_G & \exists g \in G & s \not\longleftrightarrow g \end{cases}$$

* <u>Définition 3</u> <u>Ensembles totalement liés</u>

Soient $S \subset \mathcal{S}$ et $G \subset \mathcal{G}$; S et G sont **totalement liés** si

$S = S_G$ (S est maximal pour G)

$G = G_S$ (G est maximal pour S)

* <u>Définition 4</u> <u>Ensembles saturés</u>

Soit $S \subset \mathcal{S}$

$G \subset \mathcal{G}$ est **saturé** par S si :

(i) $\forall s \in S \qquad G = G_s$

(ii) $\forall s \notin S \qquad G \neq G_s$

<u>Remarque</u>

Soit la relation d'équivalence :

$$s \sim s' \quad \text{si} \quad G_s = G_{s'}$$

Lorsque G est saturé par S, S coïncide avec la classe d'équivalence de tout $s \in S$. (En fait, au sens Bourbaki c'est S qui est saturé par cette relation d'équivalence).

* <u>Définition 5</u> <u>Ensembles mutuellement saturés</u>

$S \subset \mathcal{S}$ et $G \subset \mathcal{G}$ sont mutuellement saturés si :

$\forall s \in S \qquad G = G_s$

$\forall g \in G \qquad S = S_g$

Deux ensembles mutuellement saturés S et G tels que chacun de ces ensembles soit réduit à un seul élément définissent un **couple élémentaire**.

<u>Section 2</u> - Propriétés

Plaçons nous dans l'hypothèse suivante :

$$\begin{cases} \forall s \in \mathcal{S} \quad \exists g \in \mathcal{G} \quad \text{tel que} \quad s \longleftrightarrow g \\ \\ \forall g \in \mathcal{G} \quad \exists s \in \mathcal{S} \quad \text{tel que} \quad s \longleftrightarrow g \end{cases}$$

S et G désigneront des sous ensembles de $\mathcal{S}$ et $\mathcal{G}$.

* <u>Propriété 1-1</u> : *Soient* G_1 *maximal pour* S_1 *et* G_2 *maximal pour* S_2. *Alors* $G = G_1 \cap G_2$ *est maximal pour* $S = S_1 \cup S_2$.

Démonstration :

G_1 est maximal pour S_1 : $G_1 = \bigcap_{s \in S_1} G_s$

G_2 est maximal pour S_2 : $G_2 = \bigcap_{s \in S_2} G_s$

D'où $G_1 \cap G_2 = \bigcap_{s \in S_1 \cup S_2} G_s$. $G_1 \cap G_2$ est donc maximal pour $S_1 \cup S_2$.

* Propriété 1-2 : *Soient* G_1 *maximal pour* S_1 *et* G_2 *maximal pour* S_2, *tels que* $S_1 \subset S_2$ *; on a alors* $G_2 \subset G_1$.

$$G_2 = \bigcap_{s \in S_2} G_s$$

Imposons à s de ne parcourir que S_1 (qui est inclus dans S_2); on a alors $G_2 \subset G_1$.

— · —

* **Construction d'ensembles totalement liés**

Soit $\hat{g} \in G$

Posons $S = S_{\hat{g}}$ (S est maximal pour $\hat{g}$)

$G = G_S$ (G est maximal pour S)

Montrons que les ensembles G et S ainsi construits sont totalement liés.

Par construction G est maximal pour S ; il suffit donc de montrer que S est maximal pour G.

Soit $\bar{S} = \bigcap_{g \in G} S_g$ ($\bar{S}$ est maximal pour G)

Comme $\hat{g} \in G$, la propriété 2 donne :

$\{\hat{g}\} \subset G \Rightarrow \bar{S} \subset S_{\hat{g}} = S$

L'inclusion $\bar{S} \subset S$ n'est pas stricte.

Supposons que ce soit le cas : il existe donc $s^* \in S$ tel que $s^* \notin \bar{S}$.

$\bar{S}$ est tel que :

$$\begin{cases} \text{(i)} & \forall s \in \bar{S} \quad \forall g \in G \quad s \leftrightarrow g \\ \text{(ii)} & \forall s \notin \bar{S} \quad \exists g \in G \quad s \not\leftrightarrow g \end{cases}$$

La propriété (ii) appliquée à s^* implique : $\exists\ g^* \in G$ tel que

$$s^* \not\leftrightarrow g^*$$

Ceci est en contradiction avec le fait que G est maximal pour S.

On a donc $\bar{S} = S$. S est maximal pour G.

Remarques

1) Cette construction fournit par la même occasion la possibilité de construire un ensemble saturé.

En reprenant les notations précédentes ($\hat{g} \in \mathcal{G}$, $S = S_{\hat{g}}$, $G = G_S$), notons $\hat{G}$ l'ensemble des $g \in G$ tels que :

$$S_g = S_{\hat{g}}$$

S est saturé par $\hat{G}$.

(On est en effet certain que $\hat{G} \subset G$, car $\forall\ g \notin G\ \exists\ s \in S$ tel que $s \not\leftrightarrow g$, ce qui revient à dire que si $g \notin G$ $S_g \neq S_{\hat{g}}$).

2) Etant donné deux ensembles $\mathcal{S}$ et $\mathcal{G}$, il n'existe pas forcément de sous ensembles S et G mutuellement saturés ; on ne peut donc fournir de construction d'ensembles mutuellement saturés.

* Propriété 1-3 : *Si* G *est saturé par* S, *il est maximal pour* S.

G est saturé par S, donc :

$\forall\ s \in S \quad G = G_S$ c'est à dire $\forall\ g \in G \quad s \leftrightarrow g$

$\qquad\qquad\qquad\qquad\qquad\qquad\qquad\quad \forall\ g \notin G \quad s \not\leftrightarrow g$

On a donc

(i) $\forall\ g \in G \quad \forall\ s \in S \quad s \leftrightarrow g$

$\qquad\qquad\Rightarrow$ G est maximal pour S

(ii) $\forall\ g \notin G \quad \forall\ s \in S \quad s \not\leftrightarrow g$

(la propriété (ii) implique $\forall\ g \notin G,\ \exists\ s \in S \quad s \not\leftrightarrow g$)

Corollaire Si S et G sont mutuellement saturés, ils sont totalement liés.

— • —

* Dans l'ensemble produit $S \times G$, soit $R(s, g)$ la relation initiale ("s est lié à g") et définissons la relation d'équivalence entre deux couples :

$$(s, g) \sim (s', g') \quad \text{si} \quad G_s = G_{s'} \quad \text{et} \quad S_g = S_{g'}.$$

* Propriété 1-4 : $R(s, g)$ *est* **compatible** *avec la relation d'équivalence, c'est à dire :* $s \leftrightarrow g$ et $(s', g') \sim (s, g) \Rightarrow s' \leftrightarrow g'$.

Démonstration :

Soit $s \leftrightarrow g$ et $(s', g') \sim (s, g)$

Supposons $s' \not\leftrightarrow g'$; donc $g' \notin G_{s'}$, c'est à dire $g' \notin G_s$ d'où $g' \not\leftrightarrow s$

Or $s \leftrightarrow g \Rightarrow s \in S_g = S_{g'} \Leftrightarrow s \leftrightarrow g'$.

D'où la contradiction. On a donc $s' \leftrightarrow g'$.

* Soit $\mathfrak{S} \times \mathfrak{G}$ l'ensemble quotient de $S \times G$ par la relation d'équivalence. Notons $[s]$ et $[g]$ les éléments de $\mathfrak{S}$ et $\mathfrak{G}$; (ce sont les classes d'équivalence de $s \in S$ et $g \in G$).

L'ensemble $S \times G$ étant muni de la relation $R(a, b)$, (notée $\leftrightarrow$), munissons $\mathfrak{S} \times \mathfrak{G}$ de la relation $\mathfrak{R}([a], [b])$ déduite de $R(a, b)$ par passage au quotient ; notons cette relation $\vdash\!\dashv$

$$[a] \vdash\!\dashv [b] \quad \Leftrightarrow \quad \begin{matrix} [a] \in \mathfrak{S} \\ [b] \in \mathfrak{G} \end{matrix} \quad \text{et} \quad \forall (s, g) \in ([a], [b]) \quad s \leftrightarrow g$$

* Propriété 1-5 : *Soit* $\mathrm{S} \subset S$ *saturé par* $\mathrm{G} \subset G$. *Notons* $[\mathrm{S}]$ *l'ensemble déduit de* S *par remplacement de chaque élément de* S *par sa classe d'équivalence (et de même pour* $[\mathrm{G}]$*). Dans* $\mathfrak{S} \times \mathfrak{G}$, $[\mathrm{S}]$ *est saturé par* $[\mathrm{G}]$ *et* $[\mathrm{G}]$ *est réduit à un seul élément.*

Démonstration :

S est saturé par G : $\forall g \in \mathrm{G} \quad \mathrm{S} = S_g$

Soit $\mathfrak{g}$ un élément particulier de G

$[\mathrm{G}] = [\mathfrak{g}]$ ($[\mathrm{G}]$ est donc réduit à un seul élément).

Il suffit de montrer que [S] est maximal pour [$\hat{g}$].

Soit $s \in S$ et $s' \sim s$.

D'après la propriété 4 :

$$\left.\begin{array}{l} (s', \hat{g}) \sim (s, \hat{g}) \\ s \Leftrightarrow \hat{g} \end{array}\right\} \Rightarrow s' \Leftrightarrow \hat{g}$$

Ceci montre que $[s] \mid\!-\!\mid [\hat{g}]$ $(\forall\, s \in S$ c'est à dire $\forall\, [s] \in [S])$

$$\left.\begin{array}{l} \text{Soit } s \notin S \;;\; [s] \notin [S] \\ \qquad s \not\Leftrightarrow \hat{g} \end{array}\right\} \Rightarrow [s] \mid\!\ast\!\mid [\hat{g}]$$

Ceci montre que [S] est maximal pour [$\hat{g}$] ; il est donc saturé par [$\hat{g}$], qui est réduit à un seul élément.

Corollaire : Soient S et G mutuellement saturés ; [S] et [G] forment un couple élémentaire (dans $\tilde{\mathcal{S}} \times \tilde{\mathcal{G}}$).

II - Ensembles de suites et de procédés liés

Section 1

Dans cette section, nous particularisons les notions définies dans la partie précédente.

Ensemble $\mathcal{S}$ Une suite de réels convergente, sera notée $s = \{s_n\}$, de limite s^*. Nous désignerons par $\mathcal{S}$ un certain sous ensemble de (c) (ensemble des suites convergentes).

Ensemble $\mathcal{G}$ Un procédé de transformation de suites est une application de (c) dans l'ensemble des suites :

$$t = g(s)$$

Cette application sera toujours définie par une fonction continue, t_i étant calculé à partir d'un certain nombre de termes successifs de la suite s :

$s_i, s_{i+1}, \dots s_{i+p}$.

G ("ensemble des procédés") désignera un certain sous ensemble de fonctions continues.

Par la suite nous noterons avec la même lettre le **procédé** g (application de (c) dans l'ensemble des suites) et la **fonction continue** g telle que

$$t_i = g(s_i, s_{i+1}, \dots s_{i+p})$$

$$t = g(s) \iff t_i = g(s_i, \dots s_{i+p}) \ \forall i$$

<u>Relation entre S et G</u>

Soient s et g deux éléments de S et G.

s et g sont **liés pour l'accélération de la convergence** si la suite t, de terme général $t_i = g(s_i, \dots s_{i+p})$ converge plus vite que s, vers la même limite :

$$\lim_{i\to\infty} \frac{t_i - s^*}{s_i - s^*} = 0$$

Nous utiliserons la notation s <-> g et nous écrirons "g accèlère s" ou bien "s est accélérée par g".

* <u>Propriété 2-1</u> : *Soit $\hat{G}$ l'ensemble de toutes les fonctions continues ; $\hat{G}$ et* (c) *ne sont pas liés pour l'accélération de la convergence.*

Il existe en effet dans $\hat{G}$ des procédés qui n'accélèrent aucune suite (il suffit de considérer la fonction $g(x) = x$). Et étant donné une suite convergente s(de limite s^*) il existe un procédé qui n'accélère pas sa convergence (il suffit de considérer la fonction $g(x) = K$ $(K \neq s^*)$).

□

Le résultat de la propriété 2-1 n'a rien de surprenant, les ensembles $\hat{G}$ et (c) étant trop vastes.

Soit $\mathcal{C}$ l'ensemble des suites de réels convergentes, tel que si $s \in \mathcal{C}$ $s_n \neq s^* \ \forall_n$.
(Pour une suite $\in \mathcal{C}$, il est possible de définir l'accélération de la convergence, le rapport $\dfrac{g(s_n \ldots s_{n+p}) - s^*}{s_n - s^*}$ ne devenant jamais infini).

La suite de cet article est une tentative de définition de sous ensembles $\mathcal{S} \subset \mathcal{C}$ et $\mathcal{G} \subset \hat{\mathcal{G}}$ vérifiant les hypothèses :

$$\forall\, s \in \mathcal{S} \quad \exists\, g \in \mathcal{G} \quad \text{tel que } s \Longleftrightarrow g$$

$$\forall\, g \in \mathcal{G} \quad \exists\, s \in \mathcal{S} \quad \text{tel que } s \Longleftrightarrow g$$

(Ce sont les hypothèses permettant d'affirmer qu'il existe des sous ensembles S et G (de $\mathcal{S}$ et $\mathcal{G}$) totalement liés).

Section 2 Etude des procédés de type 1

Une transformation de suites est de type 1 si elle est de la forme :

$$t_i = s_i + (s_{i+1} - s_i)\, g(s_{i+1} - s_i, \ldots s_{i+p} - s_{i+p-1})$$

g étant une fonction continue séparément par rapport à chacune de ses variables.
Le calcul de t_i exige la connaissance de (p+1) éléments successifs de la suite $\{s_i\}$, et la transformation que nous venons de décrire est un procédé de type 1 à (p+1) mémoires.

Soit $\mathcal{G}_1$ l'ensemble des procédés de type 1 tels que :

$$\forall\, g \in \mathcal{G}_1 \quad \neg\exists\, s \in \mathcal{C} \quad \text{vérifiant } s \Longleftrightarrow g$$

Propriété 2-2 : *Il existe des suites de $\mathcal{C}$ qui ne sont accélérées par aucun $g \in \mathcal{G}_1$.*

Démonstration :

Soit $s = \{s_1, s_2 \ldots s_n \ldots\}$ une suite de $\mathcal{C}$.
Fabriquons à partir de s la suite t :

$$t = \{s_1, s_2, s_2, s_3, s_3, s_3, \ldots \underbrace{s_n \ldots s_n}_{n \text{ fois}} \ldots\}$$

Dans la suite t les termes successifs d'indices $1 + \frac{n(n-1)}{2}$, $2 + \frac{n(n-1)}{2}$, ...

$\frac{n(n+1)}{2}$ sont égaux à s_n. Cette suite appartient à $\hat{C}$ et n'est accélérée par aucun $g \in G_1$. Supposons que ce soit le cas ; soit g de type 1 à (p+1) mémoires, accélérant la convergence de t.

Formons la suite de terme général

$$u_i = \frac{t_i + (t_{i+1} - t_i)\, g(t_{i+1} - t_i, \dots t_{i+p} - t_{i+p-1}) - s^*}{t_i - s^*}$$

$\forall N \;\exists i > N$ tel que $t_{i+1} = t_i \Rightarrow u_i = 1$.

Ceci est en contradiction avec l'hypothèse que g accélère $\{t_i\}$ (c'est à dire $\lim_{i\to\infty} u_i = 0$).

□

La propriété 2-2 montre qu'on ne peut accélérer toute les suites de $\hat{C}$ au moyen de procédés de type 1 ; $\hat{C}$ est donc trop vaste.

Soit $S_1 \subset \hat{C}$ tel que :

$$\forall s \in S_1 \quad \exists g \in G_1 \text{ vérifiant } s \Leftrightarrow g.$$

<u>Propriété 2-3</u> : *Il n'existe aucun sous ensemble G de G_1 lié à S_1 pour l'accélération de convergence ; autrement dit, il n'existe pas de famille de procédés universelle (qui accélère la convergence de toute suite de S_1).*

<u>Démonstration</u> :

Soit S_ρ l'ensemble des suites telles que $\lim_i \frac{s_{i+1} - s^*}{s_i - s^*} = \rho$, ρ étant un réel appartenant à l'intervalle [-1 1[; Soit G_ρ l'ensemble des procédés de type 1 (ayant un nombre de mémoires supérieur ou égal à 2), défini par une fonction g telle que $g(0, 0, .. 0) = \frac{1}{1-\rho}$.

S_ρ et G_ρ sont mutuellement saturés.

En effet soit $s \in S_\rho$; tout procédé de G_ρ accélère la convergence de s car :

$$\lim_{i\to\infty} \frac{s_i + (s_{i+1} - s_i)\, g(s_{i+1} - s_i, \dots s_{i+p} - s_{i+p-1}) - s^*}{s_i - s^*} = 1 + (\rho - 1)\frac{1}{1-\rho} = 0$$

Soit g telle que $g(0, \ldots 0) \neq \frac{1}{1-\rho}$ (g définit en procédé de type 1 qui n'appartient pas à G_ρ ; s appartenant à S_ρ on a :

$$\lim_{i\to\infty} \frac{s_i + (s_{i+1} - s_i)\, g(s_{i+1} - s_i, \ldots s_{i+p} - s_{i+p-1}) - s^*}{s_i - s^*} \neq 0$$

Ceci montre que G_ρ est maximal pour s (et ceci $\forall$ s $\in S_\rho$).

Donc G_ρ est saturé par S_ρ.

Soit g $\in G_\rho$; on prouve de même que S_ρ est saturé par G_ρ.

Montrons qu'il n'existe aucun sous ensemble G de G_1 lié à S_1 pour l'accélération de convergence. Il suffit de montrer que $\forall$ g $\in G_1$ $\exists$ s $\in S_1$ tel que s <⋇> g.

g étant fixé $\exists$ ρ' $\in$ [-1 1[tel que g $\notin G_{\rho'}$ et tout s $\in S_{\rho'}$ vérifie s <⋇> g.

□

Section 3 Etude des procédés de type 2

Une transformation de type 2 est de la forme

$$t_i = s_i + g(s_{i+1} - s_i, \ldots, s_{i+p} - s_i)$$

* g : fonction continue séparément par rapport à chaque variable (g est une fonction ayant un nombre de variables ≥ 1).

* $g(0, 0, \ldots 0) = 0$

Soit G_2 l'ensemble des procédés de type 2 tels que :

$\forall$ g $\in G_2$ $\exists$ s $\in$ ĉ vérifiant s <—> g

Une démonstration analogue à celle de la propriété 2-2 montre qu'il existe des suites de ĉ qui ne sont accélérées par aucun g $\in G_2$.

Soit $S_2 \subset$ ĉ tel que

$\forall$ s $\in S_2$ $\exists$ g $\in G_2$ vérifiant s <—> g

On a les inclusions : $S_1 \subset S_2$

$G_1 \subset G_2$

* Propriété 3-3 : *Il n'existe pas de procédé appartenant à G_2 accélérant la convergence de toute suite de S_2.*

Démonstration :

Plan de la démonstration :

Soit $G_2^{(k)}$ l'ensemble des procédés de type 2 à k mémoires : transformation $t_i = s_i + g(s_{i+1} - s_i, \dots s_{i+k-1} - s_i)$.

La démonstration se fait en 3 parties :

- Partie 1 : Démonstration de la propriété pour $G_2^{(2)}$
- Partie 2 : Démonstration de la propriété pour $G_2^{(k)}$ $(k > 2)$
- Partie 3 : Démonstration de la propriété pour $G_2 = \bigcup_{k\geq 2} G_2^{(k)}$

Partie 1 :

Notons $S_2^{(2)}$ l'ensemble des suites telles qu'il existe g continue vérifiant $(g(0) = 0)$ et

$$1 + \lim_n \frac{g(s_{n+1} - s_n)}{s_n - s^*} = 0 \text{ pour toute suite } \{s_n\} \in S_2^{(2)}.$$

Supposons qu'il existe un procédé accélérant la convergence de toute suite de $S_2^{(2)}$, défini par une certaine fonction g.

1° point : g ne peut être dérivable à l'origine

Soit $S \subset S_2^{(2)}$ l'ensemble des suites $\{s_n\}$ générées par

$e_{n+1} = f(e_n)$ f continue, dérivable en 0, telle que $|f'(0)| < 1$

e_n désigne l'"erreur" $s_n - s^*$.

S est bien un sous ensemble de $S_2^{(2)}$ car

$$e_{n+1} - e_n = f(e_n) - e_n = F(e_n)$$

$F(0) \neq 0$. Donc dans un voisinage de 0, F admet une fonction réciproque ϕ ;

$e_{n+1} - e_n = F(e_n) \Leftrightarrow e_n = \phi(e_{n+1} - e_n)$.

Posons $\tilde{g}(x) = -\phi(x)$; on a :

$$1 + \frac{\tilde{g}(e_{n+1} - e_n)}{e_n} = 0, \text{ ce qui montre que } \{s_n\} \in S_2^{(2)}$$

Soient s_1 et $s_2 \in S_2^{(2)}$ définies par f_1 et f_2 telles que $f'_1(0) \neq f'_2(0)$.

Le procédé g accélère la convergence de s_1 et s_2 :

Pour s_1 :

$$1 + \lim \frac{g(e^1_{n+1} - e^1_n)}{e^1_n} = 1 + \lim \frac{g(f_1(e^1_n) - e^1_n)}{e^1_n} = 1 + g'(0)(f'_1(0) - 1)$$

Pour s_2 :

$$1 + \lim \frac{g(e^2_{n+1} - e^2_n)}{e^2_n} = \qquad = 1 + g'(0)(f'_2(0) - 1)$$

Ces deux quantités ne peuvent être simultanément nulles.

2° point : g ne peut être "non dérivable" en 0

Utilisons deux lemmes :

Lemme A

Soit $\{a_n\}$ *une suite de limite nulle telle que :*

(*i*) $\forall\, n \quad \Delta a_n \neq 0$

(*ii*) $\forall\, m, n$ *ou bien* $\Delta a_n \neq \Delta a_m$

ou bien $\Delta a_n = \Delta a_m$ *et* $a_n = a_m$

Alors $\{a_n\} \in S_2^{(2)}$

Notons $b_i = \Delta a_i$.

Il existe $\bar{g}$ continue telle que $\bar{g}(b_i) = - a_i \quad \forall\, i$

$\bar{g}(0) = 0$

Utilisons le théorème de Tietze :

Sur le fermé $\{b_i\} \cup \{0\}$ constitué par les éléments de la suite $\{b_i\}$ et de sa limite, définissons $\hat{g}$ par :

$\hat{g}(b_i) = - a_i$

$\hat{g}(0) = 0$

(L'hypothèse (ii) implique que $\hat{g}$ peut être définie ainsi : $b_i = b_j \Rightarrow a_i = a_j$; il n'y a donc aucune incompatibilité pour définir $\hat{g}$, ce qui permet de retirer de l'ensemble les b_j qui coïncident avec d'autres éléments de la suite $\{b_i\}$).

$\hat{g}$ est évidemment continue en tout point b_i.

Continuité en 0 :

Fixons ε ; $\exists$ N tel que $\forall\, n > N \Rightarrow |a_n| < \varepsilon$

Posons $\eta < \inf\limits_{i<N} |b_i|$

$|b_j| < \eta \Rightarrow N \Rightarrow |a_j| < \varepsilon$, d'où la continuité de $\hat{g}$ en 0.

Il existe $\bar{g}$ continue qui prolonge $\hat{g}$.

Donc pour la suite $\{a_n\}$: $1 + \dfrac{\bar{g}(a_{n+1} - a_n)}{a_n} = 0$

$$\{a_n\} \in S_2^{(2)}$$

$\square$

Supposons g non dérivable en 0.

Il existe $\{\xi_n\}$ de limite nulle ($\xi_n \neq 0 \quad \forall\, n$) telle que pour $n \to \infty$, $\dfrac{g(\xi_n)}{\xi_n}$ n'a pas de limite.

<u>Lemme B</u> :

Soit g *le procédé accélérant toute suite de* $S_2^{(2)}$ *(*g *non dérivable en 0) et* ξ_n *la suite telle que* $\dfrac{g(\xi_n)}{\xi_n}$.

Si $\{\xi_n\}$ *est une sous suite extraite des termes d'indice pair d'une suite* $\{a_n\}$ *vérifiant :*

(ii) $\forall\, m, n$ *ou bien* $\Delta a_n \neq \Delta a_m$

ou bien $\Delta a_n = \Delta a_m$ *et* $a_n = a_m$

(iii) $a_{2n+1} = 2\, a_{2n}$,

alors le procédé défini par g *ne peut accélérer la convergence de* $\{a_n\}$.

Transformons $\{a_n\}$:

$$t_{2n} = a_{2n} + g(a_{2n+1} - a_{2n}) = a_{2n} + g(a_{2n})$$

La condition d'accélération ($\lim \dfrac{t_n}{a_m} = 0$) ne peut être vérifiée car

$$\lim \frac{t_n}{a_n} = 0 \Rightarrow \lim \frac{t_{2n}}{a_{2n}} = 0 \Rightarrow \lim 1 + \frac{g(a_{2n})}{a_{2n}} = 0$$

$$\Rightarrow \lim \frac{g(\xi_n)}{\xi_n} \text{ existe.}$$

□

Démonstration du théorème pour $G_2^{(2)}$

A partir de $\{\xi_n\}$ il faut fabriquer $\{a_n\}$ vérifiant les conditions (i) (ii) et (iii) des lemmes A et B ($\{a_n\}$ sera accélérable d'après (A) et non accéléré par g, d'après (B)).

Posons
$$\begin{aligned} u_0 &= \xi_0 \\ u_1 &= 2\xi_0 \\ &\vdots \\ u_{2n} &= \xi_n \\ u_{2n+1} &= 2\xi_n \end{aligned}$$

$\xi_n \neq 0 \;\forall n \Rightarrow u_{2n+1} - u_{2n} \neq 0 \;\forall n$

Supposons qu'il existe n tel que $u_{2n} = u_{2n-1}$; il est possible de trouver $x \in [u_{2n-2}, u_{2n}]$ tel que dans la sous suite : $\{u_{2n-2}, u_{2n-1}, x, 2x, u_{2n}, u_{2n+1}\}$ toutes les différences premières soient différentes de 0.

Notons $\{v_n\}$ la nouvelle suite ainsi obtenue ; elle vérifie (i), (iii) et de plus $\lim v_n = 0$.

L'hypothèse (ii) est-elle vérifiée ?

Il se peut que $\Delta v_n = \Delta v_m \quad m < n$.

Faisons croitre n de 1 à l'infini et comparons Δv_n aux Δv_m précédents.

Deux cas peuvent se présenter :

<u>1° cas</u> : n pair = 2p

si m = 2q

$$\Delta v_n = \Delta v_m \Rightarrow v_n = v_m \quad \text{(et (ii) est vérifiée)}$$

si m = 2q+1, il est possible de trouver $x \in [v_m\ v_{m+1}]$ tel que dans la sous suite $\{v_m, x, 2x, v_{m+1}\}$ les différences premières vérifient :

$$\begin{cases} x - v_m \neq x \\ x - v_m \neq v_{m+1} - 2x \\ x - v_m \neq \Delta v_i \quad (i = 1 \dots n) \\ x \qquad \neq \Delta v_i \quad (i = 1 \dots n) \quad ; \text{ la propriété (ii) est alors vérifiée} \\ v_{m+1} - 2x \neq \Delta v_i \end{cases}$$

<u>2° cas</u> n impair = 2p+1 et $\Delta v_n = \Delta v_m$ (m < n)

Il est alors possible de trouver x tel que dans la sous suite $\{v_n, x, 2x, v_{n+1}\}$ les trois différences premières soient distinctes des Δv_i (i = 1 ... n)

La propriété (ii) est alors vérifiée.

Partie 2

Notons $S_2^{(k)}$ l'ensemble des suites accélérables par un procédé appartenant à $G_2^{(k)}$.

Supposons qu'il existe $g \in G_2^{(k)}$ accélérant toute suite de $S_2^{(p)}$ (p = 2 ... k).

<u>*1° point*</u> <u>g ne peut être dérivable en 0</u>

Comme $S_2^{(2)} \subset S_2^{(k)}$, le sous ensemble S défini en Partie 1 vérifie $S \subset S_2^{(k)}$

Pour $s \in S$

$$\lim \left[1 + \frac{g(e_{n+1} - e_n, \dots e_{n+k-1} - e_n)}{e_n}\right] = \lim \left[1 + \frac{g(f(e_n) - e_n, f^{[2]}(e_n) - e_n, \dots)}{e_n}\right]$$

(nous posons $f^{[j]}(x) = f(f(..(f(x)))$ composée j fois)

Cette limite est donc égale à

$$1 + [g'_1(0), g'_2(0) \dots g'_{k-1}(0)] \begin{bmatrix} f'(0) - 1 \\ (f'(0))^2 - 1 \\ (f'(0))^{k-1} - 1 \end{bmatrix}$$

Il est possible de choisir deux fonctions f (c'est à dire deux suites de S) telles que les quantités correspondantes soient distinctes.

=> Il n'existe pas de fonction g dérivable à l'origine, accélérant la convergence de toute suite de $S_2^{(p)}$ p = 2 ... k.

2° point g ne peut être "non dérivable" à l'origine

Enonçons les deux lemmes permettant la démonstration.

Lemme (A)

Soit $\{a_n\}$ *une suite de limite nulle telle que*

(*i*) $\forall n \quad B_n \neq 0$

(*ii*) $\forall m, n \quad B_m \neq B_n$

$$B_n = \begin{pmatrix} a_{n+1} - a_n \\ a_{n+2} - a_n \\ \vdots \\ a_{n+k-1} - a_n \end{pmatrix}$$

Alors $\{a_n\} \in S_2^{(k)}$

Démonstration analogue à celle du lemme A (partie 1).

Il existe $\bar{g}$ telle que $\bar{g}(B_i) = - a_i$

$\bar{g}(0) = 0$

$\bar{g}$ peut être prolongée en une application $\hat{g}$, continue de $\mathbb{R}^{k-1} \to \mathbb{R}$, et pour la suite $\{a_n\}$ on a :

$$\lim \left(1 + \frac{\hat{g}(a_{n+1} - a_n, \ \dots \ a_{n+k-1} - a_n)}{a_n}\right) = 0 \quad \Rightarrow \{a_n\} \in S_2^{(k)}$$

□

Supposons g non dérivable à l'origine. Il existe $\{\xi_n\}$, de limite nulle ($\xi_n \neq 0 \ \forall n$) telle que :

$$\frac{g(0, 0, \dots \xi_n, 0\ 0\ 0)}{\xi_n} \quad \text{n'a pas de limite pour } n \nearrow \infty$$

(ξ_n agit sur la $p^{ème}$ variable de g).

Lemme (B)

Soit g *le procédé de* $G_2^{(k)}$ *non dérivable en* 0 *et* ξ_n *la suite associée.*

Si $\{\xi_n\}$ *est une sous suite extraite de la sous suite*

$\{a_0, a_k, a_{2k}, \dots\}$ *de* $\{a_n\}$ *qui vérifie :*

Les conditions (i) et (ii) du lemme (A).

(iii) la sous suite de $\{a_n\}$ *indicée de* a_{jk} *à* a_{jk+k-1} *est :*

$$\left[\begin{array}{ll} a_{jk} = & \alpha \\ \vdots = & \alpha \\ \vdots & \vdots \\ a_{jk+p-1} = & 2\alpha \\ \quad = & \alpha \\ a_{jk+k-1} = & \alpha \end{array}\right.$$

alors g ne peut accélérer la convergence de $\{a_n\}$.

Technique de démonstration identique à celle du lemme (B) partie 1.

□

Démonstration de la partie 2 du théorème

D'une manière identique à la partie 1 formons à partir de la suite $\{\xi_n\}$ la suite

$$\left.\begin{array}{ll} u_0 = & \xi_0 \\ u_1 = & \xi_0 \\ \vdots & \\ u_{p-1} = & 2\xi_0 \\ \vdots & \\ u_{k-1} = & \xi_0 \end{array}\right\}$$
$$\begin{array}{ll} u_k = & \xi_1 \\ \vdots & \end{array}$$

Cette suite converge vers 0 et il est possible d'intercaler des termes supplémentaires de telle façon que les hypothèses (i) (ii) (iii) des lemmes (A) et (B) soient satisfaites. Cette suite est accélérable d'après (A) et non accélérée par g d'après (B).

Partie 3

Il n'existe pas de procédé appartenant à G_2 accélérant la convergence de toute suite de S_2.

Supposons qu'il en existe un ; soit g ce procédé. Il appartient à un certain $G_2^{(k)}$ et il ne peut accélérer la convergence de toute suite de S_2 d'après la partie 2.

CONCLUSION

Pour deux familles de procédés nous avons établi qu'il n'existe pas de procédé universel d'accélération de la convergence ; notre étude se poursuit actuellement dans deux directions : pour les deux familles citées, recherche de sous ensembles de suites et de procédés totalement liés ; d'autre part il est intéressant de se demander s'il n'existe pas une famille de suites plus vaste pour laquelle il y a non existence de procédé universel.

B. GERMAIN - BONNE
Université de Lille I
U.E.R. d'IEEA - Informatique, B.P. 36
F - 59650 VILLENEUVE D'ASCQ (FRANCE)

VALLEYS IN c-TABLE

Jacek GILEWICZ [*]
Alphonse MAGNUS [**]

ABSTRACT : The existence of valleys formed by level curves of quasi-equal elements in the c-table, observed at first in [1], is proved for the non-rational Stieltjes functions and for the exponential function. The same is also proved for the table of ratios of Toeplitz determinants which characterize the behaviour at the origin of the difference between a function and its Padé approximant. The optimal method of recursive computation of the c-table with blocks is presented.

1. c-TABLE ; GENERALITIES

Let c be a real sequence $(c_n)_{n\geq 0}$, C the corresponding formal series $\sum_{n\geq 0} c_n z^n$ and f the function which is a sum (in some sense) of C. The c-table associated with c, and by extension with C or f , is the infinite array of Toeplitz determinants:

$$\forall m \geq 0, \forall n > 0 : \quad C_n^m = \det(c_{m+i-j})_{i,j=1}^{n} \ ; \ c_k \equiv 0 \ \text{if} \ k < 0 \ ;$$

$$\forall m \geq 0 \quad : \quad C_0^m = 1 \tag{1}$$

* Centre de Physique Théorique, C.N.R.S. - Luminy - Case 907
F-13288 MARSEILLE CEDEX 2 (France)
et
Université de Toulon

** U.C.L., Institut de Mathématique Pure et Appliquée, Chemin du Cyclotron 2,
B-1348 Louvain-La-Neuve (Belgique).

m \ n	0	1	2	
0	1	c_0	$(c_0)^2$	...
1	1	c_1	C_2^1	...
2	1	c_2	C_2^2	...
	.	.	.	...

fig. 1 : c-table

C_n^m are the determinants of the linear system of equations defining the denominators of Padé approximants $[m/n] = P_m/Q_n$ of the series C [1]. The array of Padé approximants : Padé table, and the c-table have similar square blocks structure. We call block of type $(m,n;k+1)$ in the c-table the following square

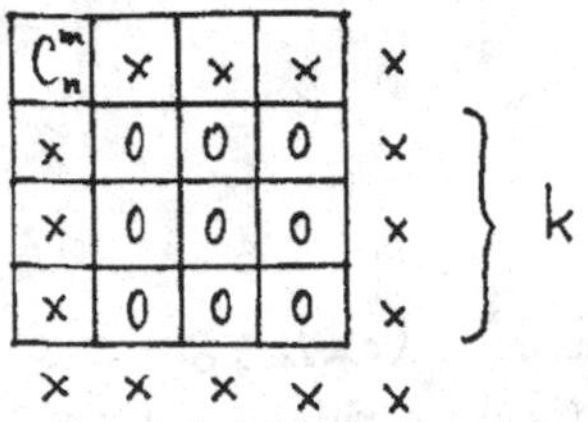

fig. 2 : Block of type $(m,n;k+1)$ in the c-table

where C_n^m and elements noted by cross are different from zero. If $k = 0$ for all m and n, we say that the series C (or c-table) is normal, and if an infinite block exists, we say that the series C is rational. In the latter case the series C is the Taylor expansion of some rational function $f = P_m/Q_n$. If an infinite block does not exist, we say that the series is non-rational.

By a permutation of columns in the Toeplitz determinant we obtain the Hankel determinant :

$$H_n^m = \det\left(c_{m+i+j}\right)_{i,j=0}^{n} \tag{2}$$

Obviously :

$$C_n^m = (-1)^{\frac{n(n-1)}{2}} H_{n-1}^{m-n+1} \tag{3}$$

In following we also use the simplified notation:

$$\mathcal{H}_n^m = H_{n-1}^{m-n+1} \tag{4}$$

In the normal case the dominant error term of the Padé approximant at the neighbourhood of the origin is given by the ratio of two Toeplitz determinants:

$$\left[\frac{f(z) - [m/n]_f(z)}{z^{m+n+1}}\right]_{z=0} = (-1)^n \frac{C_{n+1}^{m+1}}{C_n^m} \qquad (C_n^m \neq 0) \tag{5}$$

The table of these ratios is called the <u>c-ratio table</u>.

The relation (5) indicates that there is a relation between the convergence properties of the Padé approximants and the structure of the c-table or the c-ratio table. In fact, this was numerically observed in [1]. The characteristic structure of the c-table was called there valley structure.

The main result in this paper is to prove the existence of a valley structure in the c-table and the c-ratio table associated with some Stieltjes series and with exponential function. An extension of this result is suggested by new numerical observations.

2. VALLEY AND MINIMAL LINE IN A TABLE

We define the <u>minimal line</u> in a table (if it exists) as the line joining the minima of the absolute values of elements in the antidiagonals, i.e. for m+n fixed. Joining the quasi-equal absolute values of elements in a table we frequently obtain the characteristic level curves forming the <u>valley structure</u> :

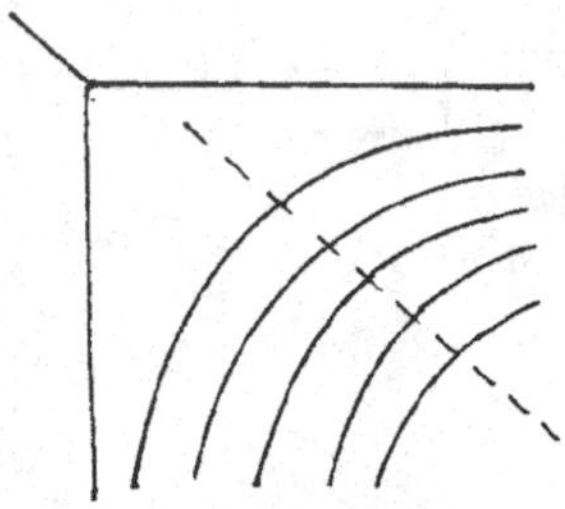

fig. 3 : Valley in a table

which is the cartographic continuous representation of the discrete structure. The existence of valleys in the c-table was observed numerically for various cases [1]. Recently we have observed the analogous valleys in the corresponding c-ratio tables. It is clear that the infinite block in the c-table represents the trivial case of a valley. In this case we have $f - [m/n]_f = 0$, and the convergence problem of Padé approximants is trivial (rational case). If numerically the elements of the blocks are not exactly zero, then the diagonals of the blocks appear as minimal lines.

In practice the analysis of a valley in a c-table allows one to choose the "best Padé approximant" [1].

The more promising result is the ascertainment that the minimal line defines the position of the best convergent chain of Padé approximants in two known cases : normal Stieltjes function and exponential function. Actually the convergence theory of Padé approximants is still very incomplete. The suggestions based on the numerical investigations of valleys can assist its advance.

3. VALLEYS IN THE STIELTJES AND EXPONENTIAL CASES

Consider the non-rational Stieltjes function :

$$z \mapsto f(z) = \int_0^{1/R} \frac{d\mu(x)}{1 - xz} \qquad (R \neq 0) \tag{6}$$

and the corresponding series of Stieltjes :

$$C = \sum_{n=0}^{\infty} s_n z^n \quad , \quad s_n = \int_0^{1/R} x^n \, d\mu(x) \tag{7}$$

where $d\mu$ is a positive measure. If $R = 1$ the Stieltjes moments s_n become the Hausdorff moments. The sequence of Hausdorff moments is frequently called the totally monotonic sequence. If, in addition it is non-rational, we write

$$s \in TM\,(NR) \tag{8}$$

All TM(NR) sequences can be <u>normalized</u> by :

$$c = \left(\frac{s_n}{s_0}\right)_{n \geq 0} \quad ; \quad c \in TM(NR) \tag{9}$$

All Stieltjes sequences (7) become TM(NR) normalized sequences by the scale transformation :

$$\frac{1}{R} = \limsup \sqrt[n]{s_n} \quad , \quad c_n = \frac{R^n s_n}{s_0} \quad , \quad c \in TM(NR) \, . \tag{10}$$

In the following we consider also the (non-Stieltjes) function $z \mapsto e^z$ and the sequence $(1/n!)_{n \geq 0}$ which is totally positive (TP). The positivity of the Hankel determinants for TM(NR) sequences ([1] , p.206) and of the Toeplitz determinants for the TP sequences ([1], p.115) :

$$\forall m \geq 0, \forall n \geq 0 : \quad \begin{array}{lll} H_n^m(c) > 0 & \text{if} & c \in TM(NR) \\ C_n^m(c) > 0 & \text{if} & c \in TP(NR) \end{array} \tag{11}$$

plays an essential role in the following proofs. For proving the existence of valleys in the c-table and c-ratio table we establish a complete set of inequalities between the elements of these tables. This is clearly represented in fig. 4, where the central element belongs to the diagonal chain :

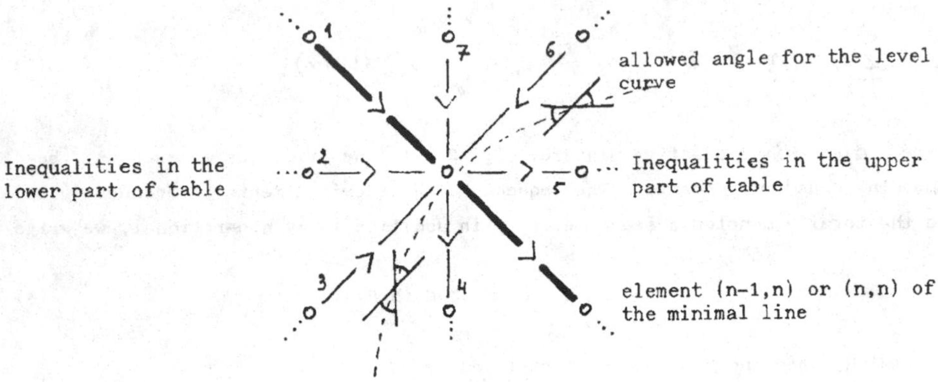

fig. 4 : Inequalities in the Theorem 1

Theorem 1

Among the following inequalities :

$$|C_{n'}^{m'}| < |C_n^m| \tag{12}$$

$$|C_{n'+1}^{m'+1} / C_{n'}^{m'}| < |C_{n+1}^{m+1} / C_n^m| \tag{13}$$

$$\forall m, m', n, n' : \begin{cases} m' \geq n'-1 \ , \ m \geq n-1 \ , \ n' \geq n \geq 1 \ , m'+n' \geq m+n & (14) \\ m' \leq n' \ , \ m \leq n \ , \ m' \geq m \geq 1 \ , m'+n' \geq m+n & (15) \end{cases}$$

i) (12) holds for the normalized TM(NR) sequences,

ii) (13) holds for all TM(NR) sequences,

iii) (12) and (13) hold, except the case $m' = n = N$, $m = n' = N+1$ where inequalities become equalities, for the sequence $(1/n!)_{n \geq 0}$.

Remark :
According to the numerical results (12) holds also for ii).

Proof : Note that we must establish only the inequalities in the directions 3, 4, 5 and 6 of fig.4, remainder inequalities are implied by this four. For instance

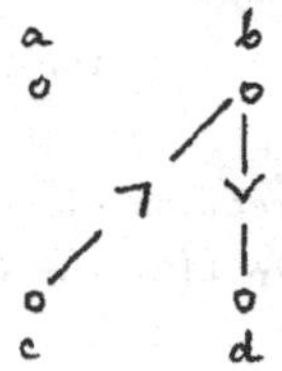

c > b and b > d implies c > d and by translation a > b , and therefore a > d.

i) : (12) is proved with help of the Schweinsian expansion [2] of the determinants :

$$\frac{\det(a_{i,j})_{i,j=2}^{m}}{\det(a_{i,j})_{i,j=1}^{m}} = \frac{1}{a_{1,1}} + \sum_{k=2}^{m} \frac{\det(a_{i,j+1})_{i,j=1}^{k-1}\,\det(a_{i+1,j})_{i,j=1}^{k-1}}{\det(a_{i,j})_{i,j=1}^{k-1}\,\det(a_{i,j})_{i,j=1}^{k}} \tag{16}$$

Accoridng to (4) and (11) we have $\mathcal{H}_n^m > 0$ for $m \geqslant n-1$ and $n \geqslant 1$. Keeping the first term in the Schweinsian expansion of $\mathcal{H}_{n-1}^{m+1} / \mathcal{H}_n^m$ we obtain

$$\mathcal{H}_n^m < c_{m-n+1}\, \mathcal{H}_{n-1}^{m+1} \qquad \text{for } m \geqslant n-1 \geqslant 1. \tag{17}$$

For the normalized TM sequences we have $c_j \leqslant 1$, then :

$$\mathcal{H}_n^m < \mathcal{H}_{n-1}^{m+1} \qquad \text{for } m \geqslant n-1 \geqslant 1 \tag{18}$$

which, according to (3), proves (12) in the direction 3. Iterating (17) we obtain

$$\mathcal{H}_n^m < c_{m-n+1}\, c_{m-n+3} \cdots c_{m+n-1} \tag{19}$$

Keeping the last term in the Schweinsian expansion of $\mathcal{H}_{n-1}^{m+1} / \mathcal{H}_n^m$ we obtain

$$\frac{\mathcal{H}_{n-1}^{m}}{\mathcal{H}_{n-1}^{m-1}} < \frac{\mathcal{H}_{n-1}^{m+1}}{\mathcal{H}_{n-1}^{m}} \tag{20}$$

i.e. the sequence $(\mathcal{H}_{n-1}^{m} / \mathcal{H}_{n-1}^{m-1})_{m \geqslant n-1}$ is increasing. According to (19) we have $\mathcal{H}_n^m < 1$, then this sequence has a limit no greater than 1, hence

$$\mathcal{H}^{m}_{m-1} < \mathcal{H}^{m-1}_{n-1} \qquad \text{for } m \geqslant n-1 \geqslant 1 \tag{21}$$

which, according to (3), proves (12) in the direction 4. Completing this by the inequalities in the directions 1 and 2 we obtain (12) for (14), i.e. all inequalities on the lower part of the c-table.

Consider the inverse series of our series of Stieltjes :

$$1 + \sum_{m=1}^{\infty} d_n z^n = \left(\sum_{n=0}^{\infty} c_n z^n \right)^{-1} \tag{22}$$

According to [1], p.207 we have :

$$(-d_m)_{m \geqslant 1} \in TM(NR) \tag{23}$$

Then, by analogy to (12) and (14) we have :

$$\mathcal{H}^{m'}_{n'}(-d) < \mathcal{H}^{m}_{n}(-d) \qquad \text{for } m' \geqslant n', m \geqslant n, n' \geqslant n \geqslant 1, m'+n' \geqslant m+n \tag{24}$$

where $m' \geqslant n'$ and $m \geqslant n$ eliminate d_0 with respect to (23). But for this range of indices the Hadmard formula ([1], p.31) and (3) give :

$$\mathcal{H}^{m}_{n}(-d) = |C^{m}_{n}(-d)| = |C^{n}_{m}(c)| \tag{25}$$

Then, interchanging the letters m and n in (24) we obtain (12) for (15).

<u>ii)</u> : We prove (13) using the Padé inequalities ([1], p.264 ; [3], p. 82-83(*)) for the non-rational Stieltjes functions in the case $R \neq 0$:

$$\forall n \geqslant 0, \forall k \geqslant 0, \forall x \in]0, R[:$$

$$0 < f(x) - [k+n+1/n+1](x) < f(x) - [k+n+2/n](x) \tag{26}$$

$$0 < f(x) - [n/n+1](x) < f(x) - [n+1/n](x) \tag{27}$$

(*) Errors in [3] :

p. 82, line 13 has to be replaced by $0 \leqslant [k/k+1] \leqslant f \leqslant [k+1/k]$

p. 83, line 6 has to be replaced by $0 \leqslant f - [k/k+1] \leqslant f - [k+1/k]$

$$0 < [m+1/n](x) < [n/m+1](x) < f(x) \tag{28}$$

$$R \geq 1: \quad 0 < f(x) - [k+n+1/n](x) < x\big(f(x) - [k+n/n](x)\big) \tag{29}$$

$$0 < f(x) - [n+1/k+n+1](x) < f(x) - [n/k+n+2](x) \tag{30}$$

$$R \geq 1: \quad 0 < f(x) - [n/k+n+1](x) < x\big(f(x) - [n/k+n](x)\big) \tag{31}$$

According to (5) ($z \to 0_+$) the inequalities (26) and (27) give (13) in the direction 3, (29) gives (13) in the direction 4, (30) gives (13) in the direction 6 and (31) in the direction 5. This is sufficient to prove all inequalities in the c-ratio table.

iii) : For the sequence $\left(\frac{1}{m!}\right)_{m \geq 0}$ we know the explicit expression for C_n^m:

$$C_n^m = \prod_{k=1}^{n} \frac{1}{k(k+1)\dots(k+m-1)} \tag{32}$$

which gives for the ratio :

$$\frac{C_{n+1}^{m+1}}{C_n^m} = \frac{m!\,n!}{(m+n)!\,(m+n+1)!} \tag{33}$$

Now we can easily verify the theorem in the case iii).

Q.E.D.

The minimal line in the c-table and the c-ratio table in the Stieltjes or exponential cases coincide with the fastest convergence chain of Padé approximants "[n/n] , [n/n+1]" . One of the arguments of this assertion for the Stieltjes case comes from the inequalities (28) : [n/n+1] Padé approximant is better than [n+1/n].

If the above connection with convergence problem is general, then the normalization of the sequence does not have any effect on the position of minimal line. In fact we observe this numerically, but we know prove only the following inequalities for the lower half part of c-table :

Theorem 2

Let c be a sequence of TM(NR) and k defined by :

$$\forall n: \quad c_n = A + a_n, \quad A \geq 0, \quad a \in TM(NR), \quad \lim_{n \to \infty} a_n = 0 \tag{34}$$

$\exists k \geq 0$: $a_k \leq 1$ and if $k \neq 0$ $a_{k-1} > 1$, then :

$$|C_{n'}^{m'}| < |C_n^m| \quad \forall m, m', n, n' : \quad m' \geq n'-1+k, m \geq n-1+k, n' \geq n, m'+n' \geq m+n \tag{35}$$

Proof : According to the Theorem 1 we have $\mathcal{H}_n^m(c/c_0) > \mathcal{H}_n^{m+1}(c/c_0)$. Multiplying both sides by c_0^n we obtain $\mathcal{H}_n^m(c) > \mathcal{H}_n^{m+1}(c)$, i.e. the inequality in the direction 4.

According to (17) the following inequality holds for the sequence a in the direction 3 :

$$\mathcal{H}_n^m(a) < \mathcal{H}_{n-1}^{m+1}(a) \qquad \text{for } m \geq n-1+k \tag{36}$$

We can easily prove the following formula :

$$\mathcal{H}_n^m(c) = A\,\mathcal{H}_{n-1}^{m-1}(\Delta^2 a) + \mathcal{H}_n^m(a) \tag{37}$$

when $\Delta^2 a$ is the sequence of second differences of a : $(\Delta^2 a)_0 =$ $= a_0 - 2a_1 + a_2$, $(\Delta^2 a)_1 = a_1 - 2a_2 + a_3$, ... The sequence $\Delta^2 a$ belongs to TM(NR) ([1], p.39), then with (36) we have :

$$A\,\mathcal{H}_{n-1}^{m-1}(\Delta^2 a) + \mathcal{H}_n^m(a) < A\,\mathcal{H}_{n-2}^{m}(\Delta^2 a) + \mathcal{H}_{n-1}^{m+1}(a)$$

and with (37) :

$$\mathcal{H}_n^m(c) < \mathcal{H}_{n-1}^{m+1}(c) \qquad \text{for } m \geq n-1+k.$$

By recurrence we complete the proof.

Q.E.D.

4. NEW NUMERICAL OBSERVATIONS

The interesting structure of both tables is observed for the Stieltjes moments for the case $R = 0$ (zero radius of convergence of the series of Stieltjes). We reproduce the results for the following moments :

$$\forall n : \quad c_n = \int_0^\infty x^n e^{-x} dx = n! \tag{38}$$

In the c-table the diagonal minimal line is competing with the minimal lines in the directions $m = 1$ and $n = 1$ and the elements on these lines increase with m and n . The c-ratio table has the following structure :

m \ n	0	1	2	3	4	5
0	1	1	3	13	71	461
1	2	2	4	14.7	73.2	
2	6	6	12	36		
3	24	24	48			
4	120	120				
5	720					

fig. 5 : c-ratio table

The minimal line is unique and corresponds also to the fastest convergence chain of Padé approximants. Opposite to the case $R \neq 0$, all inequalities are inversed except those of the directions 3 and 6 in fig. 4. Actually this is never proved for this case.

The numerical observations suggest then the columns in the c-tables of the totally monotonic sequences hide some TM properties. We have yet $\mathcal{H}_n^m > \mathcal{H}_n^{m+1}$ and the inequality (20) implies the logarithmic convexity:

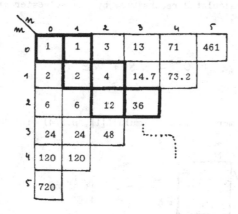

$$\log \mathcal{H}_n^m < \frac{1}{2} \left[\log \mathcal{H}_n^{m-1} + \log \mathcal{H}_n^{m+1} \right] .$$

It would be interesting to compare the structure of the c-ratio table with the convergence of Padé approximants in other non-Stieltjes cases. Actually we investigate the case of the function $z \mapsto (1+z)^\alpha$ (with complex α) for which M. Froissart elaborated the convergence thoery of Padé approximants (non published).

5. COMPUTATION OF c-TABLE

The c-table can be calculated recursively by the Sylvester crossing formula :

$$\begin{matrix} & N & \\ W & C & E \\ & S & \end{matrix} \qquad\qquad C^2 = NS + WE \tag{39}$$

and in the case of a block by more complex formula ([1], p.374) relating eight of the following elements:

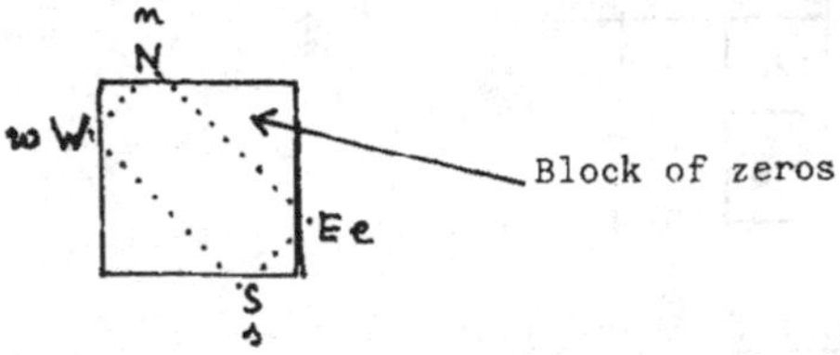

(40)

where the elements arrounding the zeros block follow the geometric progression ([1], p. 192 and 372).
One usually computes the triangular c-table "ascending" the antidiagonals and starting from the two first columns, which allow one to compute the East elements by (39) (see figs. 1 and 2). But it is also possible to compute the c-table "descending" the antidiagonals, i.e. computing the South elements by (39) , and starting from two first rows, where the second is calculated by :

$$C_m^1 = c_1 C_{m-1}^1 - c_0 c_2 C_{m-2}^1 + c_0^2 c_3 C_{m-3}^1 - \dots + (-c_0)^{m-1} c_m C_0^1 . \tag{41}$$

We notice that by the ascending algorithm (resp. descending algorithm) the region

" e " (resp. " s ") after the block can not be calculated by (39)(*) :

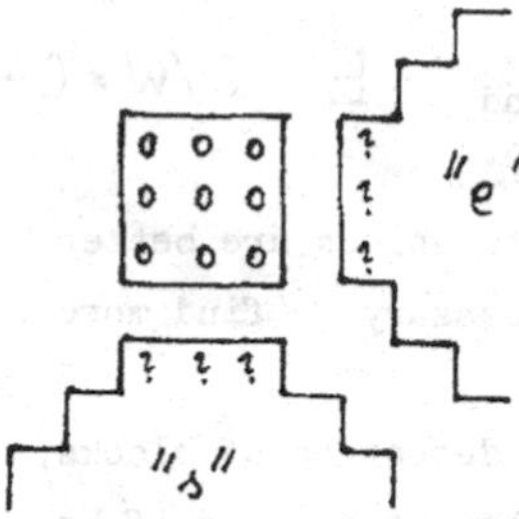

fig. 6 : Non-calculable elements in the c-table by (39)

In these cases we can use the relation (40) computing the e elements in the ascending algorithm or the s elements in the descending algorithm, afterwards we follow by (39). But it is clear that in some cases we can omit the relation (40). In fact we can complete for instance the ascending algorithm by the descending algorithm to reach the elements "?" in the " e " region, provided that these elements are not intersected by another " s " region.

Combining these algorithms we can choose the optimal algorithm with respect to the stability of computation. Firstly we can notice that the good algorithm must compute the antidiagonals going from two sides towards the minimal line. If we will minimize the "cost" of multiplications and divisions proceeding from (39), then the optimal algorithm must be the following :

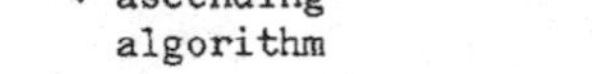

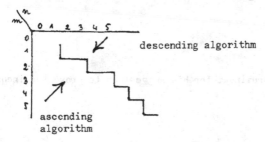

fig. 7 : Cost-optimal algorithm

(*) The scheme in [1] , p. 372 is wrong.

The inequality (19) and the Fig. 5 show that the elements of the c-tables decrease or increase rapidly. Therefore there exists some competition between the two methods of computation of the East element by (39) :

$$E = (C^2 - NS)/W \quad \text{and} \quad E = C/W * C - S/W * N$$

where in the second formula the intermediate values are better bounded. But the numerical instabilities show that it is necessary to find more stable expressions.

Another numerical problem is the detection of blocks, i.e. of numerical zeros. Guzinski [4] replaces the little element by zero if he observes a rough variation of monotonic behaviour of elements computed by the ascending algorithm. He justifies intuitively this by the smoothness of valleys in the normal c-table.

CONCLUSION

The essential problem remains open : what is the general relation between the valley structure of some tables and the convergence problem of Padé approximants? We think that the solution of this problem will greatly enhance the development of the convergence theory of Padé approximants.

ACKNOWLEDGMENTS

We wish to thank Prof. C. Brezinski for his suggestion to prove the inequalities (13) by Padé inequalities.

REFERENCES
==========

[1] GILEWICZ, J.
"Approximants de Padé",
Lecture Notes in Mathematics, 667, Springer-Verlag (1978).

[2] AITKEN, A.C.
"Determinants and Matrices",
Oliver & Boyd, Edinburgh (1946).

[3] BREZINSKI, C.
"Accélération de la convergence en analyse numérique",
Lecture Notes in Mathematics, 584, Springer-Verlag (1977).

[4] GUZINSKI, W.
"PADELIB : Library of Padé Approximation Routines",
INR 1768, Institute of Nuclear Research, Warsaw (1978).

RECURRENCE COEFFICIENTS FOR ORTHOGONAL POLYNOMIALS ON CONNECTED AND NON CONNECTED SETS.

Alphonse MAGNUS
University of Louvain.

Abstract : General methods for relating the asymptotic behaviour of the recurrence coefficients of orthogonal polynomials and properties of the corresponding weight functions are described and discussed. The cases of a function with algebraic singularities on a single bounded interval and a function positive on a finite set of bounded intervals are studied. Application to weight function reconstruction from moments is considered.

1. Introduction.

1.1. General background.

First of all, some elementary relations for orthogonal polynomials and continued fractions are recalled, merely in order to fix notations.

Let us consider a measurable set S of real numbers and a function w positive and integrable on S (such a function will be called a *weight function*). From the moments of this function, $\mu_n = \int_S t^n w(t)dt$, $n=0,1,\ldots$, it is possible to construct the corresponding orthogonal polynomials

(1.1) $B_n(z) = z^n+\ldots, \int_S B_k(t)\, B_n(t)\, w(t)dt = h_n\, \delta_{k,n}, \; k,n=0,1,\ldots,$

satisfying the recurrence relations

(1.2) $B_1(z) = z-a_0, \; B_{n+1}(z) = (z-a_n)\, B_n(z) - b_n^2\, B_{n-1}(z), \; n=1,2,\ldots$.

It is known ([19], chap. 11) that $B_n(z)$ is the denominator of the approximant $A_n(z)/B_n(z)$ of the Jacobi continued fraction expansion of the Stieltjes transform of w :

$$f(z) = \sum_0^\infty \mu_k/z^{k+1} = \int_S \frac{w(t)dt}{z-t} = \cfrac{b_0^2}{z-a_0-\cfrac{b_1^2}{z-a_1-\ldots}}.$$

Actually, $x^{-1}\, A_n(x^{-1})/B_n(x^{-1})$ is the [n-1, n] *Padé approximant* of $x^{-1}\, f(x^{-1})$, so that $f(z)-A_n(z)/B_n(z) = O(z^{-2n-1})$ when $z \to \infty$. More precisely ([1], [3] theor. 3.5),

(1.3) $f(z)-A_n(z)/B_n(z) = \varepsilon_n(z)/B_n^2(z)$, with

$$\varepsilon_n(z) = \int_S B_n^2(t)\ (z-t)^{-1}\ w(t)dt = h_n\ z^{-1}+\ldots,\ z \notin S.$$

1.2. The problem of weight function reconstruction from moments.

We now come to the following computational problem : given a finite number of moments or, equivalently, recurrence coefficients (the problem of accurate numerical determination of these coefficients from the moments is not discussed here), how to get good estimates of the values $w(t)$, $t \in S$? The Padé approximant is very poor at that : its simple fractions expansion has the form $A_N(z)/B_N(z) = \sum_0^N \mu_{k,N}\ (z-t_{k,N})^{-1}$, so that the corresponding weight function is $w_N(t) = \sum_0^N \mu_{k,N}\ \delta(t-t_{k,N})$, useless if w is known to be regular on S. In fact, all that can be said in general is that, if the moment problem is determinate (which is always true when S is bounded), $\int_{t_0}^t w_n(u)du \xrightarrow[n\to\infty]{} \int_{t_0}^t w(u)du$ [13]. Therefore, a possible way of finding w is the numerical derivation of a smoothed version of the staircase function $\int_{t_0}^t w_n(u)du$.

More accurate values can be found by a method used by people working in solid state physics ([6], [10], [16]) : it consists in evaluating the remainder

$$R_N(z) = b_N^2/(z-a_N-b_{N+1}^2/(\ldots))$$

by replacing the coefficients by asymptotic estimates. Then, as

$$(1.4)\quad w(t) = \lim_{\varepsilon\to 0,\ \varepsilon>0} -\frac{1}{\pi}\,\mathrm{Im}\, f(t+i\varepsilon) \text{ if } w \text{ is continuous at } t, \text{ and}$$

$$(1.5)\quad f(z) = \frac{A_N(z)-R_N(z)\ A_{N-1}(z)}{B_N(z)-R_N(z)\ B_{N-1}(z)},$$

w is obtained either from a closed-form for $R_{N,+}(t) = \lim_{\varepsilon\to 0,\ \varepsilon>0} R_N(t+i\varepsilon)$ if one is available (for instance, see § 4.4) ; or by smoothing and derivating the function $\int_{t_0}^t w_M(u)du$, $M \gg N$, corresponding to a large order approximant of f constructed with the given coefficients a_n, b_n for $0 \leqslant n \leqslant N$, and their estimated values for $n > N$; or by actual computations of the estimated continued fraction of $R_N(z)$ for z complex near S (the convergence is slower and slower as one approaches S ; analytic continuation could be used if w is known to be piecewise analytic).

The present paper will now proceed on establishing the asymptotic behaviour of the coefficients a_n and b_n for some families of weight functions. However, a quantitative discussion of the method just presented will be found in appendix 1.

2. General methods of establishing the asymptotic behaviour of the coefficients.

The methods which will be presented are in fact aimed at the determination of the asymptotic behaviour of the orthogonal polynomials, but information on the coefficients will always appear as a by-product. Therefore, existing theories will be recalled (by Szegö [18] if S is a bounded interval, and Nuttall and Singh [15] if S is a set of intervals), merely in order to emphasize the recurrence coefficients aspect.

2.1. The Gram matrix.

The unknown polynomial B_n is written as a linear combination of polynomials taken from a known set $\{P_k(z) = p_k z^k + p_k' z^{k-1} + \dots\}_{k=0}^{\infty}$:

$$(2.1)\quad B_n(z) = \sum_{k=0}^{n} t_{n-k,n}\, P_k(z).$$

Expressing the orthogonality of B_n and P_k with respect to w if $k < n$, and $B_n(z) = z^n + \dots = p_n^{-1} P_n(z) + \dots$, one has the *Gram system*

$$\sum_{k=0}^{n} (G_n)_{i,k}\, t_{k,n}/(p_n h_n) = \delta_{i,0},\ i=0,\dots,n, \text{ where } (G_n)_{i,k} = \int_S P_{n-i}(t)\, P_{n-k}(t) w(t) dt.$$

The $t_{k,n}/(p_n h_n)$, k=0,...,n, are therefore the elements of the first column of the inverse of the Gram matrix G_n. As $t_{0,n} = p_n^{-1}$, B_n being monic, and $h_n = b_0^2 \dots b_n^2$, from (1.1) and (1.2),

$$(2.2)\quad h_n = b_0^2 \dots b_n^2 = 1/[p_n^2 (G_n^{-1})_{1,1}],$$

which gives b_n in terms of elements of G_{n-1}^{-1} and G_n^{-1}. Again, from (1.2), $B_n(z) = z^n - (a_0 + \dots + a_{n-1}) z^{n-1} + \dots$, so that, equating the coefficients of z^{n-1} in (2.1),

$$(2.3)\quad -a_0 - \dots - a_{n-1} = t_{0,n} p_n' + t_{1,n} p_{n-1} = p_n'/p_n + (p_{n-1}/p_n)(G_n^{-1})_{2,1}/(G_n^{-1})_{1,1}.$$

To go further, one should of course know asymptotic results about the inverses of the Gram matrices G_n. The most useful choice of the set $\{P_k\}$ is a set of orthogonal polynomials for some weight function also defined on S : then, the matrix G_n is not "too far" from a diagonal matrix, and the same is true for its inverse, provided sufficient constraints have been imposed on w (see also § 2.2). Actually, this method is nothing else than the modified moments method, well known for stabilizing recurrence coefficients determination [7]. This Gram matrix method is well suited to special classes of weight functions (for instance, continuous functions with algebraic singularities, see § 3). Considering this, the class of functions concerned by the Szegö theory, outlined in what follows, is amazingly large.

2.1.1. Polynomials orthogonal on a bounded interval. The Szegö theory.

Let S be the bounded interval (d_1,d_2), $P_0(z) = 1$, $P_k(z) = 2T_k\left(\frac{2z-d_1-d_2}{d_2-d_1}\right)$, $k=1,2,\ldots$, the Chebyshev polynomials orthogonal with respect to $[(t-d_1)(d_2-t)]^{-1/2}$. Then, $p_n = 1/t_{0,n} = [4/(d_2-d_1)]^n$, $p'_n/p_n = -n(d_1+d_2)/2$, $n=0,1,\ldots$.

With $2t = d_1+d_2+(d_2-d_1)\cos\varphi$,

(2.4) $v(\varphi) = |\sin\varphi|\ w(t)$,

and using $T_k(\cos\varphi) = \cos k\varphi$, one finds

$$2(1+\delta_{k,n})(1+\delta_{m,n})(G_n)_{m,k} = (d_2-d_1)(c_{2n-m-k}+c_{m-k}),\ m,k=0,\ldots,n,$$

where $c_k=c_{-k} = \int_{-\pi}^{\pi} \cos k\varphi\ v(\varphi)d\varphi = \int_{-\pi}^{\pi} e^{-ik\varphi}\ v(\varphi)d\varphi$.

Considering now the *Toeplitz system*

$$\sum_{k=0}^{2n} c_{m-k}\ c'_{k,2n}\ c'_{0,2n} = \delta_{m,0},\ m=0,\ldots,2n,$$

one has, adding the mth and the $(2n-m)^{th}$ rows, and taking into account that $c'_{k,2n}\ c'_{0,2n}$ and $c'_{2n-k,2n}\ c'_{0,2n}$ have then the same coefficient,

$$\sum_{k=0}^{n} (G_n)_{m,k}\ (c'_{k,2n}+c'_{2n-k,2n})c'_{0,2n} = (d_2-d_1)\delta_{m,0}/2,\ m=0,\ldots,n,$$

which shows that

$$\text{(2.5)}\quad \begin{aligned} t_{k,n} &= 2p_nh_n(c'_{k,2n}+c'_{2n-k,2n})c'_{0,2n}/(d_2-d_1),\ k=0,\ldots,n \\ &= t_{0,n}(c'_{k,2n}+c'_{2n-k,2n})/(c'_{0,2n}+c'_{2n,2n}). \end{aligned}$$

This reduction of the Gram system to the simpler Toeplitz system seems peculiar to Chebyshev polynomials. From Szegö theory, sketched in appendix 2, one has the following result ([18] § 12.7), which is not new, but is worth a theorem :

Theorem 1. If w is positive and integrable on the bounded interval (d_1,d_2), and if $\int_{d_1}^{d_2} [(t-d_1)(d_2-t)]^{-1/2} \ln w(t)dt > -\infty$, then

$$a_n \xrightarrow[n\to\infty]{} a_\infty = \frac{d_1+d_2}{2}, \quad b_n \xrightarrow[n\to\infty]{} b_\infty = \frac{d_2-d_1}{4}.$$

The condition on $\ln w$ prevents w from being too small on parts of (d_1,d_2). This condition is sufficient, although not necessary. The proof uses the results recalled in appendix 2 : as

$$c'_{k,2n}+c'_{2n-k,2n} \xrightarrow[n\to\infty]{} c'_k, \quad h_n = b_0^2\ldots b_n^2 = 2b_\infty^{2n+1}/[(c'_{0,2n}+c'_{2n,2n}] \sim 2b_\infty^{2n+1}/c'^2_0 \;;$$

$$-a_0-\ldots-a_{n-1} = p'_n/p_n+(p_{n-1}/p_n)(c'_{1,2n}+c'_{2n-1,2n})/(c'_{0,2n}+c'_{2n,2n})$$

$$= -n\, a_\infty + b_\infty\, c'_1/c'_0 + o(1), \text{ hence the result.}$$

2.2. Approximations of the weight function.

One considers a sequence $\{w_N\}$ of weight functions approximating w, and the corresponding sets of orthonormal polynomials $P_{k,N}(z) = h_{k,N}^{-1/2}\, B_{k,N}(z) = h_{k,N}^{-1/2}$ $z^k - h_{k,N}^{-1/2}(a_{0,N}+\ldots+a_{k-1,N})z^{k-1}+\ldots$. To know how B_n is close to $B_{n,N}$ when w_N is close to w, one has to study the eigenvalues of the Gram matrix $G_{n,N}$ and its inverse. Considering the real quadratic forms

$\sum_{i=0}^{n}\ \sum_{k=0}^{n} x_i (G_{n,N})_{i,k}\, x_k = \int_S \left(\sum_{i=0}^{n} x_i\, P_{n-i}(t)\right)^2 \frac{w(t)}{w_N(t)}\, w_N(t)dt$, one finds easily that the eigenvalues of $G_{n,N}$ are bounded from below and from above respectively by $\inf_{t\in S} \text{ess}\ w(t)/w_N(t)$ and $\sup_{t\in S} \text{ess}\ w(t)/w_N(t)$ ([8], § 7.7). Therefore, with

$$\varepsilon_N = \sup_{t\in S} \text{ess} \left| \frac{w_N(t)}{w(t)} - 1 \right|, \text{ one has}$$

$$\text{(2.6)} \quad (G_{n,N}^{-1})_{i,k} - \delta_{i,k} = O(\varepsilon_N)$$

and, from (2.2) and (2.3),

$$h_n/h_{n,N} = 1+O(\varepsilon_N), \quad b_n/b_{n,N} = 1+O(\varepsilon_N),$$

$$a_n = a_{n,N}+b_{n+1,N}\,\varepsilon'_{n+1,N}-b_{n,N}\,\varepsilon'_{n,N}\ , \quad \varepsilon'_{n,N} = O(\varepsilon_N).$$

This method, which avoids explicit use of the Gram matrix, requires two things :

1) w must be the limit of the sequence $\{w_N\}$ in such a way that $\varepsilon_N \xrightarrow[N\to\infty]{} 0$ in (2.6),

2) asymptotic estimates of $a_{n,N}$ and $b_{n,N}$ when n and $N \to \infty$ must be known.

For instance, taking the same example as in 2.1.1., with $w_N(t) = [(t-d_1)(d_2-t)]^{-1/2}/\rho_N(t)$, where ρ_N is a polynomial of degree N, the polynomials $P_{k,N}$ are known in closed-form when $n > N/2$ ([18], § 2.6), which gives immediately

$a_{n,N} = a_\infty$ and $b_{n,N} = b_\infty$ when $n > 1+N/2$. The first condition requires $[(t-d_1)(d_2-t)]^{1/2}\, w(t)$ to be continuous and bounded from below by a positive constant on $[d_1,d_2]$, a set of conditions much stronger than the hypotheses of theorem 1.

Summing up, this method (which will be used in § 4) gives quick results, although lacking in generality. However, considering the steps of Szegö theory, extensions seem possible, but difficult.

2.3. The integral equation method.

In contrast with the two preceding paragraphs, dealing with matrix methods, a functional setting, very likely equivalent with what has been seen, is presented briefly. For polynomials orthogonal on the unit circle, it is known as the Bernstein integral equation ([18], § 12.4). Here is the Nuttall and Singh version ([15], § 6) for a real set S ; expanding B_n in terms of orthonormal polynomials $P_{k,N}$ with respect to w_N :

$$B_n(z) = \sum_{k=0} P_{k,N}(z) \int_S B_n(t)\, P_{k,N}(t)\, w_N(t)dt,$$

and using orthogonality of B_n with respect to w,

$$B_n(z) = \int_S B_n(t)\, [\sum_{k=0}^{n-1} P_{k,N}(z)\, P_{k,N}(t)]\, (w_N(t)-w(t))dt + h_{n,N}^{1/2}\, P_{n,N}(z),$$

where the kernel polynomial may be replaced by its expression from Christoffel-Darboux formula.

3. Weight function with algebraic singularities on a bounded interval (d_1,d_2).

3.1. The asymptotic behaviour of the coefficients.

This section is concerned with the asymptotic behaviour of $a_n - a_\infty$ and $b_n - b_\infty$ when w presents weak algebraic singularities (which do not destroy the continuity nor the existence of a positive lower bound). Damped oscillations have been observed and reported in [6], discussed in [6] and [16], completely described in [10] (for square root singularities).

Theorem 2.

Let w be a weight function defined on a bounded closed interval $S = [d_1,d_2]$, with algebraic singularities at d_1, d_2 and a finite number of interior points of S. At each interior singular point t^*, it must be possible to associate a finite number of functions

$$w_k(t) = \lambda_k[s_k(t-t_k)]_+^{\alpha_k-1} = \lambda_k[s_k(t-t_k)]^{\alpha_k-1} \quad \text{if } s_k(t-t_k) \geq 0$$

$$= 0 \quad \text{if } s_k(t-t_k) \leq 0$$

for $t_k = t^*$, with $s_k = +1$ or -1, $\underline{\alpha_k > 1}$, such that $w(t) - \sum_{t_k=t^*} w_k(t)$ is m times continuously differentiable in a neighbourhood of t^*, with $m \geq \alpha+2$, $\alpha = \min_k \alpha_k$.

Moreover, $[(t-d_1)(d_2-t)]^{1/2}\, w(t)$ must be *bounded from below by a positive constant* on S, and m times continuously differentiable on any closed subset of S not containing interior singular points.

Then, with $t_k = a_\infty + 2b_\infty \cos \varphi_k$, $0 < \varphi_k < \pi$,

$$(3.1a)\quad a_n - a_\infty = -\sum_k \frac{\lambda_k (b_\infty \sin \varphi_k)^{\alpha_k}\, \Gamma(\alpha_k)}{\pi\, w(t_k) n^{\alpha_k}} \sin[2n\, \varphi_k + 2\arg D(e^{i\varphi_k}) - \frac{\pi}{2} s_k \alpha_k] +$$

$$O(n^{1-2\alpha}) + O(n^{-1-\alpha})$$

$$(3.1b)\quad b_n - b_\infty = -\sum_k \frac{\lambda_k (b_\infty \sin \varphi_k)^{\alpha_k}\, \Gamma(\alpha_k)}{2\pi\, w(t_k) n^{\alpha_k}} \sin[(2n-1)\varphi_k + 2\arg D(e^{i\varphi_k}) - \frac{\pi}{2} s_k \alpha_k] +$$

$$O(n^{1-2\alpha}) + O(n^{-1-\alpha}),$$

where the complex valued function D, satisfying

$D(e^{-i\varphi})\, D(e^{i\varphi}) = 2\pi\, |\sin \varphi|\, w(a_\infty + 2b_\infty \cos \varphi)$,

is defined in appendix 2.

3.2. Remarks.

1) At the endpoints, the theorem requires an inverse square root behaviour for the weight function. Nevertheless, as already suggested by the $\sin \varphi_k$ factors in (3.1), singularities at the endpoints seem to have a small influence on the result. For instance, for all the Jacobi polynomials, the perturbation is only $O(n^{-2})$([6], § 3.4). However, a more general theorem should be needed for a proof (see also § 3.5).

2) More than one t_k may correspond to a single singular point of w : k is merely an index of ordering one-sided elementary singular functions.

3) λ_k may be positive or negative, as $w(t_k) > 0$.

4) Logarithmic functions may be introduced by the consideration of confluent set of points.

3.3. Proof of theorem 2.

From (2.5), one needs good estimates of $c'_{k,N}$, k=0,1,N-1,N,N-2n. A first estimate is $c'_{k,N} \sim c'_k$, where $\sum_0^\infty c'_k\, e^{ik\varphi} = 1/D(e^{i\varphi})$ (see (A2.7) appendix 2). In order to know the order of magnitude of the error, one needs the behaviour of the coefficients c'_k, which will be derived from the singularities of $D(e^{i\varphi})$. From (2.4),

$v(\varphi) \sim v(\varphi_k) + |\sin\varphi_k|^{\alpha_k} \sum_{\varphi_k=\varphi} {}^*\lambda_k [s_k'(\varphi-\varphi_k)]_+^{\alpha_k-1} (2b_\infty)^{\alpha_k-1}$ near a singular point φ^* (if φ^* is singular, so is $-\varphi^*$), with $s_k' = -s_k$ sign $(\sin\varphi_k)$. For the logarithm :

$\ln 2\pi\, v(\varphi) \sim \ln 2\pi\, v(\varphi_k) + \dfrac{|\sin\varphi_k|^{\alpha_k}}{v(\varphi_k)} \Sigma\, \lambda_k [s_k'(\varphi-\varphi_k)]_+^{\alpha_k-1} (2b_\infty)^{\alpha_k-1}$, powers starting with $2\alpha_k-2$ being neglected. From the hypotheses on w, and therefore on v, the following writing is valid :

$$\ln 2\pi\, v(\varphi) = \mathrm{Re}\,\{\sum_k \frac{\lambda_k |\sin\varphi_k|^{\alpha_k} (2b_\infty)^{\alpha_k-1}}{v(\varphi_k)\sin\pi\alpha_k} \lim_{r\to1,r<1} (1-re^{i(\varphi-\varphi_k)})^{\alpha_k-1} e^{-is_k'\alpha_k\pi/2}\}$$

$$+ v_1(\varphi) + v_2(\varphi),$$

where $v_1(\varphi)$ contains similar expressions with higher powers α_k, α_k+1, ..., and v_2 is m times continuously differentiable on the set of real numbers. Indeed,

$$\lim_{r\to1,r<1} (1-re^{i(\varphi-\varphi_k)})^{\alpha_k-1} = \left|2\sin\frac{\varphi-\varphi_k}{2}\right|^{\alpha_k-1} \exp\{i(\alpha_k-1)[(\varphi-\varphi_k)/2 - \pi\,\mathrm{sign}(\varphi-\varphi_k)/2]\}.$$

Taking the conjugate function, which is nothing else than 2 Im ln $D(e^{i\varphi})$ (from (A2.6) appendix 2), real parts are replaced by imaginary parts, and it is known that the conjugate of v_2 is still m-1 times continuously differentiable ([23] § 3.13). Therefore,

$$(3.2)\quad \ln D(e^{i\varphi}) \sim \ln D(e^{i\varphi_k}) + \sum_{\varphi_k=\varphi^*} \frac{\lambda_k |\sin\varphi_k|^{\alpha_k} (2b_\infty)^{\alpha_k-1}}{2v(\varphi_k)\sin\pi\alpha k} |\varphi-\varphi_k|^{\alpha_k-1}$$

$$\exp\{-i\frac{\pi}{2}[s_k'\alpha_k + (\alpha_k-1)\,\mathrm{sign}\,(\varphi-\varphi_k)]\},\ \varphi\sim\varphi^*.$$

From this, one has the behaviour of $1/D(e^{i\varphi})$ near φ^* and, using Lighthill's asymptotic formulas for the Fourier coefficients ([12], § 5.5 theorem 30 ; see also [5] § 2.8),

$$c_N' = -\sum_k \frac{\lambda_k |\sin\varphi_k|^{\alpha_k} (2b_\infty)^{\alpha_k-1} \Gamma(\alpha_k)}{2\pi D(e^{i\varphi_k})\, \check{v}(\varphi_k)\, N^{\alpha_k}} e^{-i\frac{\pi}{2}s_k'\alpha_k - iN\varphi_k} + O(N^{-1-\alpha})$$

$$+ O(N^{1-2\alpha}) + O(N^{1-m}),$$

where $N^{-1-\alpha}$ and $N^{1-2\alpha}$ come from the neglected powers, and N^{1-m} from the regular part of the function.

Now, we apply Hartwig and Fisher's result ([9], theorem 6) :

$$c_{k,N}'\, c_{0,N}' = c_k'\, c_0' + O((N-k)^{1-\alpha} N^{\alpha}),$$

which shows the validity of $c_{k,N}' \sim c_k'$ for k=O(1), but not for k=O(N), where the error has the same order of magnitude as c'_k itself. Fortunately, we have also

$$c'_{k,N}\, c'_{0,N} = c'_0 \text{ coeff. of } e^{ik\varphi} \text{ of } \frac{1}{D(e^{-i\varphi})} \left[\frac{D(e^{-i\varphi})}{D(e^{i\varphi})}\right]_N + O(k^{-\alpha} N^{2-2\alpha}),$$

where $[D(e^{-i\varphi})/D(e^{i\varphi})]_N$ means the Fourier expansion limited to powers of $e^{i\varphi}$ less or equal than N ([9] theorem 6). Actually, only $c'_{N,N}$ is needed :

$$c'_{N,N} = c'_0 \text{ coeff. of } e^{iN\varphi} \text{ of } D(e^{-i\varphi})/D(e^{i\varphi}) + O(N^{2-3\alpha}).$$

The imaginary part of (3.2) yields the behaviour of $\ln D(e^{-i\varphi})/D(e^{i\varphi}) = -2i \arg D(e^{i\varphi})$ near a singular point. Exponentiating and using Lighthill's estimates,

$$c'_{N,N} \sim - c'_0 \sum_k \frac{\lambda_k \,|\sin \varphi_k|^{\alpha_k} (2b_\infty)^{\alpha_k - 1}\, \Gamma(\alpha_k)}{2\pi\, v(\varphi_k)\, N^{\alpha_k}} \exp[-2i \arg D(e^{i\varphi_k}) - i \tfrac{\pi}{2} s'_k \alpha_k - iN\varphi_k].$$

Summing the contributions of φ_k and $-\varphi_k$, (3.1) follows from (A2.3), (A2.4), (A2.5) with m=1 (appendix 2) and (2.5).

3.4. Relation with Hodges result.

The equivalent of (3.1) is obtained by Hodges [10] (for α_k=3/2) by a much shorter argument, valid if $\sum_0^\infty |a_k - a_\infty|$ and $\sum_0^\infty |b_k - b_\infty| < \infty$ is assumed. As this has just been established, we may accept his reasoning, which makes use of $\lim_{n\to\infty} e^{in\varphi} Q_{n,+}(a_\infty + 2b_\infty \cos\varphi)$ (see (A1.2) appendix 1). The existence of this limit is a consequence of the fast convergence of the sequences of coefficients of the recurrence relation (A1.3) (appendix 1), the value is $-i(2b_\infty)^{-1/2}\, D(e^{-i\varphi})/\sin\varphi$, using Barrett's result [2]. The essence of Hodges proof is to consider $a_n - a_\infty$ and $b_n - b_\infty$, from (A1.2) (appendix 1), as approximations of Fourier coefficients of the expansion of $w(a_\infty + 2b_\infty \cos\varphi)\sin^2\varphi\,(D(e^{-i\varphi}))^{-2}$, which is nothing else than $(2\pi)^{-1} \sin\varphi\, D(e^{i\varphi})/D(e^{-i\varphi})$, showing the importance of this function.

3.5. Extension to stronger algebraic singularities.

The present proof is no more valid when $\alpha < 1$ or when $w(t_k)=0$, i.e. when $\ln v(\varphi)$ is not bounded. However, the following general technique is still applicable : what is needed is an accurate estimate of the first column x_N of the inverse of the Toeplitz matrix C_N (see appendix 2). If y_N is a guess for x_N, the residue $r_N = C_N y_N - e_N$ may be computed (e_N is the first column of I_N, the unit matrix of order N), and $y_N - x_N = C_N^{-1} r_N$, which asks for a bound of C_N^{-1}. This last problem may be solved by extreme eigenvalues considerations ([8] § 5.4, [22] § 1.4 and chap. 2). Of course, the same ideas hold for estimates of the whole matrix C_N^{-1} : if X_N is proposed,

$$C_N^{-1} = X_N + (I_N - X_N C_N) C_N^{-1} = X_N + C_N^{-1}(I_N - C_N X_N),$$

thus, if $\|I_N - C_N X_N\| = \varepsilon_N \underset{N\to\infty}{\to} 0$, $C_N^{-1} = X_N + X_N Y_N$, with $\|Y_N\| = O(\varepsilon_N)$.

This may be refined, using the symmetry about the secondary diagonal for Toeplitz matrices, a property invariant by inversion ([9], [20] § 2, see also appendix 2) :

$$C_N^{-1} = X_N + Z_N(I_N - C_N X_N) + (I_N - Z_N C_N)C_N^{-1}(I_N - C_N X_N),$$

where Z_N is the matrix X_N turned upside down and right to left.

The starting matrix used in [9] and [20] is $(X_N)_{i,j} = \sum_{k=0}^{\min(i,j)} c'_{i-k} c'_{j-k}$, $i,j \leq N$, justified for fixed i and j by the Szegö theory (appendix 2). This should be modified for large i and j when strong singularities are present. The resulting asymptotic behaviour for $a_n - a_\infty$ and $b_n - b_\infty$ will very likely exhibit an N^{-1} factor, as suggested by Hartwig and Fisher's conjecture for the determinant of C_N [9], and the example $w(t) = |t|^{\alpha-1}$, $-1 \leq t \leq 1$, $\alpha > 0$, solved by Gauss continued fraction ((89.16) of [19] with $F(\alpha/2, 1, \alpha/2+1, z^{-2})$) : one finds $a_n = 0$, $b_n = 1/2 - (-1)^n (\alpha-1)/(4n) + O(n^{-2})$.

4. System of intervals.

4.1. Functions needed for the description of the coefficients.

Let S be a collection of m bounded open intervals $(d_1, d_2), \ldots, (d_{2m-1}, d_{2m})$, $d_k < d_{k+1}$. The most important work concerning orthogonal polynomials associated with a weight function defined on such set is the one by Nuttall and Singh [15], who dealt actually with the distribution of poles of Padé approximants of functions with complex branch points. The purpose of this section is to take from this work informations on the recurrence relation coefficients.

The periodic or quasi periodic oscillating behaviour of these coefficients has been observed in [6] § 3.4, where a correct formula for the amplitude is given for m=2. It will be shown that the asymptotic behaviour is explained by special Abelian functions, the periods and amplitudes depending only on S, whereas the weight function w has an influence on the phase of the oscillations.

To introduce these special functions, one considers

$$X(z) = \prod_{k=1}^{2m} (z - d_k)$$

and $X^{1/2}(z)$, completely defined *outside* S as a continuous function, positive when $z > d_{2m}$. It is important to note that sign $X^{1/2}(z) = (-1)^{m-k}$ when $d_{2k} < z < d_{2k+1}$, $k=1,\ldots,m-1$. On S, one defines $X_+^{1/2}(t) = \lim_{\varepsilon\to 0, \varepsilon>0} X^{1/2}(t+i\varepsilon)$. The values of this function are pure imaginary and sign $i^{-1} X_+^{1/2}(t) = (-1)^{m-k}$ for $d_{2k-1} < t < d_{2k}$,

$k=1,\dots,m$.

As a set of independent periods of the Abelian (hyperelliptic) integrals of the first kind $U_r(z) = \int_{d_1}^{z} t^{r-1} X^{-1/2}(t)dt$, $r=1,\dots,m-1$, let us choose ([11] § 14)

$$K_{r,k}=2 \sum_{s=1}^{k} \int_{d_{2s}}^{d_{2s+1}} t^{r-1} X^{-1/2}(t)dt,\ iK'_{r,k}=2 \int_{d_{2k+1}}^{d_{2k+2}} t^{r-1} X_+^{-1/2}(t)dt,\ r,k=1,\dots,m-1.$$

We consider finally the following problem : for given real numbers $W_1,\dots,W_{m-1}$, to find $\alpha_1,\dots,\alpha_{m-1}$ such that

$$(4.1)\quad \sum_{k=1}^{m-1} s_k \int_{d_1, t\notin S}^{\alpha_k} t^{r-1} X^{-1/2}(t)dt = W_r + \sum_{k=1}^{m-1} M_k K_{r,k},\ r=1,\dots,m-1,$$

where each $s_k=+1$ or -1 and M_k are integers. This problem has always a unique solution $\{\alpha_k, s_k, M_k\}_{k=1}^{m-1}$; furthermore, one has $d_{2k} \leqslant \alpha_k \leqslant d_{2k+1}$, $k=1,\dots,m-1$. Indeed, (4.1) is a real numbers formulation of the Jacobi inversion problem (discussed in [17] § 4.8 and [15] § 4)

$$\sum_{k=1}^{m-1} s_k U_r(\alpha_k) = W_r + \sum_{k=1}^{m-1} M_k K_{r,k} + \sum_{k=1}^{m-1} M'_k iK'_{r,k},\ r=1,\dots,m-1,\ M_k, M'_k \in Z,$$ problem

known to have at least one solution. An expression of a solution in terms of theta functions ([11] p. 142) shows that, if the W_r are real, $\alpha(z) = \prod_{k=1}^{m-1} (z-\alpha_k)$ is real and presents alternate signs when $z=d_2, d_4,\dots,d_{2m}$. Confluence of the α_k being impossible, unicity follows ([15] lemma 4.3). Finally, the way the integrals of the left-hand side of (4.1) are written, the imaginary periods disappear.

The fact that the s_k and the M_k are completely determined by the W_r is shown clearly when $m=2$: $s \int_{d_2}^{\alpha} X^{-1/2}\, dt = W+MK$. Dividing by K, the left-hand side lies between -1/2 and 1/2, which shows that M is the closest integer to W/K ; α is found by the inversion of an elliptic integral, i.e. is an elliptic function of W. In general, any symmetric rational function of $\alpha_1,\dots,\alpha_{m-1}$ is a meromorphic periodic function of $W_1,\dots,W_{m-1}$ (Jacobi-Abel functions : [17] § 4.12).

4.2. The asymptotic behaviour of the coefficients.

Theorem 3.

If $|X(t)|^{-1/2} w(t)$ is continuous, bounded, and bounded from below by a positive constant on S, then

$$(4.2a)\qquad a_n - d_1 + \frac{X'(d_1)}{4\alpha_n(d_1)} \left[\frac{1}{\alpha_{n-1/2}(d_1)} + \frac{1}{\alpha_{n+1/2}(d_1)}\right] \underset{n\to\infty}{\to} 0,$$

(4.2b) $$b_n + X'(d_1)\,|\alpha_{n-1}(d_1)\alpha_n(d_1)|^{-1/2}\,|\alpha_{n-1/2}(d_1)|^{-1}/4 \underset{n\to\infty}{\to} 0,$$

where $\alpha_n(z) = \prod_{k=1}^{m-1} (z-\alpha_{n,k})$, $\alpha_{n,1},\ldots,\alpha_{n,m-1}$ solves the problem (4.1) with

(4.3) $$W_r = \frac{i}{\pi}\int_S X_+^{-1/2}(t)\, t^{r-1} \ln[w(t)|X_+(t)|^{-1/2}]dt + (2n+m+1)\int_{-\infty}^{d_1} X^{-1/2}(t)\, t^{r-1}dt.$$

The proof will show that (4.2) is true when the weight function is

(4.4) $$w_N(t) = i^{-1}\, X_+^{1/2}(t)/\rho_N(t),$$

where ρ_N is a polynomial. Using the method of § 2.2 with a sequence of such weight functions approximating w, and from the continuity of the Jacobi-Abel functions, the result follows.

It is here conjectured that (4.2) is still valid without the continuity and positive lower bound requirement for w, but a proof should be at least as elaborated as Szegö's proof for one interval ... (see the end of appendix 2).

To show (4.2) when (4.4) holds, one establishes first that, if ρ is a real monic polynomial of degree N-m+1 with its zeros outside S, and such that sign $\rho(t) = (-1)^{m-k}$ for $d_{2k-1} < t < d_{2k}$, k=1,...,m, then

(4.5) $$f(z) = \int_S \pi^{-1}\, i^{-1}\, X_+^{1/2}(t)\, \frac{dt}{\rho(t)(z-t)} = \frac{Y(z)-X^{1/2}(z)}{\rho(z)}, \quad z \notin S,$$

where Y is the polynomial interpolating $X^{1/2}$ at the zeros of ρ, and such that $f(z) = O(|z|^{-1})$ when $z\to\infty$. Indeed, the function is then holomorphic outside S, and can be expressed by a Cauchy integral on a contour which may shrink to S. Next, using (4.5) and the Padé property (1.3) of the continued fraction approximants,

(4.6) $$C_n(z) - B_n(z)\, X^{1/2}(z) = \rho(z)\varepsilon_n(z)/B_n(z),$$

where $C_n(z) = Y(z)B_n(z) - \rho(z)A_n(z)$.
Multiplying by

(4.7) $$C_n(z) + B_n(z)\, X^{1/2}(z) = 2B_n(z)\, X^{1/2}(z) + \rho(z)\varepsilon_n(z)/B_n(z),$$

the right-hand side of the product vanishes at the zeros of ρ and behaves at infinity like $2h_n\, z^N$ when n > N/2-m, and the left-hand side is a polynomial : therefore,

(4.8) $$C_n^2(z) - B_n^2(z)\, X(z) = 2h_n\, \rho(z)\alpha_n(z),\ n > N/2-m,$$

where α_n is a monic polynomial of degree m-1. The connection with the Jacobi-Abel functions is explained in appendix 3.

The easiest way to get the recurrence coefficients from values of the orthogonal polynomials is to consider the companion weight function $w(t)/(t-d_1)$ [4]. Writing $B_{n-1/2}$ for the corresponding orthogonal polynomial of degree n, one has, [4] § 2 :

(4.9) $B_{n+1}(z) = B_{n+1/2}(z) - u_{n+1/2}\, B_n(z)$

(4.10) $B_{n+1/2}(z) = (z-d_1)\, B_n(z) - u_n\, B_{n-1/2}(z)$

(4.11) $a_n = d_1 + u_{n+1/2} + u_n,\ b_n^2 = u_n\, u_{n-1/2},\ n=0,1,\ldots,$

(4.12) $u_n = h_n/h_{n-1/2},\ n=0,1/2,1,3/2,\ldots,$

In these relations, B_n may be replaced by A_n or C_n.
Now, we just have to take (4.8) with $z=d_1$:

(4.13) $C_n^2(d_1) = 2h_n\, \rho(d_1)\alpha_n(d_1)$ for integer $n > N/2-m$; for half-integer n, $\rho(z)$ must be replaced by $(z-d_1)\,\rho(z)$ and $C_n(d_1) = 0$ but, dividing (4.8) by $z-d_1$,

(4.14) $-B_n^2(d_1)\, X'(d_1) = 2h_n\, \rho(d_1)\alpha_n(d_1)$ for half-integer $n > N/2-m$.

From (4.9) and (4.10) :
$C_n(d_1) = (-1)^n\, u_{n-1/2}\, u_{n-3/2} \cdots u_{1/2}\, Y(d_1),\ B_{n-1/2}(d_1) = (-1)^n\, u_{n-1}\, u_{n-2} \cdots u_0,$ $n=0,1,\ldots$.
Comparing with (4.13), (4.14), using $h_n = u_n\, u_{n-1/2}\, u_{n-1} \cdots U_0\, h_{-1/2}$ from (4.12) and $Y(d_1) = -\rho(d_1)\, h_{-1/2}$ from the interpolation properties of Y, one has finally

$$u_n = -\frac{X'(d_1)}{4\alpha_{n-1/2}(d_1)\alpha_n(d_1)}$$

for integer and half-integer values of $n > (N+1)/2-m$, and (4.2) follows from (4.11).

4.3. Recurrence relations for $\alpha_n(d_1)$ when m=2.

General recurrence relations for the values $\alpha_n(d_1)$, ready for computational use, can very likely be constructed, either from the theory of the Jacobi-Abel functions ([17] § 4.12, theorems 3 and 5), or directly from (4.8) and (4.9).

For m=2, let us define the even elliptic function F by

$$\pm\int_{d_2}^{F(x)} X^{-1/2}(t)\,dt = x + MK_1 + iM'K_1'.$$

$F(x_n) = \alpha_{n,1}$ for $x_n = c^t + 2nJ$, $J = \int_{-\infty}^{d_1} X^{-1/2}(t)\,dt$. The poles of F are $\pm(J + iK_1'/2)$.

As $(F(x-J)-d_1)^{-1} + (F(x+J)-d_1)^{-1}$ and $(F(x-J)-d_1)^{-1}\,(F(x+J)-d_1)^{-1}$ are even elliptic functions with double poles at the same points, they are quadratic polynomials of

F(x). One has finally

$$(\alpha_{n-1/2,1}-d_1)^{-1} + (\alpha_{n+1/2,1}-d_1)^{-1} = \frac{4}{X'(d_1)}(\alpha_{n,1}-d_1)(\alpha_{n,1}-d_1-c_1)$$

$$(\alpha_{n-1/2,1}-d_1)^{-1}(\alpha_{n+1/2,1}-d_1)^{-1} = \frac{4}{(X'(d_1))^2}(\alpha_{n,1}-d_1)(c_2(\alpha_{n,1}-d_1) - X'(d_1)),$$

with $c_1 = (d_2+d_3+d_4-3d_1)/2$ and $c_2 = (\frac{d_4-d_1+d_3-d_2}{2})^2 - (d_3-d_1)(d_4-d_1)$, which give simpler forms for a_n and b_n :

$$a_n \sim 2d_1+c_1-\alpha_{n,1}, \quad b_n^2 \sim \frac{c_2}{4} - \frac{X'(d_1)}{4(\alpha_{n-1/2,1}-d_1)}.$$

As $d_2 \leqslant \alpha_{n,1} \leqslant d_3$, $(d_1+d_2-d_3+d_4)/2 \leqslant a_n \leqslant (d_1-d_2+d_3+d_4)$ and $(d_4-d_1-d_3+d_2)/4 \leqslant b_n \leqslant (d_4-d_1+d_3-d_2)/4$, confirming the bounds given in [6] § 3.4.

4.4. Weight function reconstruction.

The function $R_n(z) = b_n^2/(z-a_n-b_{n+1}^2/(\ldots))$, where b_k and a_k, $k \geqslant n$, are replaced by estimates based on (4.2), corresponds to a function of the form $f(z) = [Y(z) - X^{1/2}(z)]/\rho(z)$, Y interpolating $X^{1/2}$ at the zeros of the polynomial ρ. From the continued fraction identity (1.5), one has

$$R_n(z) = \frac{C_n(z)-B_n(z)\,X^{1/2}(z)}{C_{n-1}(z)-B_{n-1}(z)\,X^{1/2}(z)} = \frac{C_n(z)\,C_{n-1}(z)-B_n(z)\,B_{n-1}(z)\,X(z) + (C_n(z)\,B_{n-1}(z) - C_{n-1}(z)\,B_n(z))\,X^{1/2}(z)}{C_{n-1}^2(z)-X(z)\,B_{n-1}^2(z)}.$$

The denominator is replaced by (4.8), $C_n(z)\,B_{n-1}(z)-C_{n-1}(z)\,B_n(z) = b_{n-1}^2\ldots b_0^2\,A_{-1}(z)\rho(z) = -h_{n-1}\,\rho(z)$ from (1.2), and $C_n\,C_{n-1} - B_n\,B_{n-1}\,X$ vanishes at the zeros of ρ, from the definition $C_n = YB_n - \rho A_n$ and the interpolation properties of Y. Therefore,

$$R_n(z) = \frac{Y_n(z)-X^{1/2}(z)}{2\alpha_{n-1}(z)},$$

where Y_n is a polynomial of degree m such that $R_n(z) = O(|z|^{-1})$ when $z\to\infty$, but which *need not interpolate* $X^{1/2}$ at the zeros of α_{n-1}. From this, it is possible to compute in closed-form $R_{n,+}(t) = \lim_{\varepsilon\to 0,\varepsilon>0} R_n(t+i\varepsilon) = (Y_n(t)-X_+^{1/2}(t))/(2\alpha_{n-1}(t))$ and $w(t) = -\pi^{-1}\,\mathrm{Im}[b_0^2/(t-a_0-b_1^2/\ldots\, t-a_{n-1}-R_{n,+}(t))]$, where a_k and b_k, $0 \leqslant k \leqslant n-1$, are the original values of the coefficients.

For m=2, the knowledge of a_{n-1} and b_n yields

$$R_n(z) = \frac{z^2 - \frac{d_1+d_2+d_3+d_4}{2} z + \sum_{i<j} d_i d_j + 2b_n^2 - X^{1/2}(z)}{2(z + a_{n-1} - \frac{1}{2} \sum_1^4 d_k)} .$$

4.5. Numerical example.

The preceding formula has been used to reconstruct the weight function defined by $w(t) = 1$ for $1 \leqslant t \leqslant 2$ and $w(t) = (t-4)^{1/2}$ for $4 \leqslant t \leqslant 8$. The true recurrence coefficients are

n	0	1	2	3	4	5	..	9	10
a_n	5.63	3.43	5.24	4.83	3.64	5.40		4.27	4.07
b_n^2	6.33	4.13	3.61	1.66	4.44	2.76		1.88	5.05

Some values of the reconstructed w with n=5 and n=10 are :

t	0.25	0.5	0.75	4.25	6	7	7.5	7.75
w n=5	0.98	0.97	1.07	0.48	1.40	1.75	1.84	1.92
w n=10	1.00	1.00	0.98	0.50	1.41	1.73	1.86	1.95
w exact	1.00	1.00	1.00	0.50	1.41	1.73	1.87	1.94

A Gibbs phenomenon appears at 1,2 and 8 where $w(t)\, X^{-1/2}(t)$ is singular.

To test the validity of (4.2), the integrals of (4.3) have been computed :

$\int_{-\infty}^{d_1} X^{-1/2}(t)dt = 0.681$, $\int_{d_2}^{d_3} X^{-1/2}(t)dt = -1.041$, $\frac{i}{\pi} \int_S \ln \frac{w(t)}{|X_+(t)|^{1/2}} X_+^{-1/2}(t)dt =$ -0.389, and one finds $\alpha_{-1/2,1} = 3.976$, $\alpha_{0,1} = 2.552$, $\alpha_{1/2,1} = 2.203$, $\alpha_{1,1} = 3.938$, ..., and the estimates for a_n and b_n :

n	0	1	2	3	4	5	..	9	10
a_n	4.95	3.56	5.35	4.73	3.69	5.44		4.24	4.10
b_n^2	1.58	4.17	2.99	1.63	4.59	2.60		1.90	5.05.

4.6. Further remarks on the asymptotic behaviour of the coefficients.

A more complete asymptotic behaviour may be obtained in a fast way by the equivalent of Hodges method, already discussed in § 3.4 : the relation (A1.2) (appendix 1) assuming $Q_{n,+}(t) \sim Q_n^{(1)}(t) = q_n(t)e^{-in\varphi}$ (from the end of appendix 3, with $Z=e^{i\varphi}$ for $t \in S$), appears very roughly as a Fourier series. Therefore, if w is singular at t_0, corresponding to φ_0, a factor $\sin(2n\varphi_0 + c^t)$ may be expected in the behaviour of $a_n - a'_n$ and $b_n - b'_n$ (a'_n and b'_n being the estimates of (4.2)), but more quantitative results need a closer study.

Appendix 1. Influence of the coefficients on the weight function.

Let $f(z) = b_0^2/(z-a_0-b_1^2/\dots)$ and $f_1(z) = b_0'^2/(z-a_0'-b_1'^2/\dots)$ correspond to two weight functions w and w_1 defined on S. If the a_n and a_n', b_n and b_n' are close together, what can be said on $w(t)-w_1(t)$? First, we consider z complex. Then, $f(z)$ may be written as the head element (first row, first column) of the inverse of a symmetric tridiagonal infinite matrix ([19] § 60, [13]) :

$$f(z) = b_0^2[(zI-T)^{-1}]_{1,1},$$

with $(T)_{k,l} = a_{k-1}$ if $k=l$, b_k if $l=k+1$ or $k=l+1$, 0 otherwise. For z complex, this inverse is bounded (if the moment problem is determinate, which is always assumed), usual matrix algebra holds, and one has (if $b_0'=b_0$)

$$f(z)-f_1(z) = b_0^2[(zI-T)^{-1}\ (T-T_1)(zI-T_1)^{-1}]_{1,1}.$$

As the elements of the first row and the first column of $(zI-T)^{-1}$ are $b_0^{-1}(P_{k-1}(z)\ f(z)-N_{k-1}(z)) = b_0^{-1}\ Q_{k-1}(z)$, $k=1,2,\dots$, where P_k is the *orthonormal* polynomial of degree k with respect to w, and N_k the corresponding numerator of the kth approximant of f ([19]§ 60), one has

$$(A1.1)\ f(z)-f_1(z) = \sum_{k=0}^{\infty} (a_k-a_k')Q_k(z)Q_k^{(1)}(z) + \sum_{k=1}^{\infty} (b_k-b_k')(Q_{k-1}(z)Q_k^{(1)}(z) + Q_{k-1}^{(1)}(z)\ Q_k(z)).$$

From (1.4), with $Q_{k,+}(t) = \lim_{\varepsilon\to 0,\varepsilon>0} Q_k(t+i\varepsilon)$, if these limits exist :

$$(A1.2)\ w(t)-w_1(t) = -\pi\ \mathrm{Im}\ \{\sum_{k=0}^{\infty} (a_k-a_k')Q_{k,+}(t)Q_{k,+}^{(1)}(t) + \sum_{k=0}^{\infty} (b_k-b_k')(Q_{k-1,+}(t)\ Q_{k,+}^{(1)}(t) + Q_{k-1,+}^{(1)}(t)Q_{k,+}(t))\},$$

if, for instance, the series (A1.1) converges uniformly on a set containing S or a subinterval of S.

If bounds for the functions $Q_{k,+}$ and $Q_{k,+}^{(1)}$ are known, it is thus possible to estimate the error on w from the errors on the coefficients. Moreover, as $Q_{k,+}(t)$ behaves often like $e^{-ik\varphi}$ when $k\gg 1$ (see [2], recalled in § 3.4, and the end of appendix 3), *Fourier series methods* may be used to discuss (A1.2).

The behaviour of the functions $P_k(z)$ and $Q_k(z)$ is related to the coefficients by the recurrence relations

(A1.3) $b_n\ R_n(z) = (z-a_{n-1})R_{n-1}(z)-b_{n-1}\ R_{n-2}(z)$, $n \geqslant 2$, with $R_n=P_n$ or Q_n, or, in matrix notation, $S_n=M_n\ S_{n-1}$:

$$\begin{bmatrix} R_{n-1} \\ R_n \end{bmatrix} = \begin{bmatrix} 0 & 1 \\ -b_{n-1}/b_n & (z-a_{n-1})/b_n \end{bmatrix} \begin{bmatrix} R_{n-2} \\ R_{n-1} \end{bmatrix}.$$

Comparison with known functions $P_n^{(1)}$ and $Q_n^{(1)}$ is made with the matrix

$$W_n^{(1)} = \begin{bmatrix} P_{n-1}^{(1)} & Q_{n-1}^{(1)} \\ P_n^{(1)} & Q_n^{(1)} \end{bmatrix} :$$

$$W_n^{(1)-1} S_n = (W_n^{(1)-1} M_n W_{n-1}^{(1)}) W_{n-1}^{(1)-1} S_{n-1}$$

$$= \prod_{k=2}^{n} (W_k^{(1)-1} M_k W_{k-1}^{(1)}) W_1^{(1)-1} S_1.$$

If $M_n - M_n^{(1)}$ decreases fast enough, the product converges, yielding asymptotic estimates (see [14] for similar techniques).

Appendix 2. The inverse of a Toeplitz matrix and the Szegö theory.

One considers a function v even, positive and integrable on $[-\pi,\pi]$ and the associate Toeplitz matrix C_N of order N+1 $(C_N)_{k,m} = c_{k-m}$, k,m=1,...,N+1, with $c_k = c_{-k} = \int_{-\pi}^{\pi} e^{-ik\varphi} v(\varphi)d\varphi = 2\int_0^{\pi} \cos k\varphi\, v(\varphi)d\varphi$. C_N is real, symmetric, and may be shown to be positive definite : indeed, if $\sum_0^N x_k^2 = 1$, $\sum_{k=0}^{N}\sum_{m=0}^{N} x_k c_{k-m} x_m = \int_{-\pi}^{\pi} |\sum_0^N x_k e^{ik\varphi}|^2 v(\varphi)d\varphi$,

which shows also that the eigenvalues of C_N lie between $2\pi \operatorname{inf\,ess}_{-\pi\leqslant\varphi\leqslant\pi} v(\varphi)$ and $2\pi \operatorname{sup\,ess}_{-\pi\leqslant\varphi\leqslant\pi} v(\varphi)$ ([8] chap. 5, [22] § 1.4).

As the first element of the first column of C_N^{-1} is a ratio of two positive determinants, it is the square of a real number $c'_{0,N}$, chosen to be positive. The elements of the first column of C_N^{-1} are written $c'_{k,N}\, c'_{0,N}$, k=0,...,N. One considers now the lower triangular matrix L_N of entries $c'_{k-m,N-m}$ at row k+1 and column m+1 ; k,m=0,...,N, $k \geqslant m$. It is then easy to show that $C_N L_N$ is an upper triangular matrix with diagonal elements $c'^{-1}_{0,N},\dots,c'^{-1}_{0,0}$. Therefore, as C_N and C_N^{-1} are symmetric,

(A2.1) $C_N^{-1} = L_N L'_N$,

the Cholesky factorisation of C_N^{-1}. Finally, using the symmetry of C_N about its secondary diagonal,

(A2.2) $C_N^{-1} = P_N C_N^{-1} P_N = (P_N L_N P_N)(P_N L'_N P_N)$,

where P_N is the matrix whose sole nonzero elements are ones on the secondary diagonal.

Important identities follow from the comparison of (A2.1) and (A2.2) :

(A2.3) $c'^{2}_{0,N} = \sum_{m=0}^{N} c'^{2}_{m,m}$,

which shows that $c'_{0,N}$ increases with N,

(A2.4) $c'_{m,N}\, c'_{0,N} = \sum_{s=m}^{N} c'_{s,s}\, c'_{s-m,s} = c'_{m,N-1}\, c'_{0,N-1} + c'_{N,N}\, c'_{N-m,N}$, or

(A2.5) $\dfrac{c'_{m,N} + c'_{N-m,N}}{c'_{0,N} + c'_{N,N}} = \dfrac{c'_{m,N-1} + c'_{N-m,N-1}}{c'_{0,N-1}}$, $m=1,\ldots,N-1$.

We investigate now the asymptotic properties of the $c_{m,N}$ for fixed m and $N\to\infty$. First of all, we consider the infinite set of equations $\sum_{m=0}^{\infty} c_{k-m}\, c'_m\, c'_0 = \delta_{k,0}$, $k=0,1,\ldots$. If these series converge, one has $2\pi\, v(\varphi)\, (D(e^{i\varphi}))^{-1}\, c'_0 = 1+$ negative powers of $e^{i\varphi}$, with

$$(D(e^{i\varphi}))^{-1} = \sum_{m=0}^{\infty} c'_m\, e^{im\varphi}.$$

As v is even, this means

(A2.6) $2\pi\, v(\varphi) = D(e^{i\varphi})\, D(e^{-i\varphi}) = |D(e^{i\varphi})|^2$.

Actually, such a function D, analytic and without zero in the open unit disk, square integrable on the boundary, exists and is unique ($c'_0 > 0$) when v is positive, integrable and has an integrable logarithm on $[-\pi,\pi]$ ([8] § 1.14, [18] chap. 10). Then, the Szegö theory shows that

(A2.7) $c'_{m,N} \underset{N\to\infty}{\to} c'_m$, $m=0,1,\ldots$.

This may be proved by standard arguments of boundedness of the infinite matrices C and C^{-1} when v is continuous and bounded from below by a positive constant on $[-\pi,\pi]$ [21]. The general case is more difficult to handle and is solved in several steps ([8] chap. 3, [18] § 12.3) : first, as the inverse of the head element of A^{-1}, for any symmetric positive definite matrix A, is the minimum of the quadratic form $\sum_{k=0}^{N}\sum_{m=0}^{N} a_{k,m}\, x_k\, x_m$ with $x_0=1$, one has, in the Toeplitz case $c'^{-2}_{0,N} = \min_{x_1,\ldots,x_N} \int_{-\pi}^{\pi} \left| 1+\sum_{1}^{N} x_k\, e^{ik\varphi}\right|^2 v(\varphi)d\varphi$; if one succeeds in proving (A2.7) for m=0, then it is true for any m : indeed, using (A2.6), $c'^{-2}_{0,N} = (2\pi)^{-1}\, c'^{-2}_0 \int_{-\pi}^{\pi} \left| \dfrac{1+\sum_{1}^{N} (c'_{k,N}/c'_{0,N})e^{ik\varphi}}{1+\sum_{1}^{\infty} (c'_k/c'_0)e^{ik\varphi}}\right|^2 d\varphi$,

where the integral will remain larger than 2π unless (A2.7) holds for all m ; one has already (A2.7) when $1/v(\varphi)$ is a positive trigonometric polynomial of degree M, as $c'_{m,N} = c'_m$ provided $N > M/2$; next, (A2.7) holds if v has a positive lower bound, as $1/v$ may then be approximated in the L^1 norm by a trigonometric polynomial, such that the error on c'_0 is arbitraryly small ; finally, the positive lower bound condition on v may be dropped, as the c'_0 corresponding to $v(\varphi)+\varepsilon$ tends to c'_0 when $\varepsilon \to 0$... .

Appendix 3. Orthogonal polynomials on a system of intervals and Jacobi-Abel functions.

From (4.6), (4.7) and (4.8), $C_n+B_nX^{1/2}$ and $C_n-B_nX^{1/2}$ are analytic functions outside S, with zeros at the zeros of α_n (distributed in some way between the two functions) and at the zeros of ρ (for the second function only), behaving at infinity respectively like $2z^{n+m}$ and h_nz^{N-n-m}. The logarithms will be defined as continuous functions outside a system of rectilinear cuts joining d_1 and these zeros, including $(-\infty,d_1]$, such that the logarithms are real for large positive values of z.

We consider now the integral of $[\ln(C_n(t)+B_n(t)X^{1/2}(t))]X^{-1/2}(t)(z-t)^{-1}$ on a contour starting at $t=d_1-i\varepsilon, \varepsilon > 0$, continuing with $t=d_1-x-i\varepsilon$, $0 \leqslant x \leqslant R$, $t=Re^{i\theta}$, $-\pi < \theta < \pi$, $t=d_1-R+x+i\varepsilon$, $0 \leqslant x \leqslant R$. As the logarithms differ by $2\pi\,(n+m)i$ on the two sides of $[-R,d_1)$, the value of the integral is $2\pi\,(n+m)i \int_{-\infty}^{d_1} X^{-1/2}(t)(z-t)^{-1}\,dt$ when $R \to \infty$. On the other hand, if the contour shrinks up to the system of cuts, one finds

$$-2\pi i \sum_{z'} \int_{d_1,\,t\notin S}^{z'} X^{-1/2}(t)(z-t)^{-1}dt \;-\; 2\pi i\, X^{-1/2}(z)\ln(C_n(z)+B_n(z)X^{1/2}(z))$$

$$C_n+B_nX^{1/2} = 0 \text{ at } z'$$

$$-\int_S \ln(C_n^2(t)-B_N^2(t)X(t))X_+^{-1/2}(t)(z-t)^{-1}dt \;-\; 2\pi i \sum_{k=1}^{m-1} M_{n,k} \sum_{s=1}^{k} \int_{d_{2s}}^{d_{2s+1}} X^{-1/2}(t)(z-t)^{-1}dt,$$

where the first sum comes from the increase of $2\pi i$ of the logarithm when one circles a zero of $C_n+B_nX^{1/2}$, the second term comes from the residue at z, the third term from the part of the contour along S, summing complex conjugate values on the two sides ; in the last term, $\pi M_{n,k}$ is the increase of argument of $C_n(t)+B_n(t)X_+^{1/2}(t)$ between d_{2k+1} and d_{2k}, and is very likely related to the number of zeros of B_n in $[d_{2k},d_{2k+1}]$. Therefore,

(A3.1) $X^{-1/2}(z)\ln(C_n(z)+B_n(z)X^{1/2}(z)) =$

$$-(n+m)\int_{-\infty}^{d_1} X^{-1/2}(t)(z-t)^{-1}dt - \sum_{k=1}^{m-1} M_{n,k} \sum_{s=1}^{k}\int_{d_{2s}}^{d_{2s+1}} X^{-1/2}(t)(z-t)^{-1}dt$$

$$-\sum_{\substack{C_n+B_n X^{1/2}=0 \text{ at } z'}} \int_{d_1, t\notin S}^{z'} X^{-1/2}(t)(z-t)^{-1}dt - (2\pi i)^{-1}\int_S \ln(C_n^2(t)+B_n^2(t)\; X(t))\; X_+^{-1/2}(t)(z-t)^{-1}dt.$$

Similarly, for $C_n - B_n X^{1/2}$

(A3.2) $X^{-1/2}(z)\ln(C_n(z)-B_n(z)X^{1/2}(z)) =$

$$(n+m-N)\int_{-\infty}^{d_1} X^{-1/2}(t)(z-t)^{-1}dt + \sum_{k=1}^{m-1} M_{n,k}\sum_{s=1}^{k}\int_{d_{2s}}^{d_{2s+1}} X^{-1/2}(t)(z-t)^{-1}dt$$

$$-\sum_{\substack{C_n-B_n X^{1/2}=0 \text{ at } z'}} \int_{d_1, t\notin S}^{z'} X^{-1/2}(t)(z-t)^{-1}dt - (2\pi i)^{-1}\int_S \ln(C_n^2(t)+B_n^2(t)X(t))\; X_+^{-1/2}(t)(z-t)^{-1}dt.$$

Subtracting (A3.1) from (A3.2), and expanding in series of z^{-1}, one has, for the coefficient of z^{-r} :

$$\text{(A3.3)}\quad \sum_{\substack{C_n^2-B_n^2X=0 \text{ at } z'_k}} s_k\int_{d_1, t\notin S}^{z'_k} X^{-1/2}(t)t^{r-1}dt=(2n+2m-N)\int_{-\infty}^{d_1} X^{-1/2}(t)t^{r-1}dt + \sum_{k=1}^{m-1} M_{n,k}\,K_{r,k},$$

$$r=1,\ldots,m-1,$$

where $s_k=+1$ for the zeros of $C_n-B_nX^{1/2}$, -1 for the zeros of $C_n+B_nX^{1/2}$.

This result could have been obtained in a shorter but less elementary way by Abel's theorem, considering the zeros and poles of $C_n+B_n X^{1/2}$ on the Riemann surface of $X^{1/2}$ ([17] § 4.7, theorem 1).

The contribution of the N-m+1 zeros of the polynomial ρ, which are a part of the zeros of $C_n-B_nX^{1/2}$ may be written in (A3.3) as

$$\sum_{\text{zeros of } \rho}\int_{d_1, t\notin S}^{z'} X^{-1/2}(t)t^{r-1}dt = -(N-m+1)\int_{-\infty}^{d_1} X^{-1/2}(t)t^{r-1}dt + \frac{1}{\pi}\int_S [\ln|\rho(t)|]$$

$$X_+^{-1/2}(t)t^{r-1}dt,$$

which shows (4.3), if w is given by (4.4).

An important consequence of (A3.3) is $M_{n,k}=n\mu_k + O(1)$ when $n\to\infty$, where the μ_k are solutions of $2\int_{-\infty}^{d_1} X^{-1/2}(t)t^{r-1}dt + \sum_{k=1}^{m-1} \mu_k K_{r,k} = 0$, $r=1,\ldots,m-1$. From this, it is possible to show that the main factors appearing in the asymptotic expressions of $C_n+B_nX^{1/2}$ and $C_n-B_nX^{1/2}$ are respectively Z^n and Z^{-n}, where Z(z) maps the exterior of S on the exterior of the unit circle, with $Z(z)=c^t z + O(1)$ when $z\to\infty$([15], lemma 6.7). Therefore, φ, with $Z(z)=e^{i\varphi}$ for $z\in S$, is the natural variable describing conveniently the orthogonal polynomials and the related functions

$Q_n = B_n f - A = (C_n - B_n X^{1/2})/\rho$ on S.

REFERENCES.

[1] ALLEN, G.D., CHUI, C.K., MADYCH, W.R., NARCOWICH, F.J., SMITH, P.W. : Padé approximation of Stieltjes series, *J. Approx. Theory*, *14* (1975), 302-316.

[2] BARRETT, W. : An asymptotic formula relating to orthogonal polynomials. *J. London Math. Soc.* (2) *6* (1973), 701-704.

[3] CHUI, C.K. : Recent results on Padé approximants and related problems, pp 79-115 in *Approximation Theory II*, edited by LORENTZ, G.G., CHUI, C.K., SCHUMAKER, L.L. ; A.P., N.Y. 1976.

[4] DANLOY, B. : Numerical construction of Gaussian quadrature formulas for $\int_0^1 (-\text{Log } x).x^{\alpha}.f(x).dx$ and $\int_0^{\infty} E_m(x).f(x).dx$, *Math. Comp.*, *27* (1973), 861-869.

[5] ERDELYI, A. : *Asymptotic Expansions*, Dovor, N.Y. 1956.

[6] GASPARD, J.P., CYROT-LACKMANN, F. : Density of states from moments. Application to the impurity band. *J. of Physics C : Solid State Phys.*, *6* (1973), 3077-3096.

[7] GAUTSCHI, W. : On the construction of Gaussian quadrature rules from modified moments, *Math. Comp.*, *24* (1970) 242-260.

[8] GRENANDER, U., SZEGÖ, G. : *Toeplitz Forms and their Applications*, University of California Press, Berkeley, 1958.

[9] HARTWIG, R.E., FISHER, M.E. : Asymptotic behavior of Toeplitz matrices and determinants, *Arch. Rat. Mech. Anal.*, *32* (1969) 190-225.

[10] HODGES, C.H. : Van Hove singularities and continued fraction coefficients, *J. Physique Lett.* *38* (1977) L187-L189.

[11] LANDFRIEDT, E. : *Thetafunktionen und hyperelliptische Funktionen*, G.J. Göschensche Verlagshandlung, Leipzig 1902.

[12] LIGHTHILL, M.J. : *Introduction to Fourier Analysis and Generalized Functions*, Cambridge U.P., 1958.

[13] MASSON, D. : Padé approximants and Hilbert spaces, pp 41-52 in *Padé Approximants and their Applications*, edited by GRAVES-MORRIS, P.R. ; A.P. ; N.Y. 1973.

[14] MATTHEIJ, R.M.M. : Accurate estimates of solutions of second order recursions, *Linear Algebra and Appl.*, *12* (1975) 29-54.

[15] NUTTALL, J., SINGH, S.R. : Orthogonal polynomials and Padé approximants associated with a system of arcs, *J. Approx. Theory*, *21* (1977) 1-42.

[16] POTTIER, N. : *Etude de la densité d'états électroniques de quelques modèles de systèmes désordonnés.* Thèse, Paris VI, 1976.

[17] SIEGEL, C.L. : *Topics in Complex Function Theory, vol. II : Automorphic Functions and Abelian Integrals*, Wiley - Interscience, N.Y. 1971.

[18] SZEGÖ, G. : *Orthogonal Polynomials*, A.M.S., Providence, 1939.

[19] WALL, H.S. : *The Analytic Theory of Continued Fractions*, Van Nostrand, Princeton, 1948.

[20] WIDOM, H. : Asymptotic inversion of convolution operators, *Publications mathématiques I.H.E.S.*, n° 44 (1974) 191-240.

[21] WIDOM, H. : Toeplitz matrices and Toeplitz operators, pp 319-341 in *Complex Analysis and its Applications*, vol. 1, I.A.E.A., Vienna 1976.

[22] WILF, H.S. : *Finite Sections of Some Classical Inequalities*, Springer, Berlin, 1970.

[23] ZYGMUND, A. : *Trigonometric Series*, vol. I, Cambridge U.P., 1959.

Added references :

KAILATH, T., VIEIRA, A., MORF, M. : Inverses of Toeplitz operators, innovations, and orthogonal polynomials. *SIAM Review*, *20* (1978) 106-119.

NEVAI, P.G. : On orthogonal polynomials. *J. Approximation Theory*, *25* (1979) 34-37.

Alphonse MAGNUS
Institut mathématique U.C.L.
Chemin du Cyclotron, 2
B-1348 LOUVAIN-LA-NEUVE (Belgique).

Orthogonal Expansions in Indefinite Inner Product Spaces

H. van Rossum

Summary

We derive an expansion of a holomorphic function in terms of totally positive polynomials and interpret the result as an orthogonal expansion in a Krein space. As a special case, expansions in terms of Bessel polynomials are considered.

1. INTRODUCTION

We introduce some notations and definitions concerning a real sequence $\gamma = (c_n)_{n=0}^{\infty}$. We define the following determinants connected with γ :

$$D_m^{(n)}(\gamma) = \begin{vmatrix} c_m & c_{m-1} & \cdots & c_{m-n+1} \\ c_{m+1} & c_m & \cdots & c_{m-n+2} \\ \cdot & \cdot & & \cdot \\ c_{m+n-1} & c_{m+n-2} & \cdots & c_m \end{vmatrix}, \quad (m = 0,1,\ldots;\ n = 1,2,\ldots), \quad D_m^{(0)}(\gamma) = 1 .$$

$$\Delta_{m,n}(\gamma) = \begin{vmatrix} c_{m-n} & c_{m-n+1} & \cdots & c_m \\ c_{m-n+1} & c_{m-n+2} & \cdots & c_{m+1} \\ \cdot & \cdot & & \cdot \\ c_m & c_{m+1} & \cdots & c_{m+n} \end{vmatrix}, \quad (m,n = 0,1,\ldots) .$$

In both determinants, $c_l = 0$ if $l < 0$. Obviously

$$D_m^{(n+1)}(\gamma) = (-1)^{\frac{n(n+1)}{2}} \Delta_{m,n}(\gamma) , \quad (m,n = 0,1,\ldots) .$$

DEFINITION 1.1 The sequence γ is called <u>quasinormal</u> iff $\Delta_{n,n}(\gamma) \neq 0$, $(n=0,1,\ldots)$.

With QN we denote the set of all quasinormal sequences.
Important subsets of QN are:

1) The set H of all non-rational Hamburger sequences; $\gamma \in H$ iff $\Delta_{n,n}(\gamma) > 0$ $(n=0,1,\ldots)$.
2) The set S of all non-rational Stieltjes sequences; $\gamma \in S$ iff $\Delta_{n,n}(\gamma) > 0$ and $\Delta_{n+1,n}(\gamma) > 0$, $(n=0,1,\ldots)$.
3) The set T of all non-rational totally positive sequences; $\gamma \in T$ iff $D_m^{(n)}(\gamma) > 0$ $(m,n=0,1,\ldots)$.

It is clear that $S \subset H$ and $H \cap T = \emptyset$.

Let $P[x]$ denote the polynomial ring over $\mathbb{R}$, where x is an indeterminate. $P[x]$ can be considered as a real vector space. We define a linear functional Ω on $P[x]$ by

$$(1.1)\qquad \Omega(x^n) = c_n\,,\ (n=0,1,\dots),\ (c_n)_{n=0}^{\infty} \in QN\,.$$

An inner product on $P[x]$, called an inner product based on $(c_n)_{n=0}^{\infty}$, is introduced as follows:

$$\forall p,q \in P[x],\ \langle p,q\rangle := \Omega[p(x).q(x)]\,,$$

where $p.q$ denotes the ring product of p and q.
In general the inner product above is indefinite.

We apply the Gram-Schmidt orthogonalization process to the sequence $1,x,\dots,x^n,\dots$ in $P[x]$ to obtain an orthogonal sequence of monic polynomials $(q_n(x))_{n=0}^{\infty}$ $(\deg q_n = n, (n=0,1,\dots))$ and

$$q_n(x) = \frac{1}{\Delta_{n-1,n-1}}\begin{vmatrix} c_0 & c_1 & \dots & c_n \\ c_1 & c_2 & \dots & c_{n+1} \\ \cdot & \cdot & \dots & \cdot \\ c_{n-1} & c_n & \dots & c_{2n-1} \\ 1 & x & \dots & x^n \end{vmatrix}\,,\ (n=0,1,\dots),\ \Delta_{-1,-1} = 1\,.$$

The orthogonality relations are

$$(1.2)\qquad \langle q_n,q_m\rangle = \delta_{n,m}\frac{\Delta_{n,n}}{\Delta_{n-1,n-1}}\,,\ (n,m=0,1,\dots),\ \Delta_{-1,-1} = 1$$ [1)]

Owing to the quasi normality of the sequences $(c_n)_{n=0}^{\infty}$, non of the q_n's is neutral.

From here we can go in several directions. If, for instance, $(c_n)_{n=0}^{\infty} \in H$, there exists a bounded non-decreasing weight function $\psi : \mathbb{R} \to \mathbb{R}$ such that the functional in (1.1) has the representation as a Stieltjes-Lebesque integral,

$$\int_{-\infty}^{\infty} x^n d\psi(x) = c_n\,,\ (n=0,1,\dots).$$

1) We drop γ in $\Delta_{n,n}(\gamma)$ etc., from now on.

The orthogonality relations in (1.2) then take the form

$$\int_{-\infty}^{\infty} q_n(x)q_m(x)d\psi(x) = \delta_{n,m} \frac{\Delta_{n,n}}{\Delta_{n-1,n-1}} ,$$

where $\Delta_{n,n} > 0$, $(n = 0,1,\ldots)$ and $\Delta_{-1,-1} = 1$.

The polynomials q_n $(n = 0,1,\ldots)$ form a basis for the Hilbert space $L^2_\psi(-\infty,\infty)$.

We are here interested in that indefinite case obtained by taking $(c_n)_{n=0}^{\infty}$ totally positive. This implies $c_n > 0$ $(n = 0,1,\ldots)$.

In this case

$$\limsup \sqrt[n]{c_n} = \rho < \infty .$$

We consider the ring $P[z]$ over R, where z is a complex variable. Put

$$\psi(z) = \sum_{n=1}^{\infty} c_{n-1} z^{-n} .$$

Then (1.1) can be written as follows:

$$(1.3) \qquad \frac{1}{2\pi i} \oint_{|z|=r} z^n \psi(z)dz = c_n \quad (i = \sqrt{-1} ;\ r > \rho),\ (n = 0,1,\ldots)$$

and the orthogonality relations in (1.2) take the form

$$(1.4) \quad \frac{1}{2\pi i} \oint_{|z|=r} q_n(z)q_m(z)\psi(z)dz = \delta_{n,m} \frac{\Delta_{n,n}}{\Delta_{n-1,n-1}} \quad (n = 0,1,\ldots),\ (\Delta_{-1,-1} = 1) .$$

We want to extend the inner product (in (1.4)) on $P[z]$, to the linear space V (over $\mathbb{C}$) of functions holomorphic on a disc $D = \{z \in \mathbb{C} \mid |z| \le R\}$. We will assume that $R > \rho$. We define this inner product $[.,.]$ as follows [1)]

$$\forall f,g \in V,\ [f,g] := \frac{1}{2\pi i} \oint_{|z|=R} f(z)\overline{g(\bar{z})}\psi(z)dz ,$$

where the bar denotes complex conjugation and $\psi(z) = \sum_{n=1}^{\infty} c_{n-1} z^{-n}$ and $\gamma = (c_n)_{n=0}^{\infty} \in T$.

1) The use of this inner product is a suggestion of Dr de Bruin, University of Amsterdam.

We remark that $\overline{\psi(\bar{z})} = \psi(z)$, $\forall z \in D$. It is clear that we need only to check the Hermitian symmetry.

$$[g,f] = \frac{1}{2\pi i} \oint_{|z|=R} g(z)\overline{f(\bar{z})}\psi(z)dz = \frac{1}{2\pi i} \oint_{|z|=R} g(\bar{z})\overline{f(z)\psi(\bar{z})d\bar{z}} =$$

$$= \frac{1}{2\pi i} \oint_{|z|=R} \overline{f(z)\overline{g(\bar{z})}\psi(z)d\bar{z}} = \frac{-1}{2\pi i} \oint_{|z|=R} \overline{f(z)\overline{g(\bar{z})}\psi(z)dz} =$$

$$= \overline{\frac{1}{2\pi i} \oint_{|z|=R} f(z)\overline{g(\bar{z})}\psi(z)dz} = \overline{[f,g]} .$$

The new inner product coincides with the old one on $P[z]$, i.e.,

$$[p,q] = \frac{1}{2\pi i} \oint_{|z|=R} p(z)\overline{q(\bar{z})}\psi(z)dz = \frac{1}{2\pi i} \oint_{|z|=R} p(z)q(z)\psi(z)dz = \langle p,q \rangle ,$$

due to the fact that $\forall q \in P[z] \left(\overline{q(\bar{z})} = q(z)\right)$.

2. Let T_1 be the set of all sequences $\gamma \in T$, with $\limsup \sqrt[n]{c_n} = \rho < 1$. The set T_1 is non-void, since, for instance, $({}^1/n!)_{n=0}^{\infty} \in T_1$.
We consider the inner product space V_D of complex functions holomorphic on the disc $D = \{z \in \mathbb{C} \mid |z| \le R\}$ where R satisfies the inequality $\rho < 1 < R < \rho^{-1}$ and where $\rho = \limsup \sqrt[n]{c_n}$.
The inner product on V_D is based on $\gamma \in T_1$. We use the notation $\langle \cdot,\cdot \rangle$ instead of $[\cdot,\cdot]$ from now on, e.g.,

$$\forall f,g \in V_D , \langle f,g \rangle = \frac{1}{2\pi i} \oint_{|z|=R} f(z)\overline{g(\bar{z})}\psi(z)dz ,$$

with $\psi(z) = \sum_{n=0}^{\infty} (-1)^n c_n z^{-n-1}$.

The orthogonal polynomials obtained by applying the Gram-Schmidt orthogonalization process to $1,z,z^2,\ldots$ with respect to the inner product above, are called totally positive polynomials. They were introduced by the author in 1964 [8], by means of the Padé table for the generating function of a totally positive sequence. This table was introduced and studied by Arms and Edrei [1] and Edrei [5]. We mention here only a few properties of this table we will use in the sequel.

Let

$$(2.1)\qquad f(z) = Ce^{\gamma z}\,\frac{\prod_{\nu=1}^{\infty}(1+\alpha_\nu z)}{\prod_{\nu=1}^{\infty}(1-\beta_\nu z)} = \sum_{n=0}^{\infty} c_n z^n ,$$

where $C \geq 0$; $\gamma > 0$; $\alpha_\nu > 0$, $\beta_\nu > 0$ $(\nu = 1,2,\ldots)$ and $\sum_{\nu=1}^{\infty} (\alpha_\nu + \beta_\nu) < \infty$.

If f does not reduce to a rational function, $(c_n)_{n=0}^{\infty} \in T$.
Conversely, if $(c_n)_{n=0}^{\infty} \in T$, then $\Sigma_{n=0}^{\infty} c_n z^n$ has the representation (2.1).
For a treatment of this Padé table we refer the reader also to Baker's book [2], p. 252-260.
If $U_{m,n}(z) / V_{m,n}(z)$ is the (m,n)-Padé approximant to f, then $V_{m,n}(-z)$ has all its coefficients positive. It was proved by the author in [8] that the sequence of polynomials

$$(z^m V_{m,k+m}(-z^{-1}))_{m=0}^{\infty} \quad (k \text{ a fixed integer} \geq 0)$$

is orthogonal with respect to the sequence $((-1)^{k+m+1} c_{k+m+1})_{m=0}^{\infty}$, or equivalently

$$\frac{1}{2\pi i} \oint_{|z|=\rho+\varepsilon} z^m V_{m,k+m}(-z^{-1}) z^n V_{n,k+n}(-z^{-1}) \psi(z)dz = \begin{cases} 0 & , m \neq n \\ \dfrac{\Delta_{k+n+1,n}}{\Delta_{k+n,n-1}} & , m = n \end{cases}$$

with $\psi(z) = (-1)^{k+1} c_{k+1} z^{-1} + (-1)^{k+2} c_{k+2} z^{-2} + \ldots$, $(\rho = \limsup \sqrt[n]{c_n}\ ;\ \varepsilon > 0)$.

Obviously, if $(c_n)_{n=0}^{\infty} \in T$, then all sequences $(c_{n+k})_{n=0}^{\infty}$ $k = 0,1,\ldots$ belong to T.
Another important result reads: $\lim_{n\to\infty} V_{n,k+n}(z)$ exists (k a fixed integer ≥ 0).
See Baker [2] Th. 18.2, p. 257-259 for a more complete result in this direction.
We extract the following two properties from the preceding:

<u>Property I</u> All coefficients of the totally positive polynomials are positive.
<u>Property II</u> The set of coefficients of any sequence of totally positive polynomials is bounded.

We introduce some notations, following Whittaker's book [11].
Let the power series $f(z) = \Sigma_{n=0}^{\infty} a_n z^n$ have a radius of convergence $R > 0$.
Furthermore let $(q_n(z))_{n=0}^{\infty}$ be a sequence of polynomials in z with degree

$q_n = n \quad (n=0,1,\dots)$. We put

$$z^n = \sum_k \pi_{n,k} q_k(z)\ , \quad k \text{ finite}, \ (n=0,1,\dots) \tag{2.2}$$

and

$$f(z) = \sum_{n=0}^{\infty} a_n \sum_k \pi_{n,k} q_k(z).$$

So we have the formal expansion

$$f(z) = \sum_{n=0}^{\infty} \alpha_n q_n(z), \quad \text{with} \quad \alpha_n = \sum_{m=0}^{\infty} \pi_{n,m} \frac{f^{(m)}(0)}{m!}\ , \quad (n=0,1,\dots). \tag{2.3}$$

Let $M_m(R) = \max\limits_{|z|=R} |q_n(z)|$, $\omega_n(R) = \sum\limits_{m=0}^{\infty} \pi_{n,m} M_m(R)$, $(n=1,2,\dots)$.

Then by Weierstrass' criterion:

$$\sum_{n=0}^{\infty} |a_n| \omega_n(R) < \infty \ \Rightarrow \ \sum_{n=0}^{\infty} \alpha_n q_n(z) \text{ converges to } f(z) \text{ uniformly on} \tag{2.4}$$

$$D = \{z \in \mathbb{C} \mid \ |z| \le R\}.$$

We use the above notations and (2.4) in the proof of:

THEOREM 2.1 *Let* $(p_n(z))_{n=0}^{\infty}$ $(z \in \mathbb{C})$ *be the sequence of polynomials orthonormal with respect to the inner product* $<.,.>$ *based on the sequence* $(c_n)_{n=0}^{\infty} \in T_1$. *Then every complex valued function* f *holomorphic on* $D = \{z \in \mathbb{C} \mid \ |z| \le R\}$ *with* $R > \rho = \limsup c_n^{1/n}$ $(\rho \ne 0)$ *can be expanded in terms of the sequence* $(p_n)_{n=0}^{\infty}$ *such that* $\Sigma_{n=0}^{\infty} \alpha_n p_n(z)$ *converges to* $f(z)$ *uniformly on* D.

PROOF. We apply (2.2) with q_n replaced by p_n $(n=0,1,\dots)$ to obtain

$$< z^n, p_m(z) > = \pi_{n,m}\,.$$

Put

$$p_m(z) = p_m^{(m)} z^m + p_{m-1}^{(m)} z^{m-1} + \dots + p_0^{(m)}\ ,$$

$$\psi(z) = \sum_{n=0}^{\infty} c_n z^{-n-1}$$

then

$$|\pi_{n,m}| = \left|\frac{1}{2\pi i} \oint_{|z|=R} z^n p_m(z)\psi(z)dz\right| = p_m^{(m)} c_{n+m} + p_{m-1}^{(m)} c_{n+m-1} + \ldots + p_0^{(m)} c_n .$$

Owing to Property II, there exists a positive constant N such that

$$p_k^{(m)} < N, \quad (m = 0,1,\ldots ;\ k \le m).$$

Let $P = \max_{|z| \le \rho^{-1} - \varepsilon} \left|\sum_{n=0}^{\infty} c_n z^n\right|$, $0 < \varepsilon < \rho^{-1}$, then we have $|c_k| \le P(\beta\rho)^k$, $(k = 1,2,\ldots)$, where β is a number between 0 and 1. This gives us the following estimate for $\pi_{n,m}$:

$$|\pi_{n,m}| < NP(\beta\rho)^n \frac{1 - (\beta\rho)^{m+1}}{1 - \beta\rho} . \tag{2.5}$$

Using Property I we get

$$M_m(R) = p_m^{(m)} R^m + p_{m-1}^{(m)} R^{m-1} + \ldots + p_0^{(m)} < N \frac{1 - R^{m+1}}{1-R} . \tag{2.6}$$

From (2.5) and (2.6) it follows (using $\pi_{n,m} = 0$, if $m > n$)

$$\omega_n(R) < \sum_{m=0}^{n} N^2 P(\beta\rho)^n \frac{1 - (\beta\rho)^{m+1}}{1 - \beta\rho} \cdot \frac{1 - R^{m+1}}{1-R} .$$

Let $\Sigma_{n=0}^{\infty} a_n z^n$ be the Taylor expansion of f on D; if

$$M = \max_{|z| \le R} |f(z)|,$$

then: $|a_n| \le MR^{-n}$, $(n = 0,1,\ldots)$. Formally we have

$$\sum_{n=0}^{\infty} |a_n| \omega_n(R) << \sum_{n=0}^{\infty} \frac{M}{R^n} \sum_{m=0}^{n} N^2 P(\beta\rho)^n \frac{1 - (\beta\rho)^{m+1}}{1 - \beta\rho} \cdot \left|\frac{1 - R^{m+1}}{1-R}\right| . \tag{2.7}$$

We will show that the double series in the right-hand member converges. We consider two cases:

1) $\rho < R < 1$.

Writing the right-hand member of (2.7) in the form

$$\frac{MN^2P}{(1-\beta\rho)(1-R)} \sum_{n=0}^{\infty} \left(\frac{\beta\rho}{R}\right)^n \sum_{m=0}^{n} \left(1-(\beta\rho)^{m+1}\right)\left(1-R^{m+1}\right),$$

we see that the second sum is bounded so, on account of $0 < \beta\rho/R < 1$, the double sum converges.

2) $R > 1$. Let A denote a constant chosen such that $R < A\rho^{-1}$. Consider the expression:

$$(\beta\rho)^n \sum_{m=0}^{n} \left(1-(\beta\rho)^{m+1}\right)\left(R^{m+1}-1\right).$$

This is dominated by

$$(\beta\rho)^n \sum_{m=0}^{n} \left(1-(\beta\rho)^{m+1}\right)\left(A\rho^{-m-1}-1\right),$$

which, in turn, is dominated by:

$$A(\beta\rho)^n(\beta+\beta^2+\ldots+\beta^{n+1}) + (n+1)(\beta\rho)^n + (\beta\rho)^n[\beta\rho+(\beta\rho)^2+\ldots+(\beta\rho)^{n+1}] +$$
$$+ A\beta^n(\rho^{n-1}+\rho^{n-2}+\ldots+1+\rho^{-1}).$$

This is bounded above by Cn where C is a suitably chosen constant. So the right-hand member in (2.7) is dominated by the convergent series

$$\frac{MN^2P}{(1-\beta\rho)(R-1)} \sum_{n=0}^{\infty} \frac{Cn}{R^n} \quad \Box$$

Remark 1. As a consequence of Theorem 2.2 we have $\alpha_n = \langle f,p_n\rangle$, $(n=0,1,\ldots)$. The span $\bigvee(p_n)_{n=0}^{\infty}$ is a non-degenerate, decomposable infinite dimensional indefinite inner product space. See Bognar [3]. It can be embedded as a dense subspace in a Krein space $K = H_1 \oplus H_2$. In this orthogonal direct sum, H_1 is a Hilbert space containing all p_n's with $\langle p_n,p_n\rangle = 1$ and H_2 the antispace of a Hilbert space containing all p_n's with $\langle p_n,p_n\rangle = -1$. Since

$$\langle p_n,p_n\rangle \begin{cases} = 1, & n \text{ even},\\ = -1, & n \text{ odd},\end{cases}$$

both H_1 and H_2 are infinite dimensional.

Remark 2. The partial sum $\Sigma_{k=0}^{n} < f,p_k > p_k(z)$ is the unique projection of f onto the subspace W of K, spanned by $p_0,p_1,\dots,p_n$. This is due to the fact that the inner product is based on a sequence $(c_n)_{n=0}^{\infty} \in T_1$, hence $(c_n)_{n=0}^{\infty} \in QN$. Compare van Rossum [10].

Remark 3. In Theorem 2.2 the case $\rho = 0$ was excluded. In this paper we will only consider a special case of $\rho = 0$, involving the ordinary Bessel polynomials.

A very special case of a sequence in T_1 is the sequence $(2^{n+1}/(n+1)!)_{n=0}^{\infty}$. In this case $\lim\sup c_n^{1/n} = 0$. The totally positive polynomials connected with the sequence $((-2)^{n+1}/(n+1)!)_{n=0}^{\infty}$ are the ordinary Bessel polynomials $b_n(z)$, $(n=0,1,\dots)$ introduced by Krall and Frink [6] in 1949, and, in a different way, by the present author [9] in 1953.
Explicit expression and orthogonality relations are respectively

$$b_n(z) = \sum_{k=0}^{n} \frac{(n+k)!}{(n-k)!k!} \left(\frac{z}{2}\right)^k ,$$

$$< b_n,b_m > = \frac{1}{2\pi i} \oint_{|z|=r} b_n(z)b_m(z)e^{-\frac{2}{z}} dz = \delta_{n,m} \frac{(-1)^{n+1}2}{2n+1} , \quad (r > 0).$$

It was shown by Nassif [7] in 1954, that any complex valued function f holomorphic on a disc $D_r = \{z \in \mathbb{C} \mid |z| \le r\}$ $(r > 0$, arbitrary) can be expanded in a series of ordinary Bessel polynomials converging to $f(z)$ uniformly on D_r. Nassif's proof does not use the orthogonality of the Bessel polynomials.
His result reads:

$$(2.8) \qquad f(z) = \sum_{n=0}^{\infty} \gamma_n b_n(z), \text{ with } \gamma_n = 2^n(2n+1) \sum_{m=0}^{\infty} \frac{(-2)^m f^{(m+n)}(0)}{m!(2n+m+1)!} \qquad (n=0,1,\dots).$$

THEOREM 2.2 *Let* K_b *be the Krein space obtained by completion of the span* $\bigvee (b_n)_{n=0}^{\infty}$, *where* b_n *is the* n-*th Bessel polynomial. If* f *is holomorphic on* $D_r = \{z \in \mathbb{C} \mid |z| \le r\}$ $r > 0$ *arbitrary, then* f *belongs to* K_b.

PROOF. Put $t_m = \dfrac{(-2)^m f^{(m+n)}(0)}{m!(2n+m+1)!}$, $(m = 0,1,\dots)$, compare (2.8).
Then, if $M = \max_{|z|=r} |f(z)|$,

$$|t_m| \le \frac{2^m(m+n)!M}{r^{m+n}m!(2n+m+1)!} < \left(\frac{2}{r}\right)^m \cdot \frac{1}{m!} \cdot \frac{M}{r^n} \cdot \frac{1}{n!} .$$

So

$$|\gamma_n| \le \frac{2^n(2n+1)}{r^n.n!} M \sum_{m=0}^{\infty} \left(\frac{2}{r}\right)^m . \frac{1}{m!} = \frac{2^n(2n+1)}{r^n.n!} e^{\frac{2}{r}} M .$$

From this estimate we see that $\sum_{n=0}^{\infty} |\gamma_n|^2 < \infty$.

Let $(\tilde{b}_n)_{n=0}^{\infty}$ denote the normalized Bessel polynomials, i.e.

$$b_n(z) = \alpha_n \tilde{b}_n(z), \quad (n = 0,1,\ldots),$$

with

$$|\alpha_n|^2 = \frac{2}{2n+1}, \quad (n = 0,1,\ldots).$$

We show that the sequence $(\sum_{k=0}^{n} \gamma_k \alpha_k \tilde{b}_k(z))_{n=0}^{\infty}$ converges to $f(z)$ with respect to the norm $\|.\|$ on K_b defined as follows:

$$\forall h \in K_b, \ \|h\|^2 = \sum_{k=0}^{\infty} |<h,\tilde{b}_k>|^2 .$$

We put

$$f(z) - s_n(z) = g_n(z) = \sum_{k=n}^{\infty} \gamma_k \alpha_k \tilde{b}_k(z), \quad (n = 0,1,\ldots)$$

$$\|g_n\|^2 = \sum_{m=0}^{\infty} |<g_n,\tilde{b}_m>|^2 = \sum_{m=0}^{\infty} \left| \frac{1}{2\pi i} \oint_{|z|=r} \left(\sum_{k=n}^{\infty} \gamma_k \alpha_k \tilde{b}_k(z)\right) \tilde{b}_m(z) e^{-\frac{2}{z}} dz \right|^2 .$$

Due to the uniform convergence in (2.8), we get

$$\|g_n\|^2 = \sum_{m=0}^{\infty} \left| \sum_{k=n}^{\infty} \frac{\gamma_k \alpha_k}{2\pi i} \oint_{|z|=r} \tilde{b}_k(z) \tilde{b}_m(z) e^{-\frac{2}{z}} dz \right|^2 = \frac{1}{4\pi^2} \sum_{m=n}^{\infty} |\gamma_m|^2 \frac{2}{2m+1} .$$

This last expression obviously can be made arbitrarily small by taking n large enough. □

Remark 4. The ordinary Bessel polynomials $b_n(z)$ are related to the Padé denominators $V_{n,n}(z)$ of the function $e^z = {}_1F_1(1;1;z)$ as follows.

Let $V_{m,n}(z) = {}_1F_1(-m;-m-n;-z)$ be the (m,n)-Padé denominator, then

$$z^n {}_1F_1(-n;-2n;z^{-1}) = \frac{n!}{(2n)!} b_n(2z) .$$

Since e^z is a special case of the generating function f in (2.1), the ordinary Bessel polynomials are a special case of the totally positive polynomials. On the other hand e^z is a special case $(c = 1)$ of ${}_1F_1(1;c;z)$.
(We will assume $c > -1$ in the following). The (m,n)-Padé denominator $(m \leq n+1)$ for ${}_1F_1(1;c;z)$ is ${}_1F_1(-m;-m-n-c+1;-z)$.
The generalized Bessel polynomials introduced by Krall and Frink are $Y_m^{(\alpha)}$ given by

$$Y_m^{(\alpha)}(z) = \sum_{k=0}^{m} \binom{m}{k} (m+\alpha+1)_k \left(\frac{z}{2}\right)^k =$$

$$= (m+\alpha+1)(m+\alpha+2)\dots(2m+\alpha) z^m {}_1F_1\left(-m;-2m-\alpha;\frac{z-1}{2}\right) .$$

So the ordinary Bessel polynomials are essentially a special case $(c = 1)$ of the generalized Bessel polynomials, but these polynomials (unless $c = 1$) are not totally positive polynomials. This is implied by a result of de Bruin [4]: if and only if $c = 1$, is ${}_1F_1(1;c;z)$ the generating function of a totally positive sequence.

REFERENCES

1. Arms, R.J. and Edrei, A., The Padé tables and continued fractions generated by totally positive sequences, Mathematical Essays dedicated to A.J. Macintyre, 1-21. Ohio Univ. Press, Athens, Ohio (1970).
2. Baker, G.A., Essentials of Padé Approximants, Acad. Press, New York (1975).
3. Bognar, J., Indefinite inner product spaces, Springer, Berlin (1974).
4. Bruin, M.G. de, Convergence in the Padé table for ${}_1F_1(1;c;x)$. Kon. Ned. Akad. v. Wet. Ser A, 79, no 5 = Indag. Math., 38, no 5, 408-418 (1976).
5. Edrei, A., Proof of a conjecture of Schoenberg on the generating function of a totally positive sequence, Can. Journ. of Math., 5, 86-94 (1953).
6. Krall, H.L. and Frink, O.A., A new class of orthogonal polynomials: the Bessel polynomials, Trans. Am. Math. Soc., 65, 100-115 (1949).
7. Nassif, M., Note on the Bessel polynomials, Trans. Am. Math. Soc., 77, 408-412 (1954).
8. Rossum, H. van, Totally positive polynomials, Kon. Ned. Akad. v. Wet. Ser A, 68 no 2 = Indag. Math., 27, no 2 (1965).
9. Rossum, H. van, A theory of orthogonal polynomials based on the Padé table, Thesis, Van Gorcum, Assen (1953).

10. Rossum, H. van, Padé approximants and indefinite inner product spaces, in: Padé and rational approximation, Theory and applications, Ed. E.B. Saff and R.S. Varga, Tampa, (1976).

11. Whittaker, J.M., Sur les séries de base de polynomes quelconques, Gauthier-Villars, Paris, (1949).

H. VAN ROSSUM
Universiteit van Amsterdam
Instituut voor Propedeutische Wiskunde
Roetersstraat 15
Amsterdam (NETHERLANDS)

SUR LE CALCUL DE CERTAINS RAPPORTS DE DETERMINANTS

Claude Brezinski

INTRODUCTION

Le but de ce travail est de donner quelques méthodes numériques nouvelles pour calculer certains rapports de déterminants qui interviennent dans les méthodes d'accélération de la convergence, dans la théorie des polynômes orthogonaux et dans celle des approximants de Padé.

Dans le premier paragraphe on étudie la transformation G. On commence par démontrer, à partir de la théorie des polynômes orthogonaux, les règles récursives qui sont utilisées pour mettre en oeuvre cette transformation. Ces règles fournissent des méthodes nouvelles pour calculer les polynômes orthogonaux adjacents ainsi que des relations concernant la transformation de Shanks. On envisage ensuite le cas vectoriel ce qui permet d'obtenir une méthode beaucoup plus économique que l'ε-algorithme topologique pour la transformation de Shanks vectorielle.

Le second paragraphe est consacré à l'étude d'algorithmes récursifs qui permettent de mettre en oeuvre différentes transformations de suites et de fonctions en en changeant simplement les initialisations. L'étude inclut la transformation de Shanks scalaire et sa forme confluente ainsi que la forme confluente du ρ-algorithme. Les algorithmes utilisés sont la relation de récurrence des déterminants de Hankel, l'ε-algorithme scalaire, le w-algorithme et la transformation G.

Le troisième paragraphe propose deux méthodes nouvelles pour le calcul des coefficients qui interviennent dans la relation de récurrence d'une famille de polynômes orthogonaux. Ces méthodes sont basées sur la connexion qui existe entre les polynômes orthogonaux et les fractions continues. Le premier algorithme est une variante de la méthode de division (algorithme d'Euclide pour calculer le p.g.c.d. de deux nombres) tandis que le second algorithme utilise la connexion avec la forme confluente du ρ-algorithme.

Un certain nombre de résultats sur les méthodes d'accélération de la convergence, les polynômes orthogonaux, les approximants de Padé et les fractions continues sont supposés connus.

On trouvera dans [6] les sous-programmes FORTRAN correspondants aux méthodes décrites ci-dessous.

1 - LA TRANSFORMATION G

Considérons les polynômes :

$$H_k^{(n)}(x) = \begin{vmatrix} c_n & \cdots\cdots\cdots & c_{n+k} \\ c_{n+1} & \cdots\cdots & c_{n+k+1} \\ \cdots & \cdots\cdots\cdots & \cdots \\ c_{n+k-1} & \cdots\cdots & c_{n+2k-1} \\ 1 & \cdots\cdots\cdots & x^k \end{vmatrix}$$

Ces polynômes sont orthogonaux par rapport à la fonctionnelle linéaire $c^{(n)}$ définie par :

$$c^{(n)}(x^i) = c_{n+i} \qquad i \geq 0$$

puisque l'on a, en effet :

$$c^{(n)}(x^i H_k^{(n)}(x)) = 0 \qquad i = 0, \ldots, k-1$$

Posons :

$$H_k^{(n)} = H_k(c_n) = \begin{vmatrix} c_n & \cdots\cdots\cdots & c_{n+k-1} \\ \cdots & \cdots\cdots\cdots & \cdots \\ c_{n+k-1} & \cdots\cdots & c_{n+2k-2} \end{vmatrix}$$

$$\bar{H}_k^{(n)} = H_k(\Delta c_n)$$

Nous supposerons que tous les déterminants de Hankel utilisés sont non nuls.

On sait [5] que les familles de polynômes orthogonaux adjacents sont reliés par :

$$H_k^{(n)} H_k^{(n+1)}(x) = H_k^{(n+1)} H_k^{(n)}(x) - H_{k+1}^{(n)} H_{k-1}^{(n+1)}(x) \qquad (1)$$

$$H_k^{(n+1)} H_{k+1}^{(n)}(x) = x\, H_{k+1}^{(n)} H_k^{(n+1)}(x) - H_{k+1}^{(n+1)} H_k^{(n)}(x) \qquad (2)$$

$$H_k^{(n+2)} H_k^{(n)}(x) = H_k^{(n+1)} H_k^{(n+1)}(x) + x\, H_{k+1}^{(n)} H_{k-1}^{(n+2)}(x) \qquad (3)$$

$$H_k^{(n+1)} H_{k+1}^{(n-1)}(x) = x\, H_{k+1}^{(n-1)} H_k^{(n+1)}(x) - H_{k+1}^{(n)} H_k^{(n)}(x) \qquad (4)$$

$$[H_k^{(n+1)}]^2 H_{k+1}^{(n-2)}(x) = [(H_{k-1}^{(n+2)} H_{k+2}^{(n-2)} - H_{k+1}^{(n-2)} H_k^{(n+2)})x - H_k^{(n+1)} H_{k+1}^{(n-1)}] H_k^{(n)}(x) \qquad (5)$$

$$- x^2 [H_{k+1}^{(n-1)}]^2 H_{k-1}^{(n+2)}(x)$$

$$[H_k^{(n)}]^2 \, H_{k+1}^{(n)}(x) = [x \, H_k^{(n)} \, H_{k+1}^{(n)} + H_{k+2}^{(n-1)} \, H_{k-1}^{(n+1)} - H_{k+1}^{(n+1)} \, H_k^{(n-1)}] \, H_k^{(n)}(x) \qquad (6)$$

$$- [H_{k+1}^{(n)}]^2 \, H_{k-1}^{(n)}(x)$$

$$x \, H_k^{(n)} \, H_{k+1}^{(n-1)} \, H_k^{(n+1)}(x) = (x \, H_k^{(n+1)} H_{k+1}^{(n-1)} + H_k^{(n)} H_{k+1}^{(n)}) H_k^{(n)}(x) - H_{k+1}^{(n)} \, H_k^{(n+1)} H_k^{(n-1)}(x) \qquad (7)$$

$$H_k^{(n)} H_k^{(n+1)} H_{k+1}^{(n-1)}(x) = (x \, H_{k+1}^{(n-1)} H_k^{(n+1)} - H_k^{(n)} H_{k+1}^{(n)}) H_k^{(n)}(x) - x \, H_{k+1}^{(n-1)} H_{k+1}^{(n)} H_{k-1}^{(n+1)}(x) \qquad (8)$$

Si nous faisons $x = 0$ dans (1) alors on trouve la relation de récurrence bien connue entre déterminants de Hankel :

$$H_k^{(n)} \, H_k^{(n+2)} = [H_k^{(n+1)}]^2 + H_{k+1}^{(n)} \, H_{k-1}^{(n+2)} \qquad (9)$$

$$H_o^{(n)} = 1 \qquad H_1^{(n)} = c_n$$

Si nous faisons $x = 1$ alors :

$$H_k^{(n)}(1) = (-1)^k \, \bar{H}_k^{(n)} \qquad (10)$$

et l'on a :

$$H_k^{(n)} \, \bar{H}_k^{(n+1)} = H_k^{(n+1)} \, \bar{H}_k^{(n)} + H_{k+1}^{(n)} \, \bar{H}_{k-1}^{(n+1)} \qquad (11)$$

$$H_k^{(n+1)} \, \bar{H}_{k+1}^{(n)} = H_{k+1}^{(n+1)} \, \bar{H}_k^{(n)} - H_{k+1}^{(n)} \, \bar{H}_k^{(n+1)} \qquad (12)$$

$$H_k^{(n+2)} \, \bar{H}_k^{(n)} = H_k^{(n+1)} \, \bar{H}_k^{(n+1)} - H_{k+1}^{(n)} \, \bar{H}_{k-1}^{(n+2)} \qquad (13)$$

$$H_k^{(n+1)} \, \bar{H}_{k+1}^{(n-1)} = H_{k+1}^{(n)} \, \bar{H}_k^{(n)} - H_{k+1}^{(n-1)} \, \bar{H}_k^{(n+1)} \qquad (14)$$

$$[H_k^{(n+1)}]^2 \, \bar{H}_{k+1}^{(n-2)} = [H_{k+1}^{(n-2)} \, H_k^{(n+2)} + H_k^{(n+1)} \, H_{k+1}^{(n-1)} - H_{k-1}^{(n+2)} \, H_{k+2}^{(n-2)}] \, \bar{H}_k^{(n)} \qquad (15)$$

$$- [H_{k+1}^{(n-1)}]^2 \, \bar{H}_{k-1}^{(n+2)}$$

$$[H_k^{(n)}]^2 \, \bar{H}_{k+1}^{(n)} = [H_{k+1}^{(n+1)} \, H_k^{(n-1)} - H_k^{(n)} \, H_{k+1}^{(n)} - H_{k+2}^{(n-1)} \, H_{k-1}^{(n+1)}] \, \bar{H}_k^{(n)} \qquad (16)$$

$$- [H_{k+1}^{(n)}]^2 \, \bar{H}_{k-1}^{(n)}$$

$$H_k^{(n)} H_{k+1}^{(n-1)} \bar{H}_k^{(n+1)} = (H_k^{(n+1)} H_{k+1}^{(n-1)} + H_{k+1}^{(n)} H_k^{(n)}) \bar{H}_k^{(n)} - H_{k+1}^{(n)} H_k^{(n+1)} \bar{H}_k^{(n-1)} \qquad (17)$$

$$H_k^{(n)} H_k^{(n+1)} \bar{H}_{k+1}^{(n-1)} = (H_k^{(n)} H_{k+1}^{(n)} - H_{k+1}^{(n-1)} H_k^{(n+1)}) \bar{H}_k^{(n)} - H_{k+1}^{(n-1)} \, H_{k+1}^{(n)} \, \bar{H}_{k-1}^{(n+1)} \qquad (18)$$

Posons :

$$r_k^{(n)} = H_k^{(n)} / \bar{H}_{k-1}^{(n)} \qquad s_k^{(n)} = \bar{H}_k^{(n)} / H_k^{(n)} \qquad (19)$$

Divisons les deux membres de (11) par $H_k^{(n+1)}\ \bar{H}_k^{(n)}$; il vient :

$$\frac{s_k^{(n+1)}}{s_k^{(n)}} = 1 + \frac{r_{k+1}^{(n)}}{r_k^{(n+1)}} \tag{20}$$

Divisons les deux membres de (12) par $H_{k+1}^{(n)}\ \bar{H}_k^{(n+1)}$; il vient :

$$\frac{r_{k+1}^{(n+1)}}{r_{k+1}^{(n)}} = 1 + \frac{s_{k+1}^{(n)}}{s_k^{(n+1)}} \tag{21}$$

Les relations (20) et (21) jointent aux conditions initiales

$$s_o^{(n)} = 1 \qquad r_1^{(n)} = c_n \tag{22}$$

permettent de calculer récursivement les quantités $r_k^{(n)}$ et $s_k^{(n)}$ pour tout k et tout n.

Les relations (20), (21) et (22) forment l'algorithme r - s utilisé par Pye et Atchison pour mettre en oeuvre la transformation G[8].

Nous venons d'en fournir une démonstration différente.

Considérons maintenant les polynômes $\bar{P}_k^{(n)}$ définis par :

$$\bar{P}_k^{(n)}\ (x) = H_k^{(n)}\ (x)\ /\ \bar{H}_k^{(n)} \tag{23}$$

Remplaçons $H_k^{(n)}$ (x) par son expression en fonction de $\bar{P}_k^{(n)}$ (x) dans les relations (1) à (8) incluse et utilisons les relations (11) à (21) et (9).

Utilisons aussi le fait que :

$$r_{k+1}^{(n)}\ s_k^{(n)} = H_{k+1}^{(n)}\ /\ H_k^{(n)} \tag{24}$$

$$r_k^{(n)}\ s_k^{(n)} = \bar{H}_k^{(n)}\ /\ \bar{H}_{k-1}^{(n)} \tag{25}$$

On trouve que :

$$\bar{P}_k^{(n+1)}\ (x) = \frac{s_k^{(n)}}{s_k^{(n+1)}}\ [\bar{P}_k^{(n)}\ (x) - \frac{r_{k+1}^{(n)}}{r_k^{(n+1)}}\ \bar{P}_{k-1}^{(n+1)}\ (x)] \tag{26}$$

$$\bar{P}_{k+1}^{(n)}\ (x) = \frac{s_k^{(n+1)}}{s_{k+1}^{(n)}}\ [x\ \bar{P}_k^{(n+1)}\ (x) - \frac{r_{k+1}^{(n+1)}}{r_{k+1}^{(n)}}\ \bar{P}_k^{(n)}\ (x)] \tag{27}$$

$$\bar{P}_k^{(n)}(x) = \left(1 + \frac{r_{k+1}^{(n)}}{r_k^{(n+2)}}\right) \bar{P}_k^{(n+1)}(x) + \frac{r_{k+1}^{(n)}}{r_k^{(n+2)}} \times \bar{P}_{k-1}^{(n+2)}(x) \qquad (28)$$

$$\bar{P}_{k+1}^{(n-1)}(x) = \frac{s_k^{(n+1)}}{s_{k+1}^{(n-1)}} \times \bar{P}_k^{(n+1)}(x) - \left(1 + \frac{s_k^{(n+1)}}{s_{k+1}^{(n-1)}}\right) \bar{P}_k^{(n)}(x) \qquad (29)$$

Les autres relations sont un peu plus pénibles à obtenir. On a :

$$\bar{P}_{k+1}^{(n-2)}(x) = (D_k^{(n)} x - E_k^{(n)}) \bar{P}_k^{(n)}(x) - F_k^{(n)} x^2 \bar{P}_{k-1}^{(n+2)}(x) \qquad (30)$$

avec :

$$D_k^{(n)} = (H_{k-1}^{(n+2)} H_{k+2}^{(n-2)} - H_{k+1}^{(n-2)} H_k^{(n+2)}) \bar{H}_k^{(n)} / [H_k^{(n+1)}]^2 \bar{H}_{k+1}^{(n-2)}$$

$$E_k^{(n)} = H_k^{(n+1)} H_{k+1}^{(n-1)} \bar{H}_k^{(n)} / [H_k^{(n+1)}]^2 \bar{H}_{k+1}^{(n-2)}$$

$$F_k^{(n)} = [H_{k+1}^{(n-1)}]^2 \bar{H}_{k-1}^{(n+2)} / [H_k^{(n+1)}]^2 \bar{H}_{k+1}^{(n-2)}$$

D'après (15) on a :

$$1 = - D_k^{(n)} + E_k^{(n)} - F_k^{(n)}$$

On a :

$$D_k^{(n)} = \left\{ H_{k-1}^{(n+2)} \frac{H_{k+2}^{(n-2)}}{\bar{H}_{k+1}^{(n-2)}} - H_k^{(n+2)} \frac{H_{k+1}^{(n-2)}}{\bar{H}_{k+1}^{(n-2)}} \right\} / \left(H_k^{(n+2)} \frac{H_k^{(n)}}{\bar{H}_k^{(n)}} - H_{k-1}^{(n+2)} \frac{H_{k+1}^{(n)}}{\bar{H}_k^{(n)}}\right)$$

$$D_k^{(n)} = (H_{k-1}^{(n+2)} r_{k+2}^{(n-2)} - H_k^{(n+2)} / s_{k+1}^{(n-2)}) / (H_k^{(n+2)} / s_k^{(n)} - H_{k-1}^{(n+2)} r_{k+1}^{(n)})$$

Divisons par $H_{k-1}^{(n+2)}$; on obtient :

$$D_k^{(n)} = \frac{r_{k+2}^{(n-2)} - \dfrac{r_k^{(n+2)} s_{k-1}^{(n+2)}}{s_{k+1}^{(n-2)}}}{\dfrac{r_k^{(n+2)} s_{k-1}^{(n+2)}}{s_k^{(n)}} - r_{k+1}^{(n)}} = \frac{r_{k+2}^{(n-2)} s_{k+1}^{(n-2)} - r_k^{(n+2)} s_{k-1}^{(n+2)}}{r_k^{(n+2)} s_{k-1}^{(n+2)} - r_{k+1}^{(n)} s_k^{(n)}} \; \frac{s_k^{(n)}}{s_{k+1}^{(n-2)}} \qquad (31)$$

On a :

$$F_k^{(n)} = \frac{H_{k+1}^{(n)} \dfrac{H_{k+1}^{(n-2)}}{\bar{H}_{k+1}^{(n-2)}} - H_k^{(n)} \dfrac{H_{k+2}^{(n-2)}}{\bar{H}_{k+1}^{(n-2)}}}{H_k^{(n)} \dfrac{H_k^{(n+2)}}{\bar{H}_{k-1}^{(n+2)}} - H_{k+1}^{(n)} \dfrac{H_{k-1}^{(n+2)}}{\bar{H}_{k-1}^{(n+2)}}} = \frac{H_{k+1}^{(n)} / s_{k+1}^{(n-2)} - H_k^{(n)} r_{k+2}^{(n-2)}}{H_k^{(n)} r_k^{(n+2)} - H_{k+1}^{(n)} / s_{k-1}^{(n+2)}}$$

Divisons par $H_k^{(n)}$; on obtient :

$$F_k^{(n)} = \frac{r_{k+1}^{(n)} s_k^{(n)} / s_{k+1}^{(n-2)} - r_{k+2}^{(n-2)}}{r_k^{(n+2)} - r_{k+1}^{(n)} s_k^{(n)} / s_{k-1}^{(n+2)}} = \frac{r_{k+1}^{(n)} s_k^{(n)} - r_{k+2}^{(n-2)} s_{k+1}^{(n-2)}}{r_k^{(n+2)} s_{k-1}^{(n+2)} - r_{k+1}^{(n)} s_k^{(n)}} \frac{s_{k-1}^{(n+2)}}{s_{k+1}^{(n-2)}} \tag{32}$$

$$E_k^{(n)} = 1 + D_k^{(n)} + F_k^{(n)} \tag{33}$$

La relation (6) s'écrit :

$$\bar{P}_{k+1}^{(n)} (x) = (I_k^{(n)} x + J_k^{(n)}) \bar{P}_k^{(n)} (x) - K_k^{(n)} \bar{P}_{k-1}^{(n)} (x) \tag{34}$$

avec

$$I_k^{(n)} = H_k^{(n)} H_{k+1}^{(n)} \bar{H}_k^{(n)} / [H_k^{(n)}]^2 \bar{H}_{k+1}^{(n)}$$

$$K_k^{(n)} = [H_{k+1}^{(n)}]^2 \bar{H}_{k-1}^{(n)} / [H_k^{(n)}]^2 \bar{H}_{k+1}^{(n)}$$

et d'après (16) on a :

$$1 = - J_k^{(n)} - I_k^{(n)} - K_k^{(n)}$$

$$I_k^{(n)} = \frac{H_{k+1}^{(n)} \bar{H}_k^{(n)}}{H_k^{(n)} \bar{H}_{k+1}^{(n)}} = \frac{s_k^{(n)}}{s_{k+1}^{(n)}} \tag{35}$$

$$K_k^{(n)} = \frac{H_{k+1}^{(n)}}{H_k^{(n)}} \frac{H_{k+1}^{(n)}}{\bar{H}_{k+1}^{(n)}} \frac{\bar{H}_{k-1}^{(n)}}{H_k^{(n)}} = \frac{r_{k+1}^{(n)} s_k^{(n)}}{r_k^{(n)} s_{k+1}^{(n)}} \tag{36}$$

$$J_k^{(n)} = - K_k^{(n)} - I_k^{(n)} - 1 \tag{37}$$

(7) donne :

$$x \bar{P}_k^{(n+1)}(x) = [\frac{s_k^{(n)}}{s_k^{(n+1)}} x + 1 + \frac{r_{k+1}^{(n)} s_k^{(n)}}{s_k^{(n+1)} r_{k+1}^{(n-1)}} - \frac{s_k^{(n)}}{s_k^{(n+1)}}] \bar{P}_k^{(n)}(x) - \frac{r_{k+1}^{(n)} s_k^{(n)}}{s_k^{(n+1)} r_{k+1}^{(n-1)}} \bar{P}_k^{(n-1)}(x) \tag{38}$$

Enfin (8) s'écrit :

$$\bar{P}_{k+1}^{(n-1)}(x)=\left[\frac{s_k^{(n)}}{s_{k+1}^{(n-1)}}\, x-1-\frac{s_k^{(n)}}{s_{k+1}^{(n-1)}}-\frac{r_{k+1}^{(n)}s_k^{(n)}}{r_k^{(n+1)}s_{k+1}^{(n-1)}}\right]\bar{P}_k^{(n)}(x)-\frac{r_{k+1}^{(n)}s_k^{(n)}}{r_k^{(n+1)}s_{k+1}^{(n-1)}}\bar{P}_{k-1}^{(n+1)}(x) \qquad (39)$$

Soit maintenant S la fonctionnelle linéaire définie par :

$$S(x^i) = S_i$$

On a :

$$S(x^n\,\bar{P}_k^{(n)}(x)) = (-1)^k \begin{vmatrix} S_n & \cdots\cdots & S_{n+k} \\ c_n & \cdots\cdots & c_{n+k} \\ \cdots & \cdots & \cdots \\ c_{n+k-1} & \cdots & c_{n+2k-1} \end{vmatrix} \Big/ \begin{vmatrix} 1 & \cdots\cdots & 1 \\ c_n & \cdots\cdots & c_{n+k} \\ \cdots & \cdots & \cdots \\ c_{n+k-1} & \cdots & c_{n+2k-1} \end{vmatrix}$$

Par conséquent :

$$S(x^n\,\bar{P}_k^{(n)}(x)) = (-1)^k\, G_k^{(n)} \qquad (40)$$

où les $G_k^{(n)}$ sont les transformés de $\{S_n\}$ par la transformation G.

Si nous multiplions (27) par x^n et si nous appliquons S nous trouvons :

$$G_{k+1}^{(n)} = -\frac{s_k^{(n+1)}}{s_{k+1}^{(n)}}\left(G_k^{(n+1)} - \frac{r_{k+1}^{(n+1)}}{r_{k+1}^{(n)}}\, G_k^{(n)}\right)$$

ou encore en utilisant (21)

$$\left(1-\frac{r_{k+1}^{(n+1)}}{r_{k+1}^{(n)}}\right) G_{k+1}^{(n)} = G_k^{(n+1)} - \frac{r_{k+1}^{(n+1)}}{r_{k+1}^{(n)}}\, G_k^{(n)} \qquad (41)$$

qui n'est autre que la relation démontrée par Pye et Atchison [8].

Remarque : Il est facile de vérifier que les algorithmes rs et qd sont reliés.

En effet on a :

$$q_{k+1}^{(n)} = \frac{H_{k+1}^{(n+1)}\; H_k^{(n)}}{H_k^{(n+1)}\; H_{k+1}^{(n)}} = \frac{r_{k+1}^{(n+1)}\; s_k^{(n+1)}}{r_{k+1}^{(n)}\; s_k^{(n)}} \qquad (42)$$

$$e_{k+1}^{(n)} = \frac{H_{k+2}^{(n)} H_k^{(n+1)}}{H_{k+1}^{(n)} H_{k+1}^{(n+1)}} = \frac{r_{k+2}^{(n)} s_{k+1}^{(n)}}{r_{k+1}^{(n+1)} s_k^{(n+1)}} \tag{43}$$

On pourra, en reportant (42) et (43) dans les règles de l'algorithme qd, vérifier que ces règles sont satisfaites.

Nous allons maintenant considérer le cas particulier où :

$$c_n = \Delta S_n$$

Dans ce cas on a :

$$c^{(n)}(x^i) = c_{n+i} = \Delta S_{n+i} = S(x^{n+i+1} - x^{n+i})$$

Par conséquent les relations d'orthogonalité s'écrivent :

$$c^{(n)}(x^i \bar{P}_k^{(n)}(x)) = S((x^{n+i+1} - x^{n+i}) \bar{P}_k^{(n)}(x)) = 0 \qquad i = 0, \ldots, k-1$$

et par conséquent :

$$S(x^{n+i} \bar{P}_k^{(n)}(x)) \text{ est constant pour } i = 0, \ldots, k$$

et l'on a :

$$S(x^n \bar{P}_k^{(n)}(x)) = (-1)^k H_{k+1}(S_n) / H_k(\Delta^2 S_n) = (-1)^k e_k(S_n) \tag{44}$$

où les $e_k(S_n)$ sont les transformées de $\{S_n\}$ par la transformation de Shanks [9]. Les relations (26) à (39) incluse nous donnent donc :

$$e_k(S_{n+1}) = \frac{s_k^{(n)}}{s_k^{(n+1)}} [e_k(S_n) + \frac{r_{k+1}^{(n)}}{r_k^{(n+1)}} e_{k-1}(S_{n+1})] \tag{45}$$

$$e_{k+1}(S_n) = \frac{s_k^{(n+1)}}{s_{k+1}^{(n)}} [\frac{r_{k+1}^{(n+1)}}{r_{k+1}^{(n)}} e_k(S_n) - e_k(S_{n+1})] \tag{46}$$

$$e_k(S_n) = (1 + \frac{r_{k+1}^{(n)}}{r_k^{(n+2)}}) e_k(S_{n+1}) - \frac{r_{k+1}^{(n)}}{r_k^{(n+2)}} e_{k-1}(S_{n+2}) \tag{47}$$

$$e_{k+1}(S_{n-1}) = (1 + \frac{s_k^{(n+1)}}{s_{k+1}^{(n-1)}}) e_k(S_n) - \frac{s_k^{(n+1)}}{s_{k+1}^{(n-1)}} e_k(S_{n+1}) \tag{48}$$

$$e_{k+1}(S_{n-2}) = (E_k^{(n)} - D_k^{(n)})\, e_k(S_n) - F_k^{(n)}\, e_{k-1}(S_{n+2}) \tag{49}$$

où $E_k^{(n)}$, $D_k^{(n)}$ et $F_k^{(n)}$ sont calculés par (31), (32) et (33)

$$e_{k+1}(S_n) = -\,(I_k^{(n)} + J_k^{(n)})\, e_k(S_n) - K_k^{(n)}\, e_{k-1}(S_n) \tag{50}$$

où $I_k^{(n)}$, $J_k^{(n)}$ et $K_k^{(n)}$ sont calculés par (35), (36) et (37).

$$e_k(S_{n+1}) = \left(1 + \frac{r_{k+1}^{(n)}\, s_k^{(n)}}{s_k^{(n+1)}\, r_{k+1}^{(n-1)}}\right) e_k(S_n) - \frac{r_{k+1}^{(n)}\, s_k^{(n)}}{s_k^{(n+1)}\, r_{k+1}^{(n-1)}}\, e_k(S_{n-1}) \tag{51}$$

$$e_{k+1}(S_{n-1}) = \left(1 + \frac{r_{k+1}^{(n)}\, s_k^{(n)}}{r_k^{(n+1)}\, s_{k+1}^{(n-1)}}\right) e_k(S_n) - \frac{r_{k+1}^{(n)}\, s_k^{(n)}}{r_k^{(n+1)}\, s_{k+1}^{(n-1)}}\, e_{k-1}(S_{n+1}) \tag{52}$$

L'algorithme rs peut donc être utilisé pour calculer les quantités $e_k(S_n)$ situées sur un chemin arbitraire du tableau ε ; il faut alors adopter l'initialisation $r_1^{(n)} = \Delta S_n$.

Compte tenu de la relation entre les algorithmes rs et qd la relation (50) peut aussi s'écrire :

$$s_{k+1}^{(n)}\, e_{k+1}(S_n) = (q_{k+1}^{(n)} + e_k^{(n)} - 1)\, s_k^{(n)}\, e_k(S_n) - q_k^{(n)}\, e_k^{(n)}\, s_{k-1}^{(n)}\, e_{k-1}(S_n) \tag{53}$$

En exprimant les quantités $e_k(S_n)$ sous forme d'un rapport de déterminants on obtient de nouvelles relations entre déterminants de Hankel.

Nous allons maintenant établir d'autres relations entre l'algorithme rs et la transformation de Shanks. On a :

$$\frac{r_{k+1}^{(n)}}{s_{k+1}^{(n)}} = \frac{[H_{k+1}^{(n)}]^2}{\bar{H}_k^{(n)}\, \bar{H}_{k+1}^{(n)}}$$

Par conséquent d'après une relation connue [2] de la transformation de Shanks on a :

$$e_{k+1}(S_n) - e_k(S_n) = -\, r_{k+1}^{(n)} / s_{k+1}^{(n)} \tag{54}$$

D'après (44) on a :

$$e_k(S_{n+1}) - e_k(S_n) = \frac{s_{k+1}^{(n)}}{s_k^{(n+1)}}\,(e_k(S_n) - e_{k+1}(S_n))$$

D'où, en utilisant (54) :

$$e_k(S_{n+1}) - e_k(S_n) = \frac{r_{k+1}^{(n)}}{s_k^{(n+1)}} = \frac{H_{k+1}^{(n)}\ H_k^{(n+1)}}{\bar{H}_k^{(n)}\ \bar{H}_k^{(n+1)}} \tag{55}$$

On a :

$$e_{k+1}(S_n) - e_k(S_{n+1}) = e_{k+1}(S_n) - e_k(S_n) + e_k(S_n) - e_k(S_{n+1})$$

$$= - \frac{r_{k+1}^{(n)}}{s_{k+1}^{(n)}} - \frac{r_{k+1}^{(n)}}{s_k^{(n+1)}}$$

et on obtient finalement, en utilisant (21) :

$$e_{k+1}(S_n) - e_k(S_{n+1}) = - \frac{r_{k+1}^{(n+1)}}{s_{k+1}^{(n)}} = - \frac{H_{k+1}^{(n+1)}\ H_{k+1}^{(n)}}{\bar{H}_k^{(n+1)}\ \bar{H}_{k+1}^{(n)}} \tag{56}$$

Enfin on a :

$$e_{k+1}(S_n) - e_k(S_{n+2}) = e_{k+1}(S_n) - e_k(S_{n+1}) + e_k(S_{n+1}) - e_k(S_{n+2})$$

$$e_{k+1}(S_n) - e_k(S_{n+2}) = - \frac{r_{k+1}^{(n+1)}}{s_{k+1}^{(n)}} - \frac{r_{k+1}^{(n+1)}}{s_k^{(n+2)}} = - \frac{[H_{k+1}^{(n+1)}]^2}{\bar{H}_{k+1}^{(n)}\ \bar{H}_k^{(n+2)}}$$

$$= - \frac{r_{k+2}^{(n)}}{s_{k+1}^{(n+1)}} - \frac{r_{k+1}^{(n+2)}}{s_{k+1}^{(n+1)}} \tag{57}$$

Toutes les relations précédentes peuvent évidemment être utilisées pour le calcul des approximants de Padé.

Nous allons maintenant considérer le cas où la suite $\{S_n\}$ est une suite d'éléments d'un e.v.t.. Il est encore possible de définir l'application linéaire S par :

$$S(x^i) = S_i$$

et la transformation G généralisée par le même rapport de déterminants (40) que précédemment où la suite auxiliaire $\{c_n\}$ est une suite de scalaires. En multipliant (27) par x^n et en appliquant S on retrouve (41) qui permet de calculer récursivement les éléments $G_k^{(n)}$ de l'e.v.t. transformés de la suite $\{S_n\}$.

Si $c_n = \langle y, \Delta S_n \rangle$ alors $G_k^{(n)}$ est identique à la généralisation de la transformation de Shanks étudiée en [1]. Les $e_k(S_n)$ peuvent être calculés soit par la transformation G généralisée soit par l'ε-algorithme topologique qui est une généralisation de l'ε-algorithme scalaire de Wynn [10].

Pour cette généralisation de la transformation de Shanks les autres relations entre polynômes orthogonaux adjacents ne fournissent pas de relations de récurrence car alors (44) n'est plus valable.
Si nous multiplions (27) par x^{n+i} et si nous appliquons S alors (41) est toujours valable avec :

$$G_k^{(n)} = \begin{vmatrix} S_{n+i} & \cdots\cdots & S_{n+i+k} \\ c_n & \cdots\cdots & c_{n+k} \\ \cdots & \cdots & \cdots \\ c_{n+k-1} & \cdots & c_{n+2k-1} \end{vmatrix} \Big/ \bar{H}_k^{(n)}$$

Si $c_n = \langle y, \Delta S_n \rangle$ alors les $G_k^{(n)}$ sont les transformées de $\{S_n\}$ (pour $i = 0, \ldots, k$) qui avaient été étudiées dans [1] mais pour lesquelles aucun algorithme de calcul récursif n'avait pu être obtenu. Pour $i = 0$ on retrouve la première généralisation de la transformation de Shanks dont nous avons parlé précédemment tandis que pour $i = k$ on retrouve la seconde généralisation qui avait été notée $\tilde{e}_k(S_n)$ et qui peut être mise en oeuvre en utilisant le second ε-algorithme topologique.

Le calcul de $e_k(S_n)$ pour k et n fixé est beaucoup plus économique en utilisant l'algorithme G qu'en utilisant l'ε-algorithme topologique.

Supposons que la suite $\{S_n\}$ soit une suite de vecteurs à p composantes. Le calcul de $e_k(S_n)$ avec l'ε-algorithme topologique nécessite :

$k(2k + 1)p$ divisions

$2k(2k + 1)p$ additions

$k(2k + 1)$ calculs de produits scalaires

soit un total de :

$k(2k+1)p$ divisions

$k(2k+1)p$ multiplications

$k(2k+1)(3p-1)$ additions

c'est-à-dire $k(2k+1)(5p-1)$ opérations arithmétiques.

Il faut également mettre en mémoire 6k vecteurs de dimensions p[4].

Avec l'algorithme G le même calcul demande :

$k(k+5)p + \frac{5k}{2}(k-1)$ additions

$\frac{k(k+1)}{2} p + k(2k-1)$ divisions

$2kp + k(2k-1)$ multiplications

c'est-à-dire $\frac{3k}{2}(k+5)p + \frac{k}{2}(13k - 9)$.

Il ne faut mettre en mémoire que k+2 vecteurs de dimension p.

Pour calculer la solution d'un système de p équations linéaires il faut $10p^3 + 3p^2 - p$ opérations avec l'ε-algorithme topologique alors que la transformation G n'en nécessite que $\frac{3}{2} p^3 + 14p^2 - \frac{9p}{2}$.

Si l'on suppose que les vecteurs $\{S_n\}$ sont générés par :

$$S_o = 0$$

$$S_{n+1} = AS_n + b$$

où A est une matrice symétrique définie positive et si l'on prend $y = b$ alors on sait que la suite $\{e_k(S_o)\}$ est identique à la suite fournie par la méthode du gradient conjugué [3]. Pour calculer la solution $e_p(S_o)$ du système on a besoin de connaitre les vecteurs $S_o, \ldots, S_{2p}$ car l'ε-algorithme topologique ne profite pas du fait que la matrice est symétrique.

Si l'on utilise la transformation G alors on a besoin de connaitre $S_o, \ldots, S_p$ pour appliquer (41) et $(y, \Delta S_o), \ldots, (y, \Delta S_{2p-1})$ pour appliquer l'algorithme rs. Cependant on a :

$$(y, \Delta S_o) = (b, b)$$

$$(y, \Delta S_k) = (b, A^k b) = (A^n b, A^m b) = (\Delta S_n, \Delta S_m)$$

quelque soient n et m tels que $n + m = k$. Par conséquent on a seulement besoin de connaitre $S_o, \ldots, S_{p+1}$ pour résoudre le système par la transformation G.

Il faut cependant remarquer que, dans ce cas, l'algorithme du gradient conjugué est encore plus économique.

La transformation G généralisée peut être utilisée pour mettre en oeuvre une méthode due à Germain-Bonne [7, propriété 12, p. 71] qui est une généralisation de la transformation de Shanks scalaire et est assez voisine de la généralisation de la transformation de Shanks que nous venons de voir.

Etant donnée une suite $\{x_n\}$ d'éléments d'un espace de Hilbert on définit la suite $\{y_n\}$ par :

$$y_n = x_n + \alpha_1 \Delta x_n + \ldots + \alpha_k \Delta x_{n+k-1}$$

où les coefficients α_i sont déterminés par les conditions

$$(y_n^{(i)} - y_n, \Delta x_n) = 0 \qquad i = 1, \ldots, k$$

avec

$$y_n^{(i)} = x_{n+i} + \alpha_1 \Delta x_{n+i} + \ldots + \alpha_k \Delta x_{n+k+i-1}$$

Ces relations s'écrivent :

$$y_n = (1-\alpha_1)\, x_n + (\alpha_1-\alpha_2)\, x_{n+1} + \ldots + (\alpha_{k-1} - \alpha_k)\, x_{n+k-1} + \alpha_k\, x_{n+k}$$

$$y_n^{(i)} = (1-\alpha_1)\, x_{n+i} + (\alpha_1 - \alpha_2)\, x_{n+i+1} + \ldots + (\alpha_{k-1} - \alpha_k) x_{n+k+i-1} + \alpha_k x_{n+k+i}$$

Si l'on pose

$$\beta_o = 1 - \alpha_1$$

$$\beta_j = \alpha_j - \alpha_{j+1} \qquad j = 1, \ldots, k-1$$

$$\beta_k = \alpha_k$$

alors on a :

$$y_n = \beta_o x_n + \dots + \beta_k x_{n+k}$$

$$y_n^{(i)} = \beta_o x_{n+i} + \dots + \beta_k x_{n+k+i} \qquad i = 1, \dots, k$$

avec $\beta_o + \dots + \beta_k = 1$

$\beta_o, \dots, \beta_k$ sont alors déterminés par le système

$$\begin{pmatrix} 1 & \dots\dots\dots\dots & 1 \\ (\Delta x_n, \Delta x_n) & \dots\dots\dots\dots & (\Delta x_{n+k}, \Delta x_n) \\ \dots\dots & \dots\dots\dots\dots & \dots\dots \\ (\Delta x_{n+k-1}, \Delta x_n) & \dots\dots\dots & (\Delta x_{n+2k-1}, \Delta x_n) \end{pmatrix} \begin{pmatrix} \beta_o \\ \beta_1 \\ \vdots \\ \beta_k \end{pmatrix} = \begin{pmatrix} 1 \\ 0 \\ \vdots \\ 0 \end{pmatrix}$$

et par conséquent :

$$y_n = \frac{\begin{vmatrix} x_n & \dots\dots\dots\dots & x_{n+k} \\ (\Delta x_n, \Delta x_n) & \dots\dots\dots\dots & (\Delta x_{n+k}, \Delta x_n) \\ \dots\dots & \dots\dots\dots\dots & \dots\dots \\ (\Delta x_{n+k-1}, \Delta x_n) & \dots\dots\dots & (\Delta x_{n+2k-1}, \Delta x_n) \end{vmatrix}}{\begin{vmatrix} 1 & \dots\dots\dots\dots & 1 \\ (\Delta x_n, \Delta x_n) & \dots\dots\dots\dots & (\Delta x_{n+k}, \Delta x_n) \\ \dots\dots & \dots\dots\dots\dots & \dots\dots \\ (\Delta x_{n+k-1}, \Delta x_n) & \dots\dots\dots & (\Delta x_{n+2k-1}, \Delta x_n) \end{vmatrix}}$$

On voit que ce rapport n'est pas exactement celui qui intervient dans la transformation G car on ne peut pas poser $c_n = (\Delta x_n, \Delta x_n)$ (en effet les secondes lignes des déterminants sont différentes de $c_n, c_{n+1}, \dots, c_{n+k}$).

On ne pourra donc pas calculer récursivement les suites $\{y_n\}$ (pour différentes valeurs de k) en utilisant l'algorithme de la transformation G. Cependant cet algorithme peut être utilisé pour calculer y_n pour n et k fixés. Il suffit pour cela d'appliquer l'algorithme de la transformation G à la suite $\{S_n\}$ et à la suite auxiliaire $\{c_n\}$ définies par :

$S_o = x_n$ $\qquad\qquad$ $c_o = (\Delta x_n, \Delta x_n)$

$S_1 = x_{n+1}$ $\qquad\qquad$ $c_1 = (\Delta x_{n+1}, \Delta x_n)$

.... $\qquad\qquad$ ....

$S_k = x_{n+k}$ $\qquad\qquad$ $c_{2k-1} = (\Delta x_{n+2k-1}, \Delta x_n)$

On aura :

$$G_k^{(0)} = y_n$$

Si l'on change de valeur de k ou de n tous les calculs sont à recommencer.

2 - LE W-ALGORITHME

Soit à calculer les rapports de déterminants :

$$E_k^{(n)} = \frac{\begin{vmatrix} a_n & a_{n+1} & \cdots\cdots & a_{n+k} \\ a_{n+1} & a_{n+2} & \cdots\cdots & a_{n+k+1} \\ \cdots & \cdots & \cdots & \cdots \\ a_{n+k} & a_{n+k+1} & \cdots\cdots & a_{n+2k} \end{vmatrix}}{\begin{vmatrix} a_{n+2} & \cdots\cdots & a_{n+k+1} \\ \cdots & \cdots & \cdots \\ a_{n+k+1} & \cdots\cdots & a_{n+2k} \end{vmatrix}} = \frac{H_{k+1}(a_n)}{H_k(a_{n+2})} \qquad (58)$$

pour différentes valeurs de k et de n.

Ce formalisme regroupe un certain nombre de transformations de suites et de fonctions que l'on rencontre en accélération de la convergence :

$$\begin{array}{lll} a_n = \Delta^n S_i & i \text{ fixé} & E_k^{(0)} = e_k(S_i) \\ a_n = f^{(n)}(t) & t \text{ fixé} & E_k^{(0)} = \varepsilon_{2k}(t) \\ a_n = f^{(n)}(t) / n\,! & t \text{ fixé} & E_k^{(0)} = \rho_{2k}(t) \end{array} \qquad (59)$$

La méthode la plus simple pour calculer les rapports $E_k^{(n)}$ est d'utiliser la relation de récurrence des déterminants de Hankel (9) avec l'initialisation :

$$H_o^{(n)} = 1 \qquad\qquad H_1^{(n)} = a_n$$

On aura

$$E_k^{(n)} = H_{k+1}^{(n)} / H_k^{(n+2)}$$

Dans le cas de la transformation de Shanks on peut ainsi calculer la diagonale $\{E_k^{(0)} = e_k(S_i)\}$ en utilisant une seule fois la relation de récurrence des déterminants de Hankel au lieu d'écrire :

$$e_k(S_i) = H_{k+1}(S_i) / H_k(\Delta^2 S_i)$$

et d'être obligé d'utiliser cette relation de récurrence une fois pour les numérateurs et une fois pour les dénominateurs. Il faut cependant remarquer que, dans ce cas, on obtiendrait les valeurs de $e_k(S_i)$ pour tout k et pour tout i au lieu de la seule diagonale k pour i fixé. On utilisera pour ce calcul le sous-programme correspondant à la forme confluente de l'ε-algorithme donné dans [4].

Ces rapports de déterminants peuvent également être calculés en utilisantle w-algorithme qui a été conçu par Wynn [11] pour mettre en oeuvre les formes confluentes de ε et ρ-algorithmes. Ces formes confluentes sont définies par un rapport de déterminants analogue à (58). Pour calculer $E_k^{(n)}$ dans le cas général il suffit donc d'en changer les initialisations. On a :

$$w_{-1}^{(n)} = 0 \qquad w_o^{(n)} = a_n \qquad n = 0, 1, \ldots$$

$$\left.\begin{aligned} w_{2k+1}^{(n)} &= w_{2k-1}^{(n+1)} + w_{2k}^{(n)} / w_{2k}^{(n+1)} \\ w_{2k+2}^{(n)} &= w_{2k}^{(n+1)} (w_{2k+1}^{(n)} - w_{2k+1}^{(n+1)}) \end{aligned}\right\} \quad n, k = 0, 1, \ldots \qquad (60)$$

Wynn a démontré que :

$$E_k^{(n)} = w_{2k}^{(n)} \qquad (61)$$

Les quantités d'indice inférieur impair sont des calculs intermédiaires qui peuvent être éliminés et l'on obtient alors l'algorithme :

$$\begin{aligned} w_o^{(n)} &= a_n \qquad n = 0, 1, \ldots \\ w_2^{(n)} &= w_o^{(n)} - [w_o^{(n+1)}]^2 / w_o^{(n+2)} \\ w_{2k+2}^{(n)} &= w_{2k}^{(n)} + \frac{[w_{2k}^{(n+1)}]^2}{w_{2k-2}^{(n+2)} w_{2k}^{(n+2)}} (w_{2k}^{(n+2)} - w_{2k-2}^{(n+2)}) \end{aligned} \qquad (62)$$

ce qui peut encore se mettre sous la forme :

$$\begin{aligned} w_{-2}^{(n)} &= \infty \qquad w_o^{(n)} = a_n \qquad n = 0, 1, \ldots \\ w_{2k+2}^{(n)} &= w_{2k}^{(n)} - [w_{2k}^{(n+1)}]^2 \left[\frac{1}{w_{2k}^{(n+2)}} - \frac{1}{w_{2k-2}^{(n+2)}}\right] \qquad n, k = 0, 1, \ldots \end{aligned} \qquad (63)$$

D'après la relation de récurrence des déterminants de Hankel on a :

$$\frac{H_{k+1}^{(n)}}{H_k^{(n+2)}} - \frac{H_k^{(n)}}{H_{k-1}^{(n+2)}} = - \frac{[H_k^{(n+1)}]^2}{H_k^{(n+2)}\ H_{k-1}^{(n+2)}}$$

et par conséquent :

$$E_k^{(n)} - E_{k-1}^{(n)} = [H_k^{(n+1)}]^2 / H_k^{(n+2)}\ H_{k-1}^{(n+2)}$$

Si on applique l'algorithme qd à la suite $\{a_n\}$ alors on voit que :

$$e_k^{(n)} = \frac{H_{k+1}^{(n)}\ H_{k-1}^{(n+1)}}{H_k^{(n)}\ H_k^{(n+1)}} \qquad q_k^{(n)} = \frac{H_k^{(n+1)}\ H_{k-1}^{(n)}}{H_k^{(n)}\ H_{k-1}^{(n+1)}}$$

d'où

$$w_{2k}^{(n)} = a_n \prod_{j=1}^{k} e_j^{(n)} / q_j^{(n+1)}$$

$$w_{2k+2}^{(n)} = w_{2k}^{(n)} \frac{e_{k+1}^{(n)}}{q_{k+1}^{(n+1)}}$$

Remarque : en tenant compte de (61), (63) s'écrit :

$$E_{-1}^{(n)} = \infty \qquad E_0^{(n)} = a_n \qquad n = 0, 1, \ldots$$

$$E_{k+1}^{(n)} = E_k^{(n)} - [E_k^{(n+1)}]^2 \left[\frac{1}{E_k^{(n+2)}} - \frac{1}{E_{k-1}^{(n+2)}}\right] \qquad n, k = 0, 1, \ldots$$

Nous venons de voir qu'un algorithme mis au point initialement pour une transformation de fonction pouvait être utilisé pour une transformation de suite.

Inversement nous allons voir que l'ε-algorithme scalaire (ou un autre algorithme servant à mettre en oeuvre la transformation de Shanks comme, par exemple, la transformation G étudiée au paragraphe précédent) peut servir pour la forme confluente de l'ε-algorithme ou pour celle du ρ-algorithme. Il suffit, pour cela, de considérer les différentes suites $\{a_n\}$ de (59).

Soit $\{S_n\}$ la suite définie par :

$$\Delta^n S_o = f^{(n)}(t) \qquad n = 0, 1, \ldots$$

Si on applique la transformation de Shanks à cette suite $\{S_n\}$ alors :

$$e_k (S_o) = \varepsilon_{2k}(t)$$

$e_k(S_o)$ peut être calculé numériquement avec l'une quelconque des méthodes de mise en oeuvre de la transformation de Shanks : ε-algorithme, transformation G, relation de récurrence des déterminants de Hankel, etc ...

Si on applique la transformation de Shanks à la suite $\{S_n\}$ définie par :

$$\Delta^n S_o = f^{(n)}(t) \; / \; n \; ! \qquad n = 0, 1, \ldots$$

alors :

$$e_k(S_o) = \rho_{2k}(t)$$

Remarque : On peut étudier les transformations de suites qui correspondent aux choix :

$$a_n = \Delta^n S_i \; / \; n \; ! \qquad i \text{ fixé}$$

$$a_n = [x_i, \ldots, x_{i+n}] \qquad i \text{ fixé}$$

$$a_n = n \; ! \; [x_i, \ldots, x_{i+n}] \qquad i \text{ fixé}$$

où $[x_i, \ldots, x_{i+n}]$ est la différence divisée d'ordre n de la fonction f telle que $f(x_i) = S_i$.

Ces deux derniers cas pourraient être considérés comme correspondants respectivement aux formes confluentes des ε et ρ-algorithmes puisque si $x_i, \ldots, x_{i+n}$ tendent vers t alors :

$$n \; ! \; [t, \ldots, t] = f^{(n)} (t)$$

La transformation G généralisée peut être utilisée pour mettre en oeuvre la forme confluente de l'ε-algorithme topologique [2]. Il est facile de voir que :

$$G_k^{(n)} = \frac{\begin{vmatrix} S_n & \Delta S_n & \cdots\cdots & \Delta^k S_n \\ c_n & \Delta c_n & \cdots\cdots & \Delta^k c_n \\ \cdots & \cdots & \cdots & \cdots \\ \Delta^{k-1} c_n & \Delta^k c_n & \cdots\cdots & \Delta^{2k-1} c_n \end{vmatrix}}{\begin{vmatrix} \Delta c_n & \cdots\cdots & \Delta^k c_n \\ \cdots & \cdots & \cdots \\ \Delta^k c_n & \cdots\cdots & \Delta^{2k-1} c_n \end{vmatrix}}$$

Si nous appliquons la transformation G à la suite $\{S_n\}$ et à la suite auxiliaire $\{c_n\}$ définies par :

$$\Delta^n S_0 = f^{(n)}(t)$$

$$\Delta^n c_0 = \langle y, f^{(n+1)}(t) \rangle$$

alors on voit que :

$$G_k^{(0)} = \varepsilon_{2k}(t)$$

3 - Calcul des polynômes orthogonaux

Nous allons proposer deux méthodes nouvelles pour calculer récursivement les polynômes orthogonaux. Ces deux méthodes sont basées sur la connexion qui existe entre les polynômes orthogonaux et les dénominateurs partiels des fractions continues associées et correspondantes à une série.

Soit $f(t) = \sum_{i=o}^{\infty} c_i t^i$ et soit D sa fraction continue associée

$$D(t) = \frac{C_1|}{|1+B_1 t} - \frac{C_2 t^2|}{|1+B_2 t} - \dots$$

On sait que les B_k et les C_k sont les coefficients de la relations de récurrence des polynômes orthogonaux par rapport à la suite $\{c_n\}$ (où par rapport à la fonctionnelle $c^{(0)}$ du premier paragraphe). On va donc proposer une méthode de calcul de ces coefficients tout à fait analogue à celle utilisée dans le cas de la fraction continue correspondante et qui est basée sur l'algorithme d'Euclide pour obtenir le p.g.c.d. de deux nombres.

On pose :

$$h_o = D(t)$$

$$h_n = 1+B_n t - \frac{C_{n+1} t^2|}{|1+B_{n+1} t} - \dots$$

Ainsi

$$h_o = C_1 / h_1$$

$$h_n = 1+B_n t - C_{n+1} t^2 / h_{n+1} \qquad n = 1, 2, \dots$$

Posons

$$h_n = u_{n-1} / u_n$$

En remplaçant dans l'équation précédente et en multipliant par u_n on obtient :

$$u_{n-1} = (1+B_{n+1} t) u_n - C_{n+1} t^2 u_{n+1} \qquad n = 1, 2, \dots$$

$$u_{-1} = C_1 u_1$$

avec $\quad u_0 = 1 \quad$ et $\quad u_{-1} = f(t)$

Mais $D(0) = C_1 = f(0) = c_0$ et donc $u_1 = f(t) / c_0$ et $u_n(0) = 1$ pour $n = 1, 2, \ldots$

On obtient donc finalement l'algorithme suivant :

$$P_{-1}(t) = 0 \qquad P_0(t) = 1$$

$$u_0 = 1 \qquad u_1 = f(t) / c_0$$

puis, pour $n = 0, 1, \ldots$

$$B_{n+1} = \left[\frac{u_n - u_{n+1}}{t}\right]_{t=0}$$

$$P_{n+1}(t) = (t + B_{n+1})\, P_n(t) - C_{n+1}\, P_{n-1}(t)$$

$$C_{n+2} = \left[\frac{(1+B_{n+1}t)\, u_{n+1} - u_n}{t^2}\right]_{t=0}$$

$$u_{n+2} = [(1+B_{n+1}t)\, u_{n+1} - u_n] / C_{n+2}t^2$$

Remarque : en dérivant la relation de récurrence qui lie les u_n on voit que l'on a également :

$$B_{n+1} = u'_n(0) - u'_{n+1}(0)$$

$$2C_{n+2} = u''_{n+1}(0) + 2B_{n+1}\, u'_{n+1}(0) - u''_n(0)$$

Considérons maintenant la fraction continue correspondante à f.

Elle peut s'écrire de deux façons :

$$C(t) = c_0 + \frac{t|}{|\beta_1} + \frac{t|}{|\beta_2} + \ldots = c_0 + \frac{\alpha_1 t|}{|1} + \frac{\alpha_2 t|}{|1} + \ldots$$

Il est facile de voir que pour que ces deux fractions continues soient équivalentes (c'est-à-dire possédent la même suite de convergents) il faut que :

$$\alpha_i = \beta_i^{-1}\, \beta_{i-1}^{-1} \qquad i = 1, 2, \ldots$$

avec

$$\beta_0 = 1$$

On sait d'autre part que les nombres β_k sont reliés à la forme confluente du ρ-algorithme par [2] :

$$\beta_k = \rho_k(0) - \rho_{k-2}(0) \qquad k = 1, 2, \ldots.$$

D'autre part on sait qu'en effectuant une contraction de la fraction continue correspondante on obtient la fraction continue associée. Ceci se traduit par les relations :

$$B_{n+1} = \alpha_{2n+1} + \alpha_{2n+2} = \frac{\beta_{2n} + \beta_{2n+2}}{\beta_{2n}\,\beta_{2n+1}\,\beta_{2n+2}}$$

$$C_{n+1} = \alpha_{2n}\,\alpha_{2n+1} = [\beta_{2n-1}\,\beta_{2n}^2\,\beta_{2n+1}]^{-1}$$

On a donc finalement l'algorithme suivant :

$$P_{-1}(t) = 0 \qquad P_o(t) = 1$$

$$\beta_{-1} = \beta_o = 1$$

$$\rho_{-1}(t) = 0 \qquad \rho_o(t) = f(t)$$

puis, pour n = 0, 1, ...

$$\rho_{2n+1}(t) = \rho_{2n-1}(t) + \frac{2n+1}{\rho'_{2n}(t)}$$

$$\beta_{2n+1} = \rho_{2n+1}(0) - \rho_{2n-1}(0)$$

$$\rho_{2n+2}(t) = \rho_{2n}(t) + \frac{2n+2}{\rho'_{2n+1}(t)}$$

$$\beta_{2n+2} = \rho_{2n+2}(0) - \rho_{2n}(0)$$

$$B_{n+1} = (\beta_{2n} + \beta_{2n+2}) \,/\, \beta_{2n}\,\beta_{2n+1}\,\beta_{2n+2}$$

$$C_{n+1} = (\beta_{2n-1}\,\beta_{2n}^2\,\beta_{2n+1})^{-1}$$

$$P_{n+1}(t) = (t + B_{n+1})\,P_n(t) - C_{n+1}\,P_{n-1}(t)$$

Le calcul des quantités $\rho_k(0)$ par l'algorithme précédent nécessite la dérivation formelle des fonctions ρ_k successives. Il est donc difficile à mettre en oeuvre. On pourra donc le remplacer avantageusement soit par la relation de récurrence des déterminants de Hankel (avec $H_1^{(n)} = f^{(n)}(0)/n!$) soit par le w-algorithme (avec $w_o^{(n)} = f^{(n)}(0) \,/\, n!$).

Dans le premier cas on aura :

$$\rho_{2k}(0) = H_{k+1}^{(0)} \;/\; H_k^{(2)} \qquad\qquad \rho_{2k+1}(0) = H_k^{(3)} \;/\; H_{k+1}^{(1)}$$

Dans le second cas on aura :

$$\rho_{2k}(0) = w_{2k}^{(0)} \qquad\qquad \rho_{2k+1}(0) = 1/w_{2k}^{(1)}$$

Il n'est pas possible d'utiliser l'ε-algorithme scalaire appliqué à la suite $\{S_n\}$ définie par $\Delta^n S_o = f^{(n)}(0) \,/\, n!$ car celui-ci ne fournira pas les valeurs de $\rho_{2k+1}(0)$.

Il est évident que ces deux méthodes récursives de calcul des polynômes orthogonaux fournissent des méthodes de calcul pour les approximants de Padé et pour la transformation de Shanks.

RÉFÉRENCES

[1] C. BREZINSKI
"Généralisations de la transformation de Shanks, de la table de Padé et de l'ε-algorithme". Calcolo, 12(1975) 317-360.

[2] C. BREZINSKI
"Accélération de la convergence en analyse numérique". Lecture Notes in Mathematics, 584, Springer-Verlag, 1977.

[3] C. BREZINSKI
"Padé approximants and orthogonal polynomials". In "Padé and rational approximation", E.B. Saff and R.S. Varga eds, Academic Press, New York, 1977.

[4] C. BREZINSKI
"Algorithmes d'accélération de la convergence. Etude numérique". Technip, Paris, 1978.

[5] C. BREZINSKI
"Padé-type approximation and general orthogonal polynomial". Birkhaüser-Verlag, 1979.

[6] C. BREZINSKI
"Programmes FORTRAN pour transformations de suites et de séries". Publication ANO-3, Laboratoire de Calcul, Université de Lille, 1979.

[7] B. GERMAIN-BONNE
"Estimation de la limite de suites et formalisation de procédés d'accélération de convergence" Thèse, Université de Lille, 1978.

[8] W.C. PYE, T.A. ATCHISON
"An algorithm for the computation of the higher order G-transformation". SIAM J. Numer. Anal., 10 (1973) 1-7.

[9] D. SHANKS
"Non linear transformations of divergent and slowly convergent series". J. Math. Phys., 34(1955) 1-42.

[10] P. WYNN
"On a device for computing the $e_m(S_n)$ transformation".
MTAC, 10 (1956) 91-96.

[11] P. WYNN
*"Upon some continuous prediction algorithms. I".*Calcolo, 9(1972)
177-234.

Claude BREZINSKI
Université de Lille I
UER d'IEEA - Informatique
B.P. 36
F - 59650 VILLENEUVE d'ASCQ
(FRANCE)

Recursive algorithms for the Padé table : two approaches.

A. Bultheel

Abstract

In [14], a relation is given between the Viskovatoff algorithm for the determination of continued fractions and the triangular factorization of Hankel matrices. In this paper this idea will be further develloped to include most of the known recursive algorithms for the computation of Padé approximants. The factorization interpretation links together the continued fraction approach and the recursive Padé computation in a natural way.

1. Introduction

Given a formal power series $f(z) = \sum_{i=0}^{\infty} f_i z^i$, then we want a rational approximation (Padé approximant)

$[m/n] = P^{[m/n]}(z)/Q^{[m/n]}(z)$

such that the residual series

$$Z^{[m/n]}(z) = Q^{[m/n]}(z)\, f(z) - P^{[m/n]}(z) \qquad (1)$$

satisfies ord $Z^{[m/n]}(z) \leq m+n+1$. Furthermore $[m/n]$ must be irreducible and some normalization must be introduced to define $P^{[m/n]}(z)$ and $Q^{[m/n]}(z)$ uniquely. We will use for $Q^{[m/n]}(z)$ a monic (i.e. $q_n^{[m/n]}=1$) or a comonic normalization ($q_0^{[m/n]}=1$).

(1) reduces to the solution of the following set of equations:

$$\begin{bmatrix} T^- \\ \hline T^c \\ \hline T^+ \end{bmatrix} Q^{[m/n]} = \begin{bmatrix} P^{[m/n]} \\ \hline 0 \\ \hline 0 \end{bmatrix} + Z^{[m/n]} \qquad (2)$$

which is a shorthand notation for

$$
\begin{bmatrix}
f_0 & f_{-1} & \cdots & f_{-n} \\
f_1 & f_0 & \ddots & \\
\vdots & \ddots & \ddots & \\
f_m & & & \\
\hline
f_{m+1} & & & f_{m-n+1} \\
\vdots & \ddots & \ddots & \\
f_{m+n} & & \ddots & f_m \\
\hline
f_{m+n+1} & & \ddots & f_{m+1} \\
f_{m+n+2} & \ddots & & f_{m+2} \\
\cdots & & & \vdots
\end{bmatrix}
\begin{bmatrix}
q_0^{[m/n]} \\ \cdot \\ \cdot \\ \cdot \\ q_n^{[m/n]}
\end{bmatrix}
=
\begin{bmatrix}
p_0^{[m/n]} \\ \cdot \\ \cdot \\ \cdot \\ p_m^{[m/n]} \\ \hline 0 \\ \cdot \\ \cdot \\ 0 \\ \hline r_0^{[m/n]} \\ r_1^{[m/n]} \\ \vdots
\end{bmatrix}
$$

The negatively indexed f_k are zero but this will not be essential in the following.

We have set: $P^{[m/n]} = [p_0^{[m/n]} \; \ldots \; p_m^{[m/n]}]^T$

$$\text{and} \quad Z^{[m/n]} = [\underbrace{0 \;\ldots\; 0}_{m+n} \mid r_0^{[m/n]} \; r_1^{[m/n]} \; \ldots]^T = [0\ldots0 \mid (R^{[m/n]})^T]^T$$

The first part of this system evaluates $P^{[m/n]}$ and the last part gives $R^{[m/n]}$ when $Q^{[m/n]}$ is given. The central part may be used to solve $Q^{[m/n]}$ from

$$T^c \, Q^{[m/n]} = 0 \qquad (3)$$

which is a Toeplitz system of n equations in n unknowns (normalization fixes $q_0^{[m/n]}$ or $q_n^{[m/n]} = 1$).

Remarks that if we add rows on top of this system, then we introduce numerator coefficients on top of the RHS and if we add rows at the bottom, then we add residual coefficients at the bottom of the RHS.

For reasons of simplicity in presentation, we will suppose that the Padé table of f(z) is normal.

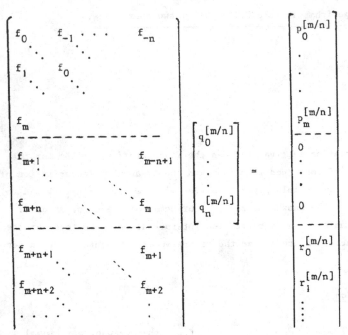

2. Hankel matrix triangularization

Let us transform now the system (3) in the following way: First add a row at the bottom to obtain

$$T^{[m/n]}\, Q^{[m/n]} \equiv \begin{bmatrix} T^c \\ \hline f_{m+n+1} \ldots f_{m+1} \end{bmatrix} Q^{[m/n]} = e_n\, r_0^{[m/n]} \tag{4}$$

e_k is the k-the unit vector (k = 0,1,2,...).

Then introduce the permutation matrix $J \equiv J_{(k+1)x(k+1)} = [\delta_{j,k-j}]_{j=0}^{k}$ and the notation $Jv = \hat{v}$ for any vector $v_{(k+1)x1}$.

Then system (4) can also be written as

$$H^{[m/n]}\, \hat{Q}^{[m/n]} \equiv (T^{[m/n]}J)\ (J\, Q^{[m/n]}) = e_n\, r_0^{[m/n]}$$

Suppose we want to compute the Padé approximants on the staircase

T_k = {[k/0], [k+1/0], [k+1,1], ..., [k+j/j], [k+j+1/j], ...} (k ⩾ + 0).

This is an interleaving of two diagonals

D_k = {[k/0], [k+1/1], ..., [k+j/j], ...} and

D_{k+1} = {[k+1/0], [k+2,1], ..., [k+j+1/j], ...}. This requires the successive solution of the systems

$$H^{[k+j/j]}\, \hat{Q}^{[k+j/j]} = e_j\, r_0^{[k+j/j]}$$

$$H^{[k+j+1/j]}\, \hat{Q}^{[k+j+1/j]} = e_j\, r_0^{[k+j+1/j]}$$

which we will write as

$$H_j\, x_j = e_j\, r_j \text{ and } \bar{H}_j\, \bar{x}_j = e_j\, \bar{r}_j \tag{5}$$

to simplify the notation. Set $x_{jj} = \bar{x}_{jj} = 1$ (comonic normalization).

We have the following nesting properties : $H_0 = h_0$;

$$H_{j+1} = \begin{bmatrix} H_j & v_j \\ \hline v_j^T & h_{2j+2} \end{bmatrix} = \begin{bmatrix} h_0 & \\ \cdot & \bar{H}_j \\ \cdot & \\ \cdot & \\ h_{j+1} & \bar{v}_j^T \end{bmatrix} = \begin{bmatrix} h_0 \ldots h_{j+1} \\ \hline \bar{H}_j \quad \bar{v}_j \end{bmatrix} \quad (j=0,1,\ldots) \tag{6}$$

$$v_j^T = [h_{j+1} \ldots h_{2j+1}];\ \bar{v}_j^T = [h_{j+2} \ldots h_{2j+2}]\ ;\ h_i = f_{k+1+i}$$

Using a bordering technique, the solution may recursively be found [4,6,9,10,14,19,23,24,28,29]

$$x_{j+1} = \begin{bmatrix} 0 \\ \bar{x}_j \end{bmatrix} - \begin{bmatrix} x_j \\ 0 \end{bmatrix} \frac{\bar{r}_j}{r_j};\ \bar{x}_{j+1} = x_{j+1} - \begin{bmatrix} \bar{x}_j \\ 0 \end{bmatrix} \frac{r_{j+1}}{\bar{r}_j} \tag{7}$$

with $r_j = [v_{j-1}^T \quad h_{2j}]\, x_j = r_0^{[k+j/j]}$

$\bar{r}_j = v_j^T \bar{x}_j = r_0^{[k+j+1/j]}$

This is another derivation of the well know Watson algorithm [10,29]. The result for the diagonal D_k is summarized in the following matrix factorization [6,14]

$$H_N^{-1} = \begin{bmatrix} 1 & & \begin{bmatrix} \\ x_i \\ \\ \end{bmatrix} & \star \\ & \ddots & & \\ & & \ddots & \\ 0 & & & 1 \end{bmatrix} \begin{bmatrix} \ddots & & & 0 \\ & r_i^{-1} & & \\ & & \ddots & \\ 0 & & & \ddots \end{bmatrix} \begin{bmatrix} 1 & & & 0 \\ \star & \ddots & & \\ & x_i^T & & \\ & \star & \ddots & 1 \end{bmatrix} = X_N D_N X_N^T \tag{8}$$

and a similar one exists for $\bar{H}_N^{-1} = \bar{X}_N \bar{D}_N \bar{X}_N^T$

The factorizations are nested as H_i is, i.e. when going over from H_i to H_{i+1} then X_i is extended with one column on the right while the other columns are left unchanged.

3. The Viskovatoff algorithm

By successively dividing power series one can obtain a continued fraction of some given formal series $f(z) = \sum_0^\infty f_k z^k$. Denote a division of two power series $R(z)$ and $S(z)$ by $\{R(z)/S(z)\}_k = (p(z), T(z))$ to mean that $p(z)$ is a polynomial consisting of the $k+1$ first terms in the formal power series of $R(z)/S(z)$ and $T(z)$ is the residual series:

$$\frac{R(z)}{S(z)} = p(z) + \frac{T(z)}{S(z)} \text{ or } T(z) = R(z) - p(z)\, S(z)$$

With this notation, the algorithm of Viskovatoff computes the staircase T_k as follows:

V: $S_{-1} = 1$

$\{f(z)/1\}_k = (P^{[k/0]}(z),\ z^{k+1}\, S_0(z))$

<u>for</u> $i = 0,1,2,\ldots$

$\{S_{i-1}/S_i\}_0 = (v_{i+1},\ z\, S_{i+1}(z))$

then it is known [7,23] that

$$f(z) = P^{[k/0]} + z^k \left(\sum_{i=1}^{n} \frac{z|}{|v_i} + \frac{zS_n|}{|S_{n-1}} \right)$$

And the n-th convergent $C^{(n)} = P^{[k/0]} + z^k \sum_{i=1}^{n} \frac{z|}{|v_i}$ is exactly

$[k + \frac{n}{2}/\frac{n}{2}]$ when n is even, and $[k + \frac{n+1}{2}/\frac{n-1}{2}]$ when n is odd up to some normalization.

This algorithm is very old and under different forms one of the most wide spread recursive algorithms for Padé approximation. It is named after Viskovatoff [21], Routh [20] and others.

However this gives neither a monic nor a comonic normalization, therefore we do an equivalence transformation to obtain a comonic normalization. When built into the V-algorithm, it runs like this [7]

C: $S_{-1} = 1$; $\{f(z)/1\}_k = (P^{[k/0]}(z),\ z^{k+1}\ S_0(z))$

<u>for</u> i = 0,1,2,...

$$c_{i+1} = S_i(0)/S_{i-1}(0)$$

$$S_{i+1} = \frac{1}{z}\,(c_{i+1}\ S_{i-1}(z) - S_i(z))$$

now $f(z) = P^{[k/0]}(z) + z^k \left(\sum_{i=1}^{n} \frac{c_i z|}{|1} + \frac{zS_n(z)|}{|S_{n-1}(z)} \right)$

The statement about the n-th convergent remains true now with a comonic normalization as one directly can see from the forward recurrence for the evaluation of this continued fraction. Equally simple it is to prove by induction that the $S_i(z)$ correspond to residual series [7]. When the n-th convergent is $C^{(n)} = P^{(n)}(z)/Q^{(n)}(z)$ and using the notation $P_i(z) = P^{(2i)}(z)$; $\bar{P}_i(z) = P^{(2i+1)}(z)$, etc. and setting $R_i(z) = S_{2i}(z)$ and $\bar{R}_i(z) = -S_{2i+1}(z)$ $i \geq 0$ one then precisely has

$$Q_i(z)\ f(z) - P_i(z) = z^{k+i+1}\ S_{2i}(z) = z^{k+i+1}\ R_i(z)$$

$$\bar{Q}_i(z)\ f(z) - \bar{P}_i(z) = -z^{k+i+2}\ S_{2i+1}(z) = z^{k+i+2}\ \bar{R}_i(z).$$

thus R_i and $\bar{R}_i$ are exactly those R introduced in (2)

$$R_i(z) = \sum_{j=0}^{\infty} r_j^{[k+i/i]} z^j;\ \bar{R}_i(z) = \sum_{j=0}^{\infty} r_j^{[k+i+1/i]} z^j \tag{9}$$

and

$$c_{2i+2} = \frac{S_{2i+1}(0)}{S_{2i}(0)} = -\frac{\bar{R}_i(0)}{R_i(0)} = -\frac{\bar{r}_i}{r_i}$$

$$c_{2i+3} = \frac{S_{2i+2}(0)}{S_{2i+1}(0)} = - \frac{R_{i+1}(0)}{R_i\ (0)} \equiv - \frac{r_{i+1}}{\bar{r}_i}$$

and thus the recursionin algorithm C stands for:

$$R_{i+1}(z) = \frac{1}{z} (\bar{R}_i(z) - \frac{\bar{r}_i}{r_i} R_i(z))$$
$$\bar{R}_{i+1}(z) = \frac{1}{z} (R_{i+1}(z) - \frac{r_{i+1}}{\bar{r}_i} \bar{R}_i(z)) \tag{10}$$

Remark that this is also the same recursion as used in the so called Thacheralgorithm as derived by Claessens [10].
The forward recursion for the denominators of the successive convergents thus coincides with the recursion (7).
The difference between (7) and the C algorithm is that now the recursion coefficients c_i that consist of leading residual coefficients are directly computed from a recursively updated residual series, rather than go around the explicit evaluation of the Padé-denominators.

4. Matrix interpretation of Viskovatoff's algorithm.

It is not difficult to see that (8) may also be written as

$$H_N X_N = X_N^{-T} D_N^{-1} = \begin{bmatrix} r_0 & & & 0 \\ & \ddots & & \\ \star & \begin{bmatrix} \\ R_i \\ \\ \end{bmatrix} & \ddots & \\ & & \star & r_N \end{bmatrix} = L_N \tag{11}$$

The lower triangular L_N now contains the first few residual coefficients as one can easily derive from (2). More precisely: in the i-th column of L_N on and below the diagonal appears $R_i = R^{[k+i/i]}$ as it was defined in (2). The columns of L_N (residuals) will of course obey the same recurrence relations as the columns of X_N (reversed denominators). Since the recursion coefficients only depend upon the diagonals of L_N and $\bar{L}_N$, it seems more natural to compute the L_N triangular factor of H_N, rather than the triangular factor X_N of H_N^{-1}. The simplest notation is obtained when we let $N \to \infty$ and write as before:

$R_j(z) = [1\ z\ z^2\ \ldots]\ R_j$, R_j as in (11) etc ..., then when multiplying (7) by H_∞ and $\bar{H}_\infty$ resp., one directly refinds the recursion (10).

Remarks

a) One can of course derive also recursions for the monic normalization. The formulas become slightly more involved, but the same principle is maintained.

b) With $\bar{x}_j$ as in (5) one sees that

$$H_j \, \bar{x}_j = e_0 \, \bar{p}_j$$

where $\bar{p}_j = p_{k+j+1}^{[k+j+1/j]}$, the highest degree coefficient in the numerator. Thus one has together with the factorizations (11) for H_N and $\bar{H}_N$ also the first column in the following triangularization:

$$H_i \begin{bmatrix} \begin{bmatrix} \\ \bar{x}_i \\ \\ \end{bmatrix} & \ddots & 0 \\ & \star & \ddots \end{bmatrix} = \begin{bmatrix} \bar{p}_i & \ddots & \star \\ & \ddots & \\ 0 & & \ddots \end{bmatrix}$$

However these factors are not nested as X_N was in (8). The other columns of thelower triangular factor in the LHS are reversed denominators of the antidiagonal trough [k+i+1/i] and the upper triangular on the RHS contains some of the corresponding numerator coefficients. This may become clear in section 6.

5. The computation of a diagonal

A contracted form of the previous algorithm is not difficult to obtain. E.g. in the V algorithm one replaces the recurrence relation by

$$\{S_{i-1}/S_i\}_1 = (v_{i+1} + w_{i+1} z,\; z^2 \, S_{i+1})$$

More explicit and with a comonic normalization one may obtain [7,19,23]

J: $\quad R_{-1} = -1; \; \{f(z)/1\}_k = (p^{[k/0]},\; z^{k+1} R_0(z))$

for $i = 0,1,2,\ldots$

$$u_{i+1} = -\frac{R_i(0)}{R_{i-1}(0)}$$

$$S_{i+1}(z) = \frac{1}{z}\,(u_{i+1}\, R_{i-1}(z) + R_i(z))$$

$$w_{i+1} = -\frac{S_{i+1}(0)}{R_i\;\;(0)}$$

$$R_{i+1}(z) = \frac{1}{z}(w_{i+1} R_i(z) + S_{i+1}(z))$$

we then have that

$$f(z) = P^{[k/0]}(z) + z^{k-1} \left(\sum_{i=1}^{n} \frac{u_i z^2|}{|1+w_i z} - \frac{z^2 R_n(z)|}{|R_{n-1}(z)} \right)$$

as can directly be verified, and if the n-th convergent is $C^{(n)} = P_n(z)/Q_n(z)$ then

$$Z_n(z) = Q_n(z) f(z) - P_n(z) = z^{k+2n+1} R_n(z)$$

which can be proved by a simple induction argument and so $C^{(n)}$ turns out to be exactly $[k+n/n]$ (see [7]), thus

$$R_n(z) = \sum_{i=0}^{\infty} r_i^{[k+n/n]} z^i$$

The matrix derivation is now as follows: Find the solution of

$$H_{i+1} x_{i+1} = e_{i+1} r_{i+1}$$

as a "linear combination" of x_i and x_{i-1}. By inspection of the nesting of the H_i, one may find that we must look for a recursion of the form

$$x_{i+1} = \begin{bmatrix} 0 \\ \\ x_i \end{bmatrix} + \begin{bmatrix} x_{i-1} \\ 0 \\ 0 \end{bmatrix} u_{i+1} + \begin{bmatrix} x_i \\ 0 \end{bmatrix} w_{i+1} \tag{12}$$

This form respects the comonic normalization and when multiplying (12) with H_{i+1} one finds that

$$u_{i+1} = -\frac{r_i}{r_{i-1}} \text{ and } w_{i+1} = \left(\frac{r'_{i-1}}{r_{i-1}} - \frac{r'_i}{r_i}\right)$$

(12')

$$\text{with} \quad \begin{bmatrix} r_i \\ \\ r'_i \end{bmatrix} = \begin{bmatrix} h_i & \cdots & h_{2i} \\ \\ h_{i+1} & \cdots & h_{2i+1} \end{bmatrix} x_i = \begin{bmatrix} v_{i-1}^T | h_{2i} \\ \hline v_i^T \end{bmatrix} x_i$$

(12) gives a recursion between three successive columns of the matrix X_N as defined in (8).

The corresponding recursion for the columns of L_N, defined in (11) can be derived as before. r_i and r'_i are the first two coefficients in the residual series $R_i(z)$ i.e. $r_i = R_i(0) = r_0^{[k+i/i]}$ and $r'_i = R'_i(0) = r_1^{[k+i/i]}$ and we obtain the recursion

$$R_{i+1}(z) = \frac{1}{z}(w_{i+1} R_i(z) + \frac{1}{z}(R_i(z) + u_{i+1} R_{i-1}(z))) \tag{13}$$

This is exactly the combination of the two recursions in the J-algorithm. A comparison delivers

$$S_{i+1}(0) = r'_i - \frac{r_i r'_{i-1}}{r_{i-1}}$$

Remark that when applying the given algorithms to $1/f(z)$, then we can obtain the other staircases and diagonals:

$T_k = \{[0/-k], [0/-k+1], \ldots, [j/-k+j], [j/-k+j+1], \ldots\}$ $k \leqslant -0$
(remark $T_{-0} \neq T_{+0}$) and

$D_k = \{[0/-k], \ldots, [j/-k+j], \ldots\}$ $k \leqslant 0$

6. Ascending staircases antidiagonals

Other paths that recieved considerable attention in the literature are the ascending staircases

$$U_k = \{[k/0], [k-1/0], [k-1/1], \ldots, [k-j/j], [k-j-1/j], \ldots, [0/k-1], [0/k]\}$$

and the antidiagonals:

$E_k = \{[k/0], [k-1/1], \ldots, [k-j/j], \ldots, [0/k]\}$

It will be convenient to transform the system (3) now as follows: add one row at the top to obtain

$$T^{[m-1/n]}\, Q^{[m/n]} = \left[\frac{f_m \;\cdots\; f_{m+n}}{T^c}\right] Q^{[m/n]} = e_0\, p_m^{[m/n]}$$

Introducing now a row permutation we obtain :

$$H_j\, x_j \equiv H^{[k-j/j]}\, Q^{[k-j/j]} \equiv (J\, T^{[k-j-1/j]})Q^{[k-j/j]} = J\, e_0\, p_{k-j}^{[k-j/j]} = e_j\, p_j$$

This is for E_k and for E_{k-1} this becomes

$$\bar{H}_j\, \bar{x}_j = H^{[k-j-1/j]}\, Q^{[k-j-1/j]} \equiv (J\, T^{[k-j-2/j]})\, Q^{[k-j-1/j]} = e_j\, p_{k-j-1}^{[k-j-1/j]} = e_j\, \bar{p}_j$$

Remark that now introducing rows on top of H_j adds residual coefficients while the numerator coefficients may be introduced at the bottom. This is opposite to the previous algorithm, but the factorization approach remains essentially the same:

Set now $h_j = f_{k-j}$, $j = 0,1,2,\ldots$ then we can use the nesting property (6) and use a monic normalization ($x_{jj} = \bar{x}_{jj} = 1$) to get the simple formula :

$$
\begin{aligned}
x_{j+1} &= \begin{bmatrix} 0 \\ \bar{x}_j \end{bmatrix} - \begin{bmatrix} x_j \\ 0 \end{bmatrix} \frac{\bar{p}_j}{p_j} \\
\bar{x}_{j+1} &= x_{j+1} - \begin{bmatrix} \bar{x}_j \\ 0 \end{bmatrix} \frac{p_{j+1}}{p_j} \\
p_j &= [h_{j+1} \ldots h_{2j+1}]\, x_j = v_j^T x_j = p_{k-j}^{[k-j/j]} \\
\bar{p}_j &= [h_{j+2} \ldots h_{2j+2}]\, \bar{x}_j = \bar{v}_j^T \bar{x}_j = p_{k-j-1}^{[k-j-1/j]}
\end{aligned}
\tag{14}
$$

These are the recursions for the Baker algorithm [2,10] (14) is exactly the same as (7), only the roles of reversed numerator and residual are interchanged. Its matrix interpretation is as in (8) with now the diagonal filled up by p_i^{-1} instead of r_i^{-1}.

We thus can also use the Viskovatoff algorithm applied on the reversed numerator series

$$\hat{P}_i(z) \equiv \sum_{j=0}^{k-i} p_j^{[k-i/i]} z^{k-i-j}$$

And this is a recursion for the columns of the RHS in the analogon of (11):

$$
H_k X_k = \begin{bmatrix} p_0 & & & & 0 \\ \vdots & \ddots & & & \\ \star & & \begin{bmatrix} \\ \hat{P}_i \\ \\ \end{bmatrix} & \ddots & \\ & & & \star & p_k \end{bmatrix}
\tag{15}
$$

containing the reversed numerators.

A contraction of the previous theory gives an antidiagonal, and again nothing essentially new is introduced : The recursion following from the Viskovatoff algorithm for the RHS of (15) is the Euclid algorithm to obtain the GCD for two polynomials [22] and is known in Padé literature as the Kronecker algorithm.

Remark that the following relation between the H_i in section 2-5 and the H_i in this section is essentially a column and row reversal for a shifted Hankel matrix.

$$H^{[m/n]} = T^{[m/n]}\, J = J\, H^{[m-1/n]}\, J$$

where the first $H^{[m/n]}$ is as in sections 2-5 and the latter $H^{[m-1/n]}$ is as defined in this section.

Left multiplication of (15) with J then justifies the remark b) at the end of section 4.

7. Toeplitz matrix factorization, rows, sawteeth, etc...

When we want to compute a row $L_k = \{[k/0], \dots [k/j], \dots\}$ $(k \geq +0)$ in a Padé table, then we have to solve

$$T^{[k/j]}\ Q^{[k/j]} = e_j\ r_0^{[k/j]}$$

which we denote in shorthand as

$$T_j\ x_j = e_j\ r_j$$

As a staircase is related to a diagonal, so is a sawtooth related to a row and we may solve in parallel with the previous family of systems also

$$\bar{T}_j\ \bar{x}_j = e_j\ \bar{r}_j$$

with $\bar{T}_j = T^{[k+1/j]}$, $\bar{x}_j = Q^{[k+1/j]}$, $\bar{r}_j = r_0^{[k+1/j]}$

to obtain the elements of the sawtooth

$$S_k^+ = \{[k/0], [k+1/0], [k/1], [k+1/1], \dots, [k/j], [k+1/j], \dots\},\ k \geq +0$$

or of a symmetrical one

$$S_k^- = \{[k+1/0], [k/0], [k+1/1], [k/1], \dots, [k+1/j], [k/j], \dots\},\ k \geq +0$$

Set $t_j = f_{k+j+1}$, then the following nesting property will be usefull:

$$T_{j+1} = \left[\begin{array}{c|c} t_0 & w_j^T \\ \hline v_j & T_j \end{array}\right] = \left[\begin{array}{c|c} T_j & \hat{w}_j \\ \hline \hat{v}_j^T & t_0 \end{array}\right] = \left[\begin{array}{c|c} \bar{w}_j^T & t_{-j} \\ \hline \bar{T}_j & \hat{\bar{w}}_j \end{array}\right] \tag{16}$$

$$w_j^T = [t_{-1} \dots t_{-j}];\ \bar{w}_j^T = [t_0 \dots t_{-j+1}];$$

$$v_j^T = [t_1 \dots t_j]\ ;\ \hat{v}_j = Jv_j;\ \hat{\bar{w}}_j = J\bar{w}_j$$

Remark that T_j is no longer symmetric as H_j was but it is persymmetric i.e. symmetric along the antidiagonal: $T_j^T = J\ T_j\ T$. This makes it possible to derive similar recursions as for the Hankel matrix.

Suppose we take a monic normalization for x_j and for $\bar{x}_j$, i.e. $x_{jj} = \bar{x}_{jj} = 1$ then it can be verified that

$$T_{j+1}\left[\begin{array}{c:c:c} 0 & & \\ & \bar{x}_j & \\ & & \bar{x}_{j+1} \\ x_j & & \\ & 0 & \end{array}\right] = \left[\begin{array}{c:c:c} p_j & \bar{p}_j & \bar{p}_{j+1} \\ 0 & 0 & 0 \\ \cdot & \cdot & \cdot \\ \cdot & \cdot & \cdot \\ r_j & \bar{r}_j & 0 \end{array}\right] \quad \text{and } \bar{T}_{j+1}\left[\begin{array}{c:c:c} 0 & & \\ & \bar{x}_j & \\ & & x_{j+1} \\ x_j & & \\ & 0 & \end{array}\right] = \left[\begin{array}{c:c:c} 0 & 0 & 0 \\ \cdot & \cdot & \cdot \\ \cdot & \cdot & \cdot \\ 0 & 0 & 0 \\ r_j & \bar{r}_j & r_{j+1} \\ r_j & \bar{r}'_j & r'_{j+1} \end{array}\right]$$

$$\text{with} \quad \begin{bmatrix} p_j \\ r_j \end{bmatrix} = \begin{bmatrix} t_{-1}\cdots t_{-j} \\ t_{j-1}\cdots t_0 \end{bmatrix} x_j = \begin{bmatrix} w_j^T \\ w_{j-1}^T | t_0 \end{bmatrix} x_j = \begin{bmatrix} p_k^{[k/j]} \\ r_0^{[k/j]} \end{bmatrix} \tag{17}$$

and

$$\begin{bmatrix} \bar{p}_j \\ \bar{r}_j \end{bmatrix} = \begin{bmatrix} t_0\cdots t_{-j+1} \\ t_j\cdots t_1 \end{bmatrix} \bar{x}_j = \begin{bmatrix} \bar{w}_j^T \\ \hat{v}_j^T \end{bmatrix} \bar{x}_j = \begin{bmatrix} p_{k+1}^{[k+1/j]} \\ r_0^{[k+1/j]} \end{bmatrix}$$

From this the recursions follow simply:

For S_k^+:

$$x_{j+1} = \begin{bmatrix} 0 \\ x_j \end{bmatrix} - \begin{bmatrix} \bar{x}_j \\ 0 \end{bmatrix} \frac{p_j}{\bar{p}_j}; \quad \bar{x}_{j+1} = x_{j+1} - \begin{bmatrix} \bar{x}_j \\ 0 \end{bmatrix} \frac{r_{j+1}}{\bar{r}_j} \tag{18a}$$

For S_k^-:

$$\bar{x}_{j+1} = \begin{bmatrix} 0 \\ x_j \end{bmatrix} - \begin{bmatrix} \bar{x}_j \\ 0 \end{bmatrix} \frac{r_j}{\bar{r}_j}; \quad x_{j+1} = \left(\begin{bmatrix} 0 \\ x_j \end{bmatrix} - \bar{x}_{j+1} R\right)/(1-R) \tag{18b}$$

with $R = p_j/\bar{p}_{j+1}$

Or a combination of the two gives two rows in parallel: $S_k^{=}$:

$$x_{j+1} = \begin{bmatrix} 0 \\ x_j \end{bmatrix} - \begin{bmatrix} \bar{x}_j \\ 0 \end{bmatrix} \frac{p_j}{\bar{p}_j}; \quad \bar{x}_{j+1} = \begin{bmatrix} 0 \\ x_j \end{bmatrix} - \begin{bmatrix} \bar{x}_j \\ 0 \end{bmatrix} \frac{r_j}{\bar{r}_j} \tag{18c}$$

Note that (17) may suggest that now four inner products should be made in each step, however two of them can be replaced by recursions:

e.g. in (18a): $r_{j+1} = r_j - \bar{r}_j p_j/\bar{p}_j$ and $\bar{p}_{j+1} = -\bar{p}_j r_{j+1}/\bar{r}_j$

in (18b): $\bar{p}_{j+1} = p_j - \bar{p}_j r_j/\bar{r}_j$ and $r_{j+1} = r_j/(1-R)$

in (18c) we can use the recursion for r_{j+1} and $\bar{p}_{j+1}$ as above.

Several other variants are possible, e.g. a blockpath

$B_k^+ = \{[k/0],[k+1/0],[k+1/1],[k/1], \ldots,[k/2j],[k+1/2j],[k+1/2j+1],[k/2j+1], \ldots$

A derivation of the corresponding algorithm may also be found from (17). To give the matrix interpretation, it is the simplest way to rewrite (18c) for a monic normalization of x_j and a comonic normalization of $\bar{x}_j$, then it reads:

$$x_{j+1} = \begin{bmatrix} 0 \\ x_j \end{bmatrix} - \begin{bmatrix} \bar{x}_j \\ 0 \end{bmatrix} \frac{p_j}{\bar{p}_j}; \quad \bar{x}_{j+1} = \begin{bmatrix} \bar{x}_j \\ 0 \end{bmatrix} - \begin{bmatrix} 0 \\ x_j \end{bmatrix} \frac{\bar{r}_j}{r_j} \qquad (19)$$

$$r_{j+1} = r_j - \bar{r}_j p_j / \bar{p}_j \qquad \bar{p}_{j+1} = p_j - p_j \bar{r}_j / r_j$$

$$p_{j+1} = w_{j+1}^T \, x_{j+1} \qquad \bar{r}_{j+1} = \tilde{v}_{j+1}^T \, \bar{x}_{j+1}$$

One can see that

$$T_n \overset{0 \;\ldots\; i \;\ldots\; N}{\begin{bmatrix} 1 & \star & \begin{bmatrix} x_i \end{bmatrix} & \star \\ & \ddots & & \\ & 0 & & \ddots \\ & & & 1 \end{bmatrix}} = \begin{bmatrix} r_0 & & & 0 \\ & \ddots & & \\ \star & & \begin{bmatrix} R_i \end{bmatrix} & \ddots \\ & & \star & r_N \end{bmatrix} \equiv T_N X_N = L_N \qquad (20a)$$

and

$$\bar{T}_N \overset{0 \;\ldots\; i \;\ldots\; N}{\begin{bmatrix} 1 & \star & \begin{bmatrix} \bar{x}_i \end{bmatrix} & \star \\ & \ddots & & \\ & 0 & & \ddots \\ & & & 1 \end{bmatrix}} = \begin{bmatrix} \bar{r}_0 & & & 0 \\ & \ddots & & \\ \star & & \begin{bmatrix} \bar{R}_i \end{bmatrix} & \ddots \\ & & \star & r_N \end{bmatrix} \equiv \bar{T}_N \bar{X}_N = \bar{L}_N \qquad (20b)$$

but also

$$T_N \underset{N \;\ldots\; i \;\ldots\; 0}{\begin{bmatrix} 1 & & & 0 \\ & \ddots & & \\ \star & & \begin{bmatrix} \bar{x}_i \end{bmatrix} & \ddots \\ & & \star & 1 \end{bmatrix}} = \begin{bmatrix} \bar{p}_N & \star & \begin{bmatrix} \bar{p}_i \end{bmatrix} & \star \\ & \ddots & & \\ & 0 & & \ddots \\ & & & \bar{p}_0 \end{bmatrix} \equiv T_N \hat{\bar{X}}_N = \bar{U}_N \qquad (20c)$$

and

$$\underline{T}_N \underset{N\ \dots\ i\ \dots\ 0}{\begin{bmatrix} 1 & & & 0 \\ & \ddots & & \\ & \star & \begin{bmatrix} x_i \end{bmatrix} & \ddots \\ & & & \star\ \ddots\ 1 \end{bmatrix}} = \begin{bmatrix} p_N & \star & & \\ & \ddots & \begin{bmatrix} p_i \end{bmatrix} & \star \\ & 0 & & \ddots \\ & & & p_0 \end{bmatrix} \equiv \underline{T}_N \hat{X}_N = U_N \qquad (20d)$$

where $\underline{T}_N$ corresponds to row k-1: $\underline{T}_N = T^{[k-1/N]}$

Using the persymmetry of T_N and T_N^{-1}, (20a) and (20c) can be combined to give

$$T_N^{-1} = \underset{X_N}{\overset{0\ \dots\ i\ \dots\ N}{\begin{bmatrix} 1 & \star & \begin{bmatrix} x_i \end{bmatrix} & \star \\ & \ddots & & \\ 0 & & & \ddots\ 1 \end{bmatrix}}} \ \underset{D_N}{\overset{0\dots i\dots N}{\begin{bmatrix} \ddots & & 0 \\ & r_i^{-1} & \\ 0 & & \ddots \end{bmatrix}}} \ \underset{(J\hat{X}_N^T J)}{\begin{bmatrix} 1 & & 0 \\ \ddots & \begin{bmatrix} \hat{x}_i^T \end{bmatrix} & \\ & \star & \ddots\ 1 \end{bmatrix}} \begin{matrix} 0 \\ \vdots \\ i \\ \vdots \\ N \end{matrix}$$

$$= \underset{\hat{\hat{X}}_N}{\underset{N\ \dots i\dots\ 0}{\begin{bmatrix} 1 & & 0 \\ \ddots & & \\ \star & \begin{bmatrix} \bar{x}_i \end{bmatrix} & \ddots \\ & & \star\ \ddots\ 1 \end{bmatrix}}} \ \underset{(JD_NJ)}{\underset{N\dots i\dots 0}{\begin{bmatrix} \ddots & & 0 \\ & \bar{p}_i^{-1} & \\ 0 & & \ddots \end{bmatrix}}} \ \underset{(JX_N^TJ)}{\begin{bmatrix} 1 & & \star \\ & \ddots & \begin{bmatrix} \hat{x}_i^T \end{bmatrix} \\ 0 & & \ddots\ \star\ 1 \end{bmatrix}} \begin{matrix} N \\ \vdots \\ i \\ \vdots \\ 0 \end{matrix}$$

from which it follows that

$$L_N = J\ \hat{\hat{X}}_N^{-T}\ D_N^{-1}\ J,\ \bar{U}_N = J\ X_N^{-T}\ D_N^{-1}\ J$$

and $r_i = \bar{p}_i$

Some variants of the above algorithms are known [5,9,25,27,31,32]

It is possible to derive a Viskovatoff-like algorithm [7].

E.g. (18a) corresponds to the denominator of the convergents of a continued fraction

$$C = P^{[k/0]} + z^{k+1} \frac{\rho_0|}{|1} + \frac{z|}{|\Pi_1} + \frac{\rho_1|}{|1} + \frac{z|}{|\Pi_2} + \dots + \frac{\rho_{j-1}|}{|1} + \frac{z|}{|\Pi_j} \qquad (21)$$

$$\rho_{j+1} = -r_{j+1}/\bar{r}_j;\ \Pi_{j+1} = -p_j/\bar{p}_j$$

The ρ_j is the constant term in the series expansion of the ratio of two residual power serie : $-R_{j+1}(z)/\bar{R}_j(z) = \rho_{j+1}+\dots$ and Π_j is introduced to monitor the degree of the numerator. The basic idea in Viškovatoff's algorithm is to compute the residuals corresponding to the successive convergents of (21), i.e. the columns of L_N and $\bar{L}_N$.

However now it is not possible to recover the necessary recursion coefficients from these alone. We now need the diagonal information from two other triangular factors e.g. U_N and $\bar{U}_N$ (numerators), or X_N and $\bar{X}_N$ (denominators) if we are willing to evaluate the inner products to obtain p_j and $\bar{p}_j$.

Using an obvious notation, the algorithm may be summarized as follows :

$$S^+: \ \{f/1\}_k = (P_0 = P^{[k/0]}, z^{k+1} R_0(z));\ \{f/1\}_{k+1} = (\bar{P}_0 = P^{[k+1/0]}, z^{k+2}\bar{R}_0(z))$$

$$\hat{P}_0(z) = \sum_{j=0}^{\infty} p_{k-j}^{[k/0]} z^j \ ;\ \hat{\bar{P}}_0(z) = \sum_{j=0}^{\infty} p_{k+1-j}^{[k+1/0]} z^j \ ;\ \rho_0 = R_0(0)$$

<u>for</u> $i = 0,1,2,\dots$

$$\Pi_{i+1} = -\hat{P}_i(0)/\hat{\bar{P}}_i(0)$$

$$R_{i+1}(z) = R_i(z) + \bar{R}_i(z)\,\Pi_{i+1};\hat{P}_{i+1}(z) = \frac{1}{z}(\hat{P}_i(z) + \hat{\bar{P}}_i(z)\,\Pi_{i+1})$$

$$\rho_{i+1} = -R_{i+1}(0)/\bar{R}_i(0)$$

$$\bar{R}_{i+1}(z) = \frac{1}{z}(R_{i+1}(z) + \bar{R}_i(z)\,\rho_{i+1});\ \hat{\bar{P}}_{i+1}(z) = z\,\hat{P}_{i+1}(z) + \hat{\bar{P}}_i\,\rho_{i+1}$$

which follows from the recursions (18) and the relations (20).

We introduced the $\hat{P}_i(z)$ as reversed numerator series and it may be supposed $p_j^{[k/0]} = 0$ for $j < 0$, but this is not essential. This makes the algorithm also applicable to Laurent Padé approximation problems [11,15,16].

Similar algorithms exist for the other paths mentioned earlier in this section.

A contraction of the previous algorithm results in a row computation. We obtain from

$$T_{j+1}\begin{bmatrix} 0 & & 0 \\ & x_j & \\ & & x_{j-1} \\ x_j & & \\ & 0 & 0 \end{bmatrix} = \begin{bmatrix} p_j & 0 & p_{j-1} \\ 0 & 0 & 0 \\ \vdots & \vdots & \vdots \\ 0 & r_j & r_{j-1} \\ r_j & r'_j & r'_{j-1} \end{bmatrix}$$

that

$$x_{j+1} = \begin{bmatrix} 0 \\ \\ x_j \end{bmatrix} - \begin{bmatrix} 0 \\ x_{j-1} \\ 0 \end{bmatrix} \frac{p_j}{p_{j-1}} + \begin{bmatrix} x_j \\ \\ 0 \end{bmatrix} \frac{p_j}{p_{j-1}} \frac{r_{j-1}}{r_j} \tag{22}$$

having a monic normalization.

The corresponding constrained division algorithm computes the continued fraction:

$$P^{[k/0]} + z^k \sum_{j=0}^{n} \frac{\Pi_j z|}{|\Pi_j \rho_j + z}$$

$$\Pi_j = -p_j/p_{j-1}; \; \rho_j = - r_{j-1}/r_j$$

by means of the recursive computation of the columns in the factors L_N and U_N in (20a) and (20d).

The algorithm goes like this: see also [7]

$$R: R_{-1} = -1; \; \{f/1\}_k = (P_0 = P^{[k/0]}, \; z^{k+1} R_0(z))$$

$$\hat{P}_{-1} = 1; \; \hat{P}_0(z) = \sum_{j=0}^{\infty} P_{k-j}^{[k/0]} z^j$$

<u>for</u> $i = 0,1,2,\ldots$

$$\Pi_i = - \hat{P}_i(0)/\hat{P}_{i-1}(0); \; \rho_i = - R_{i-1}(0)/R_i(0)$$

$$\hat{P}_{i+1}(z) = \frac{1}{z} (\hat{P}_i(z) + \Pi_i \hat{P}_{i-1}(z)) + \Pi_i \rho_i \hat{P}_i(z)$$

$$R_{i+1}(z) = \frac{1}{z} (R_{i-1}(z) + \rho_i R_i(z)) \Pi_i + R_i(z)$$

The algorithms S^+ and R are the continued fraction approaches of the problem [7], while if we would compute L_N and $\bar{L}_N$ recursively, together with X_N and $\bar{X}_N$ with the diagonal element of U_N and $\bar{U}_N$ found by a multiplication of a row in $\underline{T}_N$ (T_N) by a column in $\hat{X}_N$ ($\hat{\bar{X}}_N$) then we obtain the algorithm of Rissanen [26] when computing only X_n and $\bar{X}_N$ plus necessary inner products as in (18) or (19) we obtain algorithms in the style of Trench and Zohar [27,31,32].

Conclusion

The continued fraction algorithms like Viskovatoff's and its extensions [7] are the duals of the recursive algorithms explicitely evaluating the denominators of the Padé approximations, in that they recursively evaluate the numerators and/or the residuals. Their matrix interpretation reveals that the first correspond to the computation of triangular factors for the inverses of Hankel or Toeplitz matrices, which the latter compute triangular factors for the matrices themselves.

This way of approach may also be generalized for non normal Padé tables without considerable difficulties [5,6,7,12,13,18,22,23,25] or to the matrix Padé approximation problem [8,26] or the rational interpolation problem [3,30].

The matrix interpretation also provides a tool for the examination of the numerical stability of these algorithms [13,17].

References

1. Akaike, "Block Toeplitz matrix inversion", SIAM J. Appl. Math. 24 (1973), 234-241.

2. G.A. Baker Jr., "Recursive calculation of Padé approximants", P.R. Graves-Morris, ed. "Padé approximants and their applications", Academic Press, London, 1973, 83-91.

3. V. Belevitch, "Interpolation matrices", Philips Res. Rept. 25 (1970), 337-369.

4. C. Brezinski, "Computation of Padé approximants and continued fractions", Journ. Comp. and Appl. Math., 2, 1976, 113-123.

5. A. Bultheel, "Recursive algorithms for non-normal Padé tables", subm. for publication.

6. A. Bultheel, "Fast factorization algorithms and Padé approximation", subm. for publication.

7. A. Bultheel, "Division algorihtms for continued fractions and the Padé table", subm. for publication.

8. A. Bultheel, "Recursive algorithms for the matrix Padé problem", subm. for publication.

9. D. Bussonnais, "Tous les algorithms de calcul recurrente des approximants de Padé d'une serie. Construction des fractions continues correspondantes" Séminaire d'Analyse Numérique, n° 293, Grenoble, 1978.

10. G. Claessens, "A new look at the Padé table and the different methods for computing its elements", Journ. of Comp. and Appl. Math., 1 (1975), 141-152.

11. A. Common, "Padé-Chebyshev approximation", these proceedings.

12. B. Cordellier, "Deux algorithmes de calcul recursif des éléments d'une table de Padé non normale", Presented at the conference on Padé approximation, Lille, 1978.

13. L.S. De Jong, "Numerical aspects of recursive realization algorithm", SIAM J. Contr. and Opt., 16 (1978), 646-659.

14. W.B. Gragg, "Matrix interpretations and applications of the continued fraction algorithm", Rocky Mountain J. of Math., 4 (1974), 213-225.

15. W.B. Gragg, G.D. Johnson, "The Laurent-Padé table", IFIP Congress 74, North-Holland, 1974, 632-637.

16. W.B. Gragg, "Laurent, Fourier and Chebyshev-Padé tables", in Saff and Varga (eds.), "Padé and rational approximation, Theory and applications", Academic Press, New York, 1977, 61-72.

17. P.R. Graves-Morris, T.R. Hopkins, "Reliable rational interpolation", manuscript feb. 1978.

18. P.R. Graves-Morris, "Numerical calculation of Padé approximants", these proceedings.

19. J.F. Hart et al., "Computer approximations", John Wiley and Sons, 1968.

20. P. Henrici, "Applied and computational complex analysis, vol. II", John Wiley and Sons, revised edition, 1977.

21. A.N. Khovanskii, "The application of continued fractions and their generalization to problems in approximation theory", P. Noordhoff N.V., Groningen, 1963.

22. R.J. McEliece, J.B. Shearer, "A property of Euclid's algorithm and an application to Padé approximation", SIAM J. Appl. Math., 34 (1978), 611-616.

23. J.A. Murphy, M.P. O'Donohoe, "A class of algorithms for obtaining rational approximants to functions which are defined by power series", Journ. of Appl. Math. and Physics (ZAMP), 28 (1977), 1121-1131.

24. J.L. Philips, "The triangular decomposition of Hankel matrices", Math. of Comp., 25 (1971), 599-602.

25. J. Rissanen, "Solution of linear equations with Hankel and Toeplitz matrices", Numer. Math., 22 (1974), 361-366.

26. J. Rissanen, "Algorithms for triangular decomposition of block Hankel and Toeplitz matrices with application to factoring positive matrix polynomials", Math. of Comp., 27 (1973), 147-154.

27. W.F. Trench, "An algorithm for the inversion of finite Toeplitz matrices", SIAM J. Appl. Math. 12 (1964), 515-512.

28. W.F. Trench, "An algorithm for the inversion of finite Hankel matrices", SIAM J. Appl. Math., 13 (1965), 1102-1107.

29. P.J.S. Watson, "Algorithm for differentiation and integration", P.R. Graves-Morris (ed.), "Padé approximants and their applications", Academic Press, London, 1973, 93-98.

30. H. Werner, "Continued fractions for the numerical solution of rational interpolation", these proceedings.

31. S. Zohar, "Toeplitz matrix inversion: the algorithm of W.F. Trench", Journ. ACM, 16 (1969), 592-601.

32. S. Zohar, "The solution of a Toeplitz set of linear equations", Journ. ACM, 21 (1974), 272-276.

A. BULTHEEL
K.U.Leuven
Afdeling Toegepaste Wiskunde
en Programmatie
Celestijnenlaan 200 A
B-3030 HEVERLEE (Belgium)

THE NUMERICAL CALCULATION OF PADÉ APPROXIMANTS

P. R. Graves-Morris
Mathematical Institute
University of Kent
Canterbury, England

1. Summary and Introduction.

Ever since the spectacular successes of the Padé approximant method in estimating the critical indices in critical phenomena, the question of what is the best way of calculating Padé approximants has been asked. In this review, I seek to analyse the qualities which characterise a good method. The properties of reliability and discrimination are defined: these seem to be very important qualities of a good method. The problems of a detailed specification of an algorithm are briefly discussed. Various proposed calculational methods are categorised and analysed according to the criteria proposed. We will see that Kronecker's algorithm seems excellent when exact arithmetic is available, and that a particular matrix method seems best, at present, for floating-point computations.

We normally suppose that a function f(z) exists with the Maclaurin expansion

$$f(z) = \sum_{i=0}^{\infty} c_i z^i \tag{1.1}$$

Our notation for an [L/M] Padé approximant is given by

$$[L/M] \equiv [L/M]_f(z) \equiv \frac{a_0 + a_1 z + \dots + a_L z^L}{b_0 + b_1 z + \dots + b_M z^M} \tag{1.2}$$

Eq. (1.2) defines the [L/M] Padé approximant of f(z) provided $b_0=1$ and the Maclaurin expansion of (1.2) agrees with that of (1.1) up to and including order z^{L+M}. Details of this, Baker's modern definition, are given in [1]. We may use Padé approximants to accelerate convergence of the series (1.1) if its numerical convergence is unacceptably slow. In certain cases, it is known that suitable sequences of Padé approximants converge to the value of f(z) when the given series (1.1) is divergent. We refer to [2] for details and references. Numerous historical references are omitted from this review because they may be found in Wuytack's bibliography [3].

2. Specification of an Algorithm.

In this section, we attempt to make more precise the specification for a routine "which calculates Padé approximants". From a theoretical viewpoint, the variable z in (1.1) and (1.2) is a complex variable, whether or not the actual Padé approximant subprogram is designed to evaluate the approximant at real values of z.

The coefficients $\{ c_i, i=0,1,2,\dots \}$ in (1.1) may have various distinct specifications, leading to different approaches.

(i) $\{c_i\}$ may be integer coefficients or fractions. In this case, exact arithmetic is feasible and tests for zero are decisive. Essam [4] has already implemented an algorithm for this case. The modified Kronecker algorithm described in section 4 seems to be an ideal algorithm in this context.

(ii) $\{c_i\}$ may be real or complex coefficients. Normally the coefficients are specified to a basic fixed-point or floating-point numerical precision. Most of this article is addressed to this case. Note that in scientific applications, the actual precision of $\{c_i\}$ is unlikely to be the machine precision.

(iii) $\{c_i\}$ may be matrices with integer, fractional, real or complex elements. Such cases can be reduced (by taking elements, or real and imaginary parts, etc.) to case (ii) in principle. In practice, the subroutine of Starkand [5], which is the only published subprogram for matrix Padé approximants, may be useful.

We shall assume, unless explicitly stated to the contrary, that we are dealing with case (ii). It seems to be standard practice in numerical algorithm libraries to provide separate routines for solving the coefficient problem and for subsequent evaluation of the approximants. If one is primarily concerned with the value problem, which is to say that only values of the approximants are required, this may be inconvenient and possibly also inaccurate. The ε-algorithm of Shanks and Wynn and Bauer's η-algorithm are designed for the value problem [3]. Implementation of routines such as these, if desirable, is inconvenient in the standard framework. People familiar with Froissart's unpublished numerical experiments [6,7] know that Padé approximants can act as a noise filter. This property has yet to be "pinned down" mathematically: if it can be done, the corresponding numerical algorithms will have to solve the value problem directly. However, at present, there seem to be no cogent reasons for not solving the coefficient problem first, namely the evaluation of $a_0, a_1, \dots, a_L$, $b_1, b_2, \dots, b_M$. The value problem then becomes the relatively trivial task of evaluating (1.2).

It is often the case that a particular sequence of approximants is required. In the case of numerical evaluation of special functions [8,9,10,1,2], it is often the case that particular sequences of Padé approximants provide converging upper and lower bounds for the function values. However, in the interests of providing an algorithm of general applicability, it would be unwise to settle on an algorithm which only calculates diagonal sequences, or paradiagonal sequences, or antidiagonal sequences, or rows, or columns, or rays or even saw-tooth sequences. It seems best to select an algorithm for the calculation of a single [L/M] approximant, with L and M as prespecified integers.

In conclusion, this article is primarily addressed to the problem of calculating the coefficients of a specified [L/M] Padé approximant from given real or complex coefficients $c_0, c_1, \dots c_{L+M}$ of specified precision.

3. Methods for Padé approximation

In this section, the various methods available for Padé approximation are categorised. Comparisons between methods in the same category are relatively easy, whereas comparisons amongst the best methods of the different categories are harder.

A. Sequence to sequence transforms. The ε-algorithm is the best known of these. It is not a reliable algorithm in the sense of section 4, and its stability properties are unknown. It is possible that a device such as Cordellier's identity might be used to convert it to a reliable algorithm [11,12].

B. Iterative methods based on diagonal staircase sequences. The numerators and denominators of staircase sequences obey the Frobenius three term identities. By obtaining the constants occurring in these identities by other techniques, efficient algorithms are constructed; we refer to papers by Claessens, Gragg and Wynn [3]. The Watson algorithm [3], which exploits the accuracy-through-order condition is likely to turn out to be the most stable.

Viskovatov's algorithm is an ingenious reorganisation of the algebra involved in converting a Maclaurin series to its associated C-fraction. Rutishauser's Q.D. algorithm is verified by a comparison of neighbouring continued fraction sequences. Both these algorithms generate continued fraction representations of Padé approximants and so come into this category.

The conversion of some of these methods to become reliable algorithms was commenced by Magnus [13], continued by Claessens and Wuytack [14], and substantially improved by Bultheel [15] .

C. Iterative methods based on antidiagonal staircase sequences. The equivalent of the Q.D. algorithm is Gragg's algorithm based on comparison of neighbouring continued fraction sequences. Claessens investigated the equivalent of Watson's algorithm. In this case, the Baker algorithm, being the simplest, seems to have the advantage.

D. Iterative methods for a descending diagonal sequence. Brezinski's algorithm is the only member of this class, and Bultheel's modification for reliability [15] seems to fit naturally.

E. Iterative methods for an antidiagonal sequence. Kronecker's algorithm is the prototype of this kind, and we review its potential in detail in the next section.

F. Toeplitz and Hankel methods. These methods are fast methods of inverting the matrix of coefficients of the Padé equations. The Padé equations may be written compactly as

$$C\,\underline{b} = \underline{d} \tag{3.1}$$

where

$$C = \begin{bmatrix} c_{L-M+1} & c_{L-M+2} & \cdots & c_L \\ c_{L-M+2} & c_{L-M+3} & \cdots & c_{L+1} \\ \vdots & \vdots & & \vdots \\ c_L & c_{L+1} & \cdots & c_{L+M-1} \end{bmatrix}, \quad \underline{b} = \begin{bmatrix} b_M \\ b_{M-1} \\ \vdots \\ b_1 \end{bmatrix}, \quad \underline{d} = \begin{bmatrix} -c_{L+1} \\ -c_{L+2} \\ \vdots \\ -c_{L+M} \end{bmatrix}$$

or else as

$$T\,\underline{\tilde{b}} = \underline{d}$$

where

$$T = \begin{bmatrix} c_L & c_{L-1} & \cdots & c_{L-M+1} \\ c_{L+1} & c_L & \cdots & c_{L-M+2} \\ \vdots & \vdots & & \vdots \\ c_{L+M-1} & c_{L+M-2} & \cdots & c_L \end{bmatrix} \quad \text{and} \quad \underline{\tilde{b}} = \begin{bmatrix} b_1 \\ b_2 \\ \vdots \\ b_M \end{bmatrix}.$$

C is a Hankel matrix with a cyclic symmetry property and T is a Toeplitz matrix with the persymmetry property. The special methods for treating Hankel and Toeplitz systems are similar in spirit. Each element T_{ij} of T satisfies the equation $T_{ij} = t_{i-j}$, expressing the persymmetry. The inverse of a Toeplitz matrix is a Toeplitz matrix, which can normally be found with $O(3M^2)$ multiplicative-type operations, according to the following outline. Let $T^{(n)}$ denote an n×n Toeplitz matrix, so that $T^{(n+1)}$ has the block form

$$T^{(n+1)} = \left[\begin{array}{c|cccc} t_0 & t_{-1} & t_{-2} & \cdots & t_{-n} \\ \hline t_1 & & & & \\ t_2 & & & & \\ \vdots & & T^{(n)} & & \\ t_n & & & & \end{array}\right] = \left[\begin{array}{cccc|c} & & & & t_{-n} \\ & T^{(n)} & & & t_{-n+1} \\ & & & & \vdots \\ & & & & t_{-1} \\ \hline t_n & t_{n-1} & \cdots & t_1 & t_0 \end{array}\right]$$

We may verify that $(T^{(n)})^{-1}$ has the form

$$(T^{(n)})^{-1} = L^{(n)} D^{(n)} U^{(n)}$$

where $L^{(n)}$ is a unit lower triangular matrix with elements

$$L_{ij}^{(n)} = \ell_{i-j}^{(n-j+1)} \qquad \text{for} \quad i > j$$

and $U_{ij}^{(n)}$ is a unit upper triangular matrix with elements

$$U_{ij}^{(n)} = u_{j-i}^{(n-i+1)} \qquad \text{for} \quad i < j.$$

$D^{(n)}$ is a diagonal matrix with elements $D_{nn}^{(n)}=1$ and

$$D_{ii}^{(n)} = r_{n-i}^{-1} \;, \qquad \text{for } i = 1,2,\ldots n-1.$$

<u>Example</u>

$$(T^{(2)})^{-1} = \begin{bmatrix} 1 & 0 \\ \ell_1^{(2)} & 1 \end{bmatrix} \begin{bmatrix} r_1^{-1} & 0 \\ 0 & 1 \end{bmatrix} \begin{bmatrix} 1 & u_1^{(2)} \\ 0 & 1 \end{bmatrix}$$

and

$$(T^{(3)})^{-1} = \begin{bmatrix} 1 & 0 & 0 \\ \ell_1^{(3)} & 1 & 0 \\ \ell_2^{(3)} & \ell_1^{(2)} & 1 \end{bmatrix} \begin{bmatrix} r_2^{-1} & 0 & 0 \\ 0 & r_1^{-1} & 0 \\ 0 & 0 & 1 \end{bmatrix} \begin{bmatrix} 1 & u_1^{(3)} & u_2^{(3)} \\ 0 & 1 & u_1^{(2)} \\ 0 & 0 & 1 \end{bmatrix}$$

We note the way in which the LDU decomposition of $(T^{(2)})^{-1}$ is used to form the LDU decomposition of $(T^{(3)})^{-1}$. We refer to Trench and Zohar [3] and Bultheel [15] for details of the rapid calculation of r_n^{-1}, the first column of $L^{(n)}$ and the top row of $U^{(n)}$ in the body of the iteration. I only emphasise that an unavoidable aspect of this approach is that

$$r_{n-1} = \det(T^{(n)})/\det(T^{(n-1)}) \;.$$

This implies that the existence of inverses of all the Toeplitz submatrices $T^{(m)}$, $m = 1,2,\ldots,M-1$, is required in order to obtain the inverse of a non-singular Toeplitz matrix $T^{(M)}$. Of course, the method may be modified, [14], but it is hard to envisage a useful modification without row or column interchanges when the LDU decomposition of an intermediate stage is non-existent.

G. Matrix solution methods. Despite the proliferation of allegedly excellent algorithms for Padé approximation, the most popular method amongst users is the matrix solution method. The solution of (3.1) can be found using $O(\frac{1}{3}M^3+M^2)$ operations, ignoring the benefit of the symmetry of the coefficient matrix. As we argue in section 5, meaningful Padé calculations are necessarily low order calculations; typically $M \le 10$ in single precision. Consequently, in terms of the operational count, matrix solution is very competitive with the so-called fast methods of paragraphs A-F, which are $O(\alpha M^2)$ methods, $3 \le \alpha \le 6$. A fairly standard matrix inversion method is Gauss elimination or Gauss-Jordan elimination, each with full pivoting. A more interesting method is the Crout decomposition (row interchanges plus equilibration) supplemented by an iterative refinement. We argue in section 7 that the latter method is the best choice out of the present selection of algorithms for Padé approximation.

4. The modified Kronecker algorithm.

The material in this section is distilled from several sources: Kronecker's algorithm is defined by Warner [17], and the modification has been proposed independently by Claessens [18], McEliece and Shearer [19] and Cordellier [11,20]. The basic algorithm involves an antidiagonal sequence of interpolants shown in fig. 1 and defined for $j=0,1,2,\ldots,N$ by

$$\frac{p^{[N-j/j]}(z)}{q^{[N-j/j]}(z)} \equiv \frac{p^{(j)}(z)}{q^{(j)}(z)} . \tag{4.1}$$

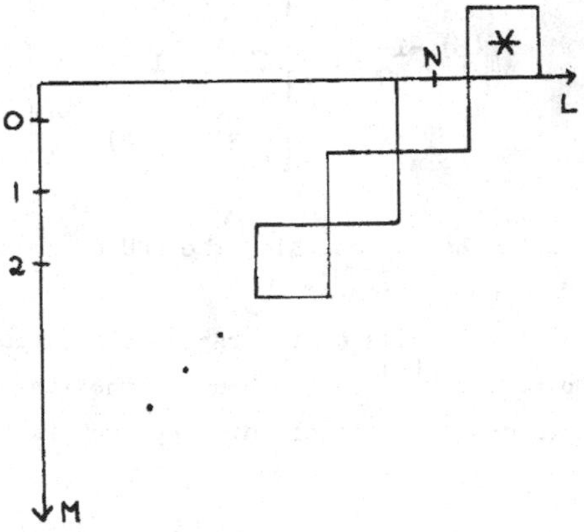

Fig. 1 An antidiagonal sequence.

The first member of the sequence of interpolants, $j=0$, is the Maclaurin series, and we define

$$p^{(0)}(z) \equiv \sum_{i=0}^{N} c_i z^i \equiv \sum_{i=0}^{N} p_i^{(0)} z^i , \tag{4.2}$$

$$q^{(0)}(z) \equiv 1 \equiv q_0^{(0)} . \tag{4.3}$$

The recurrence relations required to generate the sequence are, for $j = 0,1,2,\ldots,M-1$,

$$p^{(j+1)}(z) = (\alpha_j z + \beta_j)p^{(j)}(z) - p^{(j-1)}(z) \tag{4.4}$$

$$q^{(j+1)}(z) = (\alpha_j z + \beta_j)q^{(j)}(z) - q^{(j-1)}(z) . \tag{4.5}$$

To initialise the algorithm, it is convenient to introduce the artificial initialising values for the entry denoted by a * in fig. 1,

$$p^{(-1)}(z) = z^{N+1}, \qquad q^{(-1)} = 0 . \tag{4.6}$$

At stage (j), α_j and β_j are found so as to reduce the apparent degree of the right-hand side of (4.4) from N-j+1 to N-j-1. In fact we require that

$$\alpha_j\, p^{(j)}_{N-j} - p^{(j-1)}_{N-j+1} = 0 , \qquad (\text{coefft. of } z^{N-j+1}), \tag{4.7}$$

$$\alpha_j\, p^{(j)}_{N-j-1} + \beta_j\, p^{(j)}_{N-j} - p^{(j-1)}_{N-j} = 0 , \quad (\text{coefft. of } z^{N-j}) . \tag{4.8}$$

Provided only that $p^{(j)}_{N-j} \neq 0$, which is to say that $p^{(j)}(z)$ has full degree, α_j and β_j are uniquely determined by (4.7) and (4.8). Provided α_j is non-zero, which occurs when $p^{(j-1)}(z)$ has full degree, $q^{(j+1)}(z)$ has full degree.

<u>Verification</u>. From (4.4) and (4.5),

$$p^{(j+1)}(z) + p^{(j-1)}(z) = (\alpha_j z + \beta_j)\, p^{(j)}(z)$$

$$q^{(j+1)}(z) + q^{(j-1)}(z) = (\alpha_j z + \beta_j)\, q^{(j)}(z) .$$

Therefore,

$$\begin{vmatrix} p^{(j+1)}(z) + p^{(j-1)}(z) & p^{(j)}(z) \\ q^{(j+1)}(z) + q^{(j-1)}(z) & q^{(j)}(z) \end{vmatrix} = 0 ,$$

and

$$\begin{vmatrix} p^{(j+1)}(z) & p^{(j)}(z) \\ q^{(j+1)}(z) & q^{(j)}(z) \end{vmatrix} = \begin{vmatrix} p^{(j)}(z) & p^{(j-1)}(z) \\ q^{(j)}(z) & q^{(j-1)}(z) \end{vmatrix} \tag{4.8}$$

Hence each side of (4.8) is independent of j, and we deduce that

$$\begin{vmatrix} p^{(j+1)}(z) & p^{(j)}(z) \\ q^{(j+1)}(z) & q^{(j)}(z) \end{vmatrix} = \begin{vmatrix} p^{(0)}(z) & p^{(-1)}(z) \\ q^{(0)}(z) & q^{(-1)}(z) \end{vmatrix} = -z^{N+1} \tag{4.9}$$

Hence

$$\frac{p^{(j+1)}(z)}{q^{(j+1)}(z)} - \frac{p^{(j)}(z)}{q^{(j)}(z)} = \frac{-z^{N+1}}{q^{(j+1)}(z)\, q^{(j)}(z)} . \tag{4.10}$$

Provided $q^{(j)}(0) \neq 0$ for $j=1,2,\ldots,M$,

$$\frac{p^{(j)}(z)}{q^{(j)}(z)} = \sum_{i=0}^{N} c_i z^i + O(z^{N+1}) \quad \text{for } j=1,2,\ldots,M. \tag{4.11}$$

This algorithm encounters a block in the Padé table if and only if $p_{N-j}^{(j)} = 0$, i.e. a numerator polynomial has a lower degree than expected. It follows from (4.5) that the denominator always has full degree. Let us suppose that we are treating the first non-trivial block encountered in the antidiagonal sequence, and that the deficiency in degree of $p^{(j)}(z)$ is d, $d \geq 1$, and this means that

$$p_{N-j}^{(j)} = p_{N-j-1}^{(j)} = \ldots. = p_{N-j-d+1}^{(j)} = 0 \; ; \quad p_{N-j-d}^{(j)} \neq 0 .$$

It is then possible to find the d+2 coefficients of the polynomial $\pi_{d+1}(z)$ of order d+1 to satisfy

$$p^{(j+1)}(z) = \pi_{d+1}(z)\, p^{(j)}(z) - p^{(j-1)}(z) \tag{4.12}$$

$$q^{(j+1)}(z) = \pi_{d+1}(z)\, q^{(j)}(z) - q^{(j-1)}(z) , \tag{4.13}$$

such that $\partial\{p^{(j+1)}(z)\} = N-j-d-1$ and $\partial\{q^{(j+1)}(z)\} = j+d+1$ and the next interpolant of the sequence is uniquely defined. We see that the next interpolant occurs on the correct antidiagonal for the sequence, but is it a Padé approximant? There are two cases to consider. The previous analysis is valid, mutatis mutandis, provided $q^{(j)}(0) \neq 0$, which is designated case 1. In this case, the path of the algorithm is indicated in fig. 2, and there are no difficulties. In case 2, $q^{(j)}(0) = 0$ and we deduce from the equivalent of (4.9) that $p^{(j)}(0) = 0$ also. Hence a factor of z cancels in $p^{(j)}(z)/q^{(j)}(z)$ and the corresponding Padé approximant is non-existent [1]. Let k be the maximum power of z such that z^k cancels in $p^{(j)}(z)/q^{(j)}(z)$. Then we note that in case 2, we are dealing with a $d+k+1 \times d+k+1$ block in the Padé table. (Note that in case 1, it is neither possible nor necessary to determine the block size.) Using the previous analysis, it follows that

$$\frac{p^{(j+1)}(z)}{q^{(j+1)}(z)} - \frac{p^{(j-1)}(z)}{q^{(j+1)}(z)} = -z^{N+1}\left[\frac{1}{q^{(j+1)}(z)\, q^{(j)}(z)} + \frac{1}{q^{(j)}(z)\, q^{(j-1)}(z)}\right]$$

$$= \frac{-z^{N+1}\, \pi_{d+1}(z)}{q^{(j+1)}(z)\, q^{(j-1)}(z)} , \tag{4.14}$$

and hence we see that $p^{(j+1)}(z)/q^{(j+1)}(z)$ is the next non-degenerate Padé approximant in the antidiagonal sequence. We show it in the Padé table in fig. 3 for the case where d=1 and k=2.

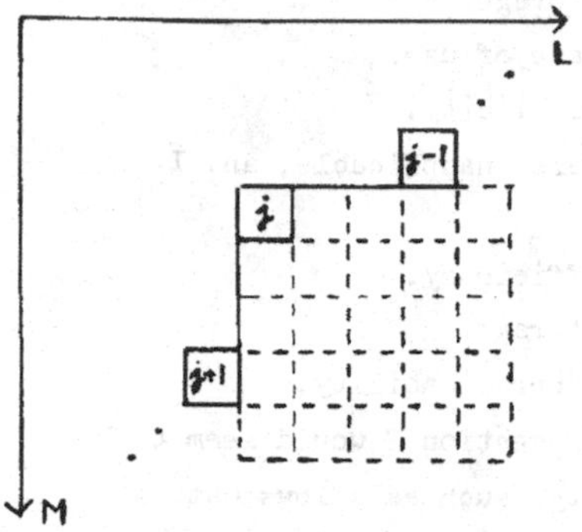

Fig. 2 Path of Kronecker's algorithm through a degenerate entry (j).

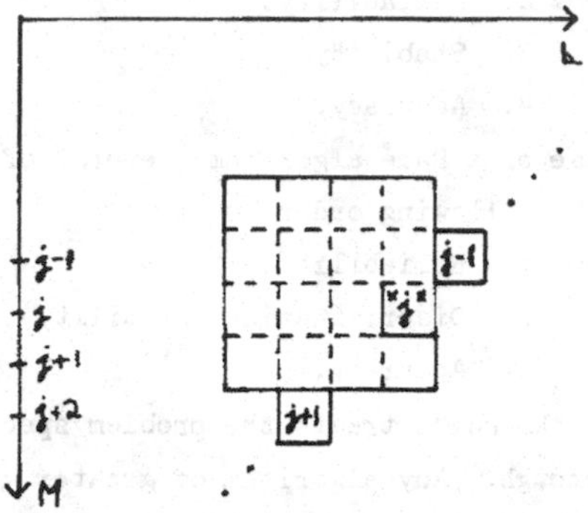

Fig. 3 Path of Kronecker's algorithm through an indeterminate entry (j).

We omit the details of the remainder of the proof which shows that the algorithm is valid whatever the block structure on the antidiagonal may be.

We have shown in this section that the definitions (4.2-4.5), modified by (4.12,4.13) if necessary, define an antidiagonal sequence of rational forms. If the numerator has anomalously low degree, the approximant is degenerate. If the denominator vanishes at z=0, the corresponding Padé approximant does not exist. All the other members of the sequence form the complete set of Padé approximants on the antidiagonal.

The modification (4.12) of the algorithm is called the Euclidean modification because of the similarity to the Euclidean algorithm for determining the common factor of two polynomials. We have elaborated this algorithm at length because it is a good example of a reliable algorithm: the algorithm finds a solution of a Padé approximation problem if and only if a solution exists. Another reason for presenting this proof, rather than one from the existing literature, is that this proof generalises immediately to the rational interpolation problem: (4.2) is replaced by the Newton interpolating polynomial and (4.11) becomes

$$\frac{p^{(j)}(z)}{q^{(j)}(z)} = \sum_{i=0}^{N} c_i z^i + r(z) \prod_{i=0}^{N} (z-z_i) \quad , \quad r(z_i) \neq \infty,\ i=0,1,\dots,N.$$

Thus we see that the modified Kronecker algorithm is also a reliable algorithm for rational interpolation.

5. Criteria for selection of a method.

A "standard" list of criteria for a good numerical method, in order of importance, is given by Moler and van Loan [21] as

1. Generality.
2. Reliability.
3. Stability.
4. Accuracy.
5. Efficiency.
6. Storage.
7. Ease of use.
8. Simplicity.

In our case of a Padé algorithm, several of these are inapplicable, and I propose the following order:

1. Reliability.
2. Discrimination (stability).
3. Accuracy.
4. Efficiency.
5. Storage.
6. Generalisability.

Any algorithm which treats the problem specified in section 2 would seem to be general enough. Any algorithm of greater generality, such as allows matrix coefficients or treats the rational interpolation problem, introduces so many extraneous considerations that it is given minimal priority. The principal problems we face are associated with blocks and allied degeneracies in the Padé table. With the specification of section 2, we ask "Is the required approximant in a block ?". If we are using a recursive procedure, such as the Q.D. algorithm, how do we navigate or circumnavigate a block ? With these thoughts in mind, we define the criteria in our list in the context of the Padé approximation problem.

We say that an algorithm is reliable if it is capable, in principle, of finding a solution of any specified Padé approximation problem when such a solution exists, and of deciding on the non-existence of a solution whenever appropriate [1,22]. The phrase "in principle" means that we may assume exact arithmetic, so that tests for zeros of real or complex numbers are unambiguous.

We expect our algorithm to be accurate in the sense that the output coefficients $a_0, a_1, \ldots, a_L, b_1, \ldots, b_M$ are determined as accurately as the data permit. Padé approximation problems are notoriously unstable: small variations in the data coefficients usually lead to substantial variation in the output coefficients. We might hope to obtain output coefficients of the same precision as the (binary) input coefficients. We should not be satisfied with output coefficients which correspond with exact arithmetic to input coefficients with greater errors than the rounding errors.

We expect our algorithm to discriminate between degenerate and non-degenerate Padé approximation problems in practice. A problem which is degenerate in principle may be rendered non-degenerate by rounding error, and vice-versa in exceptional cases. The algorithm must contain a branching instruction at which juncture the computer must decide whether the processed data might have originated from a degenerate problem (e.g. the determinant is zero within rounding error) or that the data do not correspond to a degenerate approximant.

Efficiency and economy of storage are self-explanatory virtues.

An algorithm is generalisable if the numerical method is readily adapted to similar problems, e.g. integer or matrix coefficients, the Cauchy-Jacobi problem etc..

We leave it to the reader to assess the merits of the various algorithms in the light of the criteria defined. However, the following table summarises some of the relevant features within the context of the calculation of a [M/M] approximant.

	Reliability	Discrimination	Efficiency
Sequential methods	No	nil	$O(\alpha M^2)$, $4 \le \alpha \le 6$
Toeplitz methods	No	nil	$O(3M^2)$
Modified Kronecker	Yes	yet to be analysed	$O(4M^2)$
Matrix inversion	Yes	good	$O(M^3)$

The moral is that the future for algorithm devisors lies in converting existing algorithms into reliable algorithms, and in assessing their discriminating power.

6. Instability of Padé approximation.

As an anthropological observation, it is well known that inexperienced Padé approximators underestimate the numerical accuracy needed for the Maclaurin coefficients and overestimate the order of the Padé approximants supported by their numerical data. We will show that the underlying instability is an inevitable feature of Padé approximation by considering an ideal example in which the Padé approximants converge systematically. There is no question of the occurrence of the familiar problems associated with poles and defects. We consider the Stieltjes function

$$f(z) = \int_0^1 \frac{du}{1 + zu} \quad . \tag{6.1}$$

The staircase sequence of [M/M] and [M/M+1] approximants converge with all poles on the interval (-1,0) on the negative real axis. The poles have positive residues: as $M \to \infty$, they "coalesce" to form the cut of f(z). Off the negative real axis, convergence is geometric. In short, we have selected a simple, ideal example. In fact,

$$f(z) = \frac{1}{z}\ln(1+z) = \sum_{i=0}^{\infty} \frac{(-z)^i}{i+1} , \tag{6.2}$$

and the Padé equations for $[M-1/M]_f(z)$ may be written as

$$H^{(M)}\,\underline{b}' = \underline{d}' , \tag{6.3}$$

where

$$H^{(M)} = \begin{pmatrix} 1 & \frac{1}{2} & \frac{1}{3} & \cdots & \frac{1}{M} \\ \frac{1}{2} & \frac{1}{3} & \frac{1}{4} & \cdots & \frac{1}{M+1} \\ \cdot & \cdot & \cdot & & \cdot \\ \cdot & \cdot & \cdot & & \cdot \\ \frac{1}{M} & \frac{1}{M+1} & \frac{1}{M+2} & \cdots & \frac{1}{2M-1} \end{pmatrix}, \quad \underline{b}' = \begin{pmatrix} -b_M \\ b_{M-1} \\ \cdot \\ \cdot \\ (-)^M b_1 \end{pmatrix}, \underline{d}' = (-1)^{M+1} \begin{pmatrix} \frac{1}{M+1} \\ \frac{1}{M+2} \\ \cdot \\ \cdot \\ \frac{1}{2M} \end{pmatrix}.$$

The coefficient matrix $H^{(M)}$ is a Hilbert segment, notorious for its ill-conditioning. As usual, we define the condition number as

$$\kappa(H^{(M)}) = \| H^{(M)} \|_2 \cdot \| H^{(M)-1} \|_2 . \tag{6.4}$$

Taylor has recently given a lower bound for the condition number of Gram matrices [23], and in this case the result is

$$\kappa(H^{(M)}) \geq \frac{(2M)!^2}{M!^4} \simeq \frac{16^M}{\pi M} . \tag{6.5}$$

Summarising, we note that in a situation ideal for convergence of the approximants, the <u>coefficient</u> problem is exceedingly ill-conditioned. The Padé folklore, that you lose M decimal places of accuracy in forming an [L/M] approximant is seen to be somewhat optimistic in this case. If a computer holds p bits of mantissa precision, the implication of the foregoing analysis is that one should only normally consider formation of an [L/M] Padé approximant with

$$M \leq [p/4] . \tag{6.6}$$

Indeed, even lower order approximants may be mandatory if the data coefficients $c_0, c_1, \ldots, c_{L+M}$ are known to less accuracy than the machine precision.

It is clear, and it follows from Taylor's results, that one can obtain a different estimate of the condition number by changing the example to

$$f(z) = \int_0^a \frac{du}{1+zu}, \qquad a > 0.$$

However, with equilibration of the coefficient matrix and rescaling of the variables, it would be surprising if the conclusions differed appreciably from (6.5) and (6.6).

7. Discrimination using matrix solution.

The remarks of the previous section indicate that an algorithm for calculating Padé approximants will only be satisfactory if it discriminates between ill-conditioned but soluble approximation problems and problems corresponding to degenerate Padé approximants. Rounding error, howsoever introduced, in the data coefficients is inevitably magnified in the solution of the coefficient problem. Little enough error analysis has been attempted on any of the other methods of section 3: we summarise in this section the merits of the matrix solution method, which hinge on Wilkinson's error analysis [24,25]. For Gaussian elimination, the singularity test consists of deciding whether a pivot is zero within numerical error or not. We consider the method of Crout reduction (partial pivoting with row interchanges), equilibration and residual correction. To solve the system (3.1), namely $C\underline{b} = \underline{d}$, we suppose that an LU decomposition of C has been made, and that an <u>approximate</u> inverse $(LU)^{-1}$ of C has been found, yielding an approximate solution $\underline{b}^{(1)} \equiv \underline{x}^{(1)}$. We define a sequence of residuals

$$\underline{r}^{(s)} = \underline{d} - C\underline{x}^{(s)}, \qquad s = 1,2,\ldots \quad (7.1)$$

and approximate solutions

$$\underline{x}^{(s+1)} = \underline{x}^{(s)} + (LU)^{-1}\underline{r}^{(s)}, \qquad s = 1,2,\ldots \quad (7.2)$$

From Wilkinson's analysis,

$$\underline{x}^{(s+1)} - \underline{x} = [I - (LU)^{-1}C]^s\,(\underline{x}^{(1)} - \underline{x})$$

and

$$\underline{r}^{(s+1)} = [I - (LU)^{-1}C]^s\,\underline{r}^{(1)}.$$

As a practical matter, we emphasise that the computation of (7.1) requires double precision. In principle, it follows that

$$\| \underline{r}^{(s+1)} \|_\infty \le \frac{2^{-p}}{1+2^{-p}} \| \underline{r}^{(s)} \|_\infty \tag{7.3}$$

provided that

$$\| C - LU \|_\infty \| C^{-1} \|_\infty < 2^{-p} \ . \tag{7.4}$$

Eq. (7.3) shows that the residuals converge geometrically if (7.4) is satified. In practice, geometric convergence of the residuals corresponds to non-singular but possibly ill-conditioned matrices, whereas lack of geometric convergence always corresponds to rounded singular matrices. The behaviour of the residuals provides the essential discrimination about whether or not the data correspond to a degenerate Padé approximant. The method of this section is implemented in the N.A.G. library routines EO2RAF and EO2RAA for Padé approximation.

8. Conclusions and aknowledgement.

We have stressed the necessity that a Padé approximant algorithm be reliable and discriminating. We suggest that Kronecker's algorithm for the case of integer coefficients and the method of section 7 for the case of rounded coefficients seem to be the best methods at present.

I am grateful to Dr. T.R. Hopkins, Dr. M.G. Cox and Dr. G. Hayes for discussions and collaboration leading to production of the N.A.G. routines. I am also grateful to Prof. L. Wuytack, Dr. F. Cordellier and Dr. A. Bultheel for helpful discussions and correspondence essential for an up-to-date review.

References

[1] G.A. Baker, J. Math. Anal. Applcns. 43, 498, (1973).

[2] G.A. Baker, "The Essentials of Padé Approximants", Academic Press (N.Y.,1974)
J.S.R. Chisholm, "Padé Approximants", P.R. Graves-Morris (ed.), Institute of Physics (1973), p.1.
A.C. Genz, ibid., p.112.
G.A. Baker and P.R. Graves-Morris, "Padé Approximants", Addison-Wesley, to be published.

[3] L. Wuytack, bibliography in these procedings.

[4] J. Essam, S.R.C. project at Westfield College, London, (1978).

[5] Y. Starkand, Comm. Comp. Phys. 11, 325, (1976).

[6] J.-L. Basdevant, Fortschr. Phys. 20, 282, (1972).

[7] M. Pindor, these procedings.

[8] Y. Luke, "The Special Functions and their Approximations", vols I,II, Academic Press (N.Y.,1969); "Mathematical Functions and their Approximations", Academic Press (N.Y.,1975); these procedings.

[9] P. Henrici and P. Pfluger, Num. Math. 9, 120, (1966).
P. Henrici, "Applied and Computational Complex Analysis", vol II p.615, Wiley (1977).

[10] W. Jones and W. Thron, "Continued Fractions", Addison-Wesley, to be published.

[11] F. Cordellier, Thèse, Université de Lille, to appear.

[12] J. Gilewicz, "Approximants de Padé", Springer Verlag, (1978).

[13] A. Magnus, Math. Zeit. 78, 361, (1962).
A. Magnus, Rky. Mtn. J. Math. 4, 257, (1974).

[14] G. Claessens and L. Wuytack, "On the Computation of non-normal Padé Approximants", Antwerp preprint (1977).

[15] A. Bultheel, these proceedings.

[16] L.S. de Jong, SIAM J. Control and Optimisation 16, 646, (1978).

[17] D. Warner, Thesis, University of San Diego, (1974).

[18] G. Claessens, Thesis, University of Antwerp, (1976).

[19] R.J. McEliece and J.B. Shearer, SIAM J. Appl. Math. 34, 611, (1978).

[20] F. Cordellier, lecture at the Padé Symposium, University of Lille,(1978), unpublished.

[21] C. Moler and C. van Loan, SIAM review, 20, 801, (1978).

[22] P.R. Graves-Morris and T.R. Hopkins, "Reliable Rational Interpolation" Kent preprint (1978), submitted to Num. Math.

[23] J.M. Taylor, Proc. Roy. Soc. Edin. 80A, 45, (1978).

[24] J.H. Wilkinson, "Rounding Error in Algebraic Processes", Notes in Applied Science no. 32, HMSO. Chap. 3, (1963).

[25] J.H. Wilkinson, J. Assoc. Comp. Mach. 8, 281, (1961).

SUR LE CALCUL DE L'EXPONENTIELLE D'UNE MATRICE

J.R. ROCHE

Laboratoire de Mathématiques Appliquées
B.P. 53
38041 GRENOBLE - CEDEX

Introduction :

Notre problème est de chercher les facteurs qui determinent la convergence des approximants de Padé de l'exponentielle d'une matrice carré d'ordre N.

On étudie d'abord quelle est la nature de l'erreur dans le cas "scalaire".

Ensuite on calcule une majoration de l'erreur théorique commise quand on approche l'exponentielle d'une matrice par la méthode de Padé.

Notations :

On appele g_{pq} l'approximant de Padé de l'exponentielle :

$$g_{pq}(z) = n_{pq}(z) / d_{pq}(z) \quad \text{où}$$

$$n_{pq}(z) = \sum_{k=0}^{q} \frac{(p+q-k)!\, q!\, (z)^k}{(p+q)!\, k!\, (q-k)!}$$

$$\text{et} \quad d_{pq}(z) = \sum_{k=0}^{p} \frac{(p+q-k)!\, p!\, (-z)^k}{(p+q)!\, k!\, (p-k)!}$$

1 - CALCUL DE L'ERREUR THEORIQUE DANS LE CAS "SCALAIRE"

M. PADE [3] a remarqué d'abord que si F (t) est un polynôme en t de degré m et,

$$\mathcal{F}(t) = \frac{F(t)}{z} + \frac{F^{(1)}(t)}{z^2} + \ldots + \frac{F^{(m)}(t)}{z^{m+1}}$$

alors

$$\int e^{-tz} F(t)\, dt = -e^{-tz} \mathcal{F}(t)$$

donc on obtient

$$\emptyset_0(z)\, e^z - \emptyset_1(z) = z^{m+1} e^z \int_0^1 e^{-zt} F(t)dt$$

où

$$\emptyset_0(z) = F(0)\, z^m + F^{(1)}(0)\, z^{m-1} + \ldots + F^{(m)}(0)$$

et

$$\emptyset_1(z) = F(1)\, z^m + F^{(1)}(1)\, z^{m-1} + \ldots + F^{(m)}(1)$$

Pour être dans le cas des approximants de PADE il faut et il suffit que :

1) Le degré de $\emptyset_0$ soit p

2) Le degré de $\emptyset_1$ soit q

3) m soit au moins égal à p + q

Donc, il faut et il suffit que :

$F^{(l)}_{(0)} = 0$ pour $l = 0,\ldots,m-p-1$

$F^{(l)}_{(1)} = 0$ pour $l = 0,\ldots,m-q-1$

Et, pour que le polynôme F(t) y satisfasse, il faut et il suffit que l'on ait :

$F(t) = t^{m-p}\, (t-1)^{m-q}\, G(t)$ où

G(t) est un polynôme en t de degré p+q-m

Mais m doit être au moins égal à p + q donc nécessairement, m = p + q et G est un constante.

Il en résulte, finalement :

$F(t) = t^q (t-1)^p = (-1)^p\, t^q (1-t)^p$ et

$\emptyset_0(z) = F^{(p+q)}(0) + F^{(p+q-1)} z + \ldots + F^{(q)}_{(0)}\, z^p$

$\emptyset_1(z) = F^{(p+q)}(1) + F^{(p+q-1)} z + \ldots + F^{(p)}(1)\, z^q$

Mais $\emptyset_0(z) = (p+q)!\, d_{pq}(z)$ et $\emptyset_1(z) = (p+q)!\, n_{pq}(z)$

On en déduit le résultat suivant :

$$R_{pq}(z) = e^z - \frac{n_{pq}(z)}{d_{pq}(z)} = \frac{(-1)^p\, z^{p+q+1}}{(p+q)!\, d_{pq}(z)} \int_0^1 e^{z(1-t)} (1-t)^p t^q dt$$

2 - UNE BORNE DE L'ERREUR ET SES DERIVEES

M. VARGA [1], M. UNDERHILL et WRAGG [2] ont borné l'erreur calculée par PADE.

Maintenant, soit x la partie réelle de z alors :

$$\left| \int_0^1 e^{z(1-t)} t^q (1-t)^p dt \right| \leq e^{|x|} \left| \int_0^1 t^q (1-t)^p dt \right|$$

donc, dans le cercle $|z| < R_0$

$$e^{|x|} \left| \int_0^1 t^q (1-t)^p dt \right| \leq e^{R_0} \frac{p!\, q!}{(p+q+1)!}$$

Underhill a aussi démontré que les zéros z_i du dénominateur de g_{pq} ont la propriété d'être en module, plus grands que r (p+q), où r est la racine réelle positive de l'équation : $1 + r + \mathrm{Ln}(r) = 0$

C'est à dire, la racine réelle positive de l'équation $r\, e^{r+1} = 1$. VARGA en a calculé une approximation numérique ; 0.278465.

Donc, si $z_1, \ldots, z_p$ sont les p racines de $d_{pq}(z)$ on obtient :

$$d_{pq}(z) = \frac{(-1)^p\, q!}{(p+q)!} \prod_{i=1}^{p} (z - z_i)$$

Si de plus p et q verifient $r\,(p+q) > R_0$ alors $|z-z_i| > r(p+q) - R_0 > 0$ pour $i = 1, \ldots, p$.

Cette remarque permet d'obtenir très aisement une borne d'erreur car :

$$|1 / d_{pq}(z)| \leq \frac{(p+q)!}{q!\, (r(p+q) - R_0)^p}$$

pour $|z| < R_0$ et p,q suffisament grands, donc :

$$|R_{pq}(z)| \leq \frac{e^{R_0}\, R_0^{p+q+1}\, p!}{(p+q+1)!\, (r(p+q) - R_0)^p}$$

Pour obtenir tous les éléments dont nous avons besoin pour les calculs ultérieurs, on va suivre l'article dû à UNDERHILL et WRAGG.

Soit maintenant :

$$R_{pq}^{(j)}(z) = e^z - g_{pq}^{(j)}(z) = (e^z - g_{pq}(z))^{(j)}$$

Nous pouvons choisir p et q tels que $R_{pq}(z)$ soit analytique, pour module de z plus petit que R_0 ; donc $R_{pq}^{(j)}(z)$ le sera aussi. Donc le maximum de $R_{pq}(z)$ sur le cercle fermé $|z| \leq R_0$ est obtenu sur la frontière, c'est à dire dans un point $Z_0 = R_0\, e^{i\theta_0}$ où θ_0 appartient à R

Alors si $|z| < R_0$ on obtient :

$$|R_{pq}^{(j)}(z)| \leq |R_{pq}^{(j)}(R_0 e^{i\theta_0})| \leq \frac{m(h)\; j!}{h^j}$$

où m(h) est une borne de $R_{pq}(z)$ dans le cercle $|z| < R_0+h < r(p+q)$; $h > 0$

Donc, on obtient la borne suivante :

$$|R_{pq}^{(j)}(z)| \leq \frac{e^{R_0 th}\quad (R_0+h)^{p+q+1}\, p!\; j!}{(p+q+1)!\; (r(p+q)-R_0)^p\, h^j}$$

où $|z| < R_0+h < r(p+q)$; $h > 0$ et p,q suffisament grands.

Si maintenant on considère p égal à q on obtient :

$$|R_{pp}^{(j)}(z)| \leq \frac{e^{R_0 th}\quad (R_0 th)^{2p+1}\; p!\; j!}{(p+1)!\; (2rp - R_0 - h)^p\, h^j}$$

$j = 0,1,2,\ldots$; p et q suffisament grands. Cette borne tends vers 0 lorsque p tends vers l'infini.

3 - MAJORATION DE L'ERREUR THEORIQUE DANS LE CAS MATRICIEL

Nous allons maintenant calculer une majoration de l'erreur théorique commise quand on approche l'exponentielle d'une matrice d'ordre N par la méthode de PADE.

On notera Exp (A) l'exponentielle d'une matrice d'ordre N en considérand :

$$\mathrm{Exp}\,(A) = \sum_{i=0}^{\infty} \frac{A^i}{i!}$$

On introduit un théorème classique avec l'intention d'expliciter le facteur qu'intervient dans ce calcul.

Théorème

Soit $L_k(\lambda)$ la matrice carrée d'ordre k définie par :

$$L_1(\lambda) = \lambda \text{ et } L_k(\lambda) = \begin{bmatrix} \lambda & 1 & & 0 \\ & \ddots & \ddots & \\ & & \ddots & 1 \\ 0 & & & \lambda \end{bmatrix} \qquad k = 2,\ldots$$

Alors pour A matrice carrée d'ordre n, il existe T, matrice carrée d'ordre n, non singulière telle que :

$$T^{-1} AT = B = \text{diag} \{L_{ki} (\lambda_i) \; ; \; i = 1,\ldots,r\}$$

où $\sum_{i=1}^{r} k_i = n$ et les λ_i pour $i = 1,\ldots,r$ sont les valeurs propres de A, non nécessairement differentes.

D'après ce théorème, on peut dire que

$$A = T B T^{-1}$$

d'où il en suit :

$$\text{Exp} (A) = T \; \text{Exp} (B) \; T^{-1}$$

où

$$\text{Exp} (B) = \text{diag} \{ \text{Exp} (L_{ki} (\lambda_i))\}$$

Si k_i est égal à un alors :

$$\text{Exp} (L_{ki} (\lambda_i)) = e^{\lambda i}$$

sinon

$$\text{Exp} (L_{ki}(\lambda_i)) = e^{\lambda i} \; \text{Exp} (M_{ki}) \text{ où } M_{ki} = \delta_{i,j-1}$$

On remarque que $M_{ki}^{ki} = \bar{0}$ et $M_{ki}^{k} = \delta_{i,j-k}$ si k est plus petit que k_i, tout en étant plus grand que zéro.

Donc on peut espèrer donner une forme explicite des termes de Exp (A) à partir de la connaissance de λ_i, T et T^{-1}.

Soit maintenant $T = (t_{ij})_{i,j=1}^{n}$ et $T^{-1} = (p_{ij})_{i,j=1}^{n}$

$$A = (a_{ij})_{i,j=1}^{n}$$

Alors si tous les k_i sont égaux à un, on obtient :

$$a_{ij} = \sum_{k=1}^{n} \lambda_k \, t_{ik} \, p_{kj} \quad \text{et} \quad (\text{Exp } (A))_{ij} = \sum_{k=1}^{n} e^{\lambda_k} \, t_{ik} \, p_{kj}$$

Dans le cas où il y a un k_i plus grand que un on a :

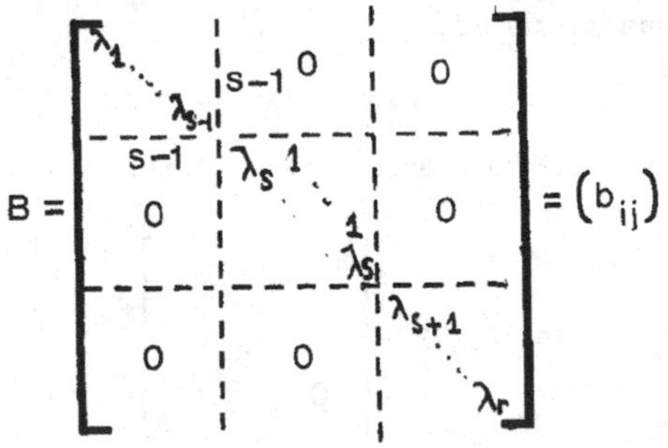

ce qui entraine :

$$a_{ij} = \sum_{k=1}^{n} \lambda_k t_{ik} p_{kj} + \sum_{k=s+1}^{s+k_s-1} t_{i,k-1} p_{k,j}$$

On peut aussi calculer directement l'Exp (A)

$$\text{Exp}(A) = T \ \text{Exp}(B)\, T^{-1} = (c_{ij})_{i,j=1}^{n} \quad \text{où}$$

$$c_{ij} = \sum_{k=1}^{n} e^{\lambda_k} t_{ik} p_{kj} + e^{\lambda_s} \sum_{k=s+1}^{s+ks-1} \left(\sum_{l=s}^{k-1} \frac{1}{(k-l)!} t_{il} p_{kj} \right)$$

Si on considère le cas général on aura autant de termes du deuxième type dans l'adition que de k_i plus grands que un.

Maintenant on considère g_{pp} les approximants de PADE diagonaux de e^x, de type p,p ; on appele l'approximant de PADE g_{pp} (A) de Exp (A) a :

$$g_{pp}(A) = d_{pp}(A)^{-1} n_{pp}(A) = n_{pp}(-A)^{-1} n_{pp}(A)$$

On remarque très aisément que :

$$g_{pp}(A) = T\, g_{pp}(B)\, T^{-1} \quad \text{où}$$

$$g_{pp}(B) = \text{diag}\,\{ g_{pp}(l_{ki}(\lambda_i)),\ i = 1,\dots,r \}$$

Si k_i est égal à un on aura :

$$g_{pp}(l_{ki}(\lambda_i)) = n_{pp}(-\lambda_i)^{-} n_{pp}(\lambda_i) \text{ et}$$

si k est plus grand que un alors :

$$g_{pp}(l_{ki}(\lambda_i)) = n_{pp}(-l_{ki}(\lambda_i))^{-1} n_{pp}(l_{ki}(\lambda_i))$$

où

$$n_{pp}(l_{ki}(\lambda_i)) = \sum_{l=0}^{p} c_l \sum_{k=0}^{l} \frac{l!}{k!(l-k)!} \lambda_i^{l-k} \quad \delta_{i,j-k}$$

où c_l sont les coefficients de n_{pp} (z).

On en déduit, d'après ce qui précède

$$\text{Exp}(A) - g_{pp}(A) = T\,(\text{Exp}(B) - g_{pp}(B))T^{-1}$$

$$= \quad T\,\text{diag}\,[\,\text{Exp}(l_{ki}(\lambda_i)) - g_{pp}(l_{ki}(\lambda_i))]\ T^{-1}$$

Mais si k_i est égal à 1 alors pour tout ε plus grand que zéro il existe p tel que :

$$|e^{\lambda i} - g_{pp}(\lambda_i)| < \varepsilon$$

et si k_i est plus grand que un alors :

$$\operatorname{Exp}(1_{ki}(\lambda_i)) - g_{pp}(1_{ki}(\lambda_i)) = \sum_{l=0}^{k_i-1} (e^{\lambda i} - g_{pp}^{(l)}(\lambda_i)) \frac{M_{ki}^{l}}{l!}$$

et pour tout ε plus grand que zéro il existe p tel que :

$$| e^{\lambda i} - g_{pp}^{(l)}(\lambda_i) | < \varepsilon /_{ki}$$

pour tout l entre zéro et ki.

Alors on obtient :

$$\operatorname{Exp}(A) - g_{pp}(A) = T \operatorname{diag}\{(\sum_{l=0}^{ki-1} (e^{\lambda i} - g_{pp}^{(l)}(\lambda_i)) \frac{M_{ki}^{l}}{l!})\} T^{-1}$$

et

$$|\operatorname{Exp}(A) - g_{pp}(A)| \leq \varepsilon\, |T| \operatorname{diag}(\sum_{l=0}^{k_i-1} \frac{M_{ki}^{l}}{l!})\} |T^{-1}|$$

Soit $\bar{M} = \operatorname{diag}\{M_{ki} ; i = 1,\ldots,r\}$

alors

$$\operatorname{diag}(\sum_{l=0}^{k_i-1} \frac{M_{ki}^{l}}{l!}) = \sum_{l=0}^{(\operatorname{Max}(ki))-1} \frac{\bar{M}^{l}}{l!}$$

Donc, on obtient que pour tout ε plus grand que zéro il existe p tel que :

$$| \operatorname{Exp}(A) - g_{pp}(A) |_{ij} \leq \varepsilon(\sum_{l=0}^{(\operatorname{Max}(ki))-1} | T | \frac{\bar{M}^{l}}{l!} | T^{-1} |)$$

Dans le cas où tous les ki sont égaux à un, on a :

$$| \operatorname{Exp}(A) - g_{pp}(A) |_{ij} \leq \varepsilon \sum_{k=}^{n} |t_{ik}| \; |p_{kj}|$$

Donc du point de vue théorique on peut toujours choisir p de façon que l'approche faite par PADE ait une précision voulue.

Si maintenant on considère p fixe et les valeurs propres de la matrice A dans une boule du plan complexe de rayon R alors,

$$| \operatorname{Exp}(A) - g_{pp}(A)|_{ij} \leq \max_{\substack{|\lambda| < R_0 \\ 0 \leq l \leq \operatorname{Max}(ki)}} |e^{\lambda} - g_{pp}^{(l)}(\lambda)| \times (\sum_{k=1}^{n} |t_{ik}| |p_{kj}| + \sum_{l=1}^{\operatorname{Max}(ki)-1} (|T| \frac{\bar{M}^{l}}{l!} |T^{-1}|)_{ij})$$

Ce qui met en évidence le rôle du produit $|T|\ |T^{-1}|$, dans l'erreur commise quand on approche l'exponentielle d'une matrice par la méthode de Padé.

On a donc remarqué du point de vue théorique deux facteurs qui déter--minent l'erreur, la taille des valeurs propres et celle des éléments du produit $|T|\ |T^{-1}|$.

<u>EXEMPLES</u>:

Pour comprendre la signification de la majoration obtenue, on présente une comparaison entre l'algorithme de WARD pour calculer les approximants de Padé de l'exponentielle d'une matrice, et un algorithme qui consiste a bloc diagonali--ser la matrice et effectuer un calcul approprié de l'exponentielle de chaque bloc.

L'algorithme pour bloc diagonaliser la matrice A a été programmé par G. Stewart et C. A. Barely [6] .

Soit A la matrice suivante:

-8.467999999999999	-2.149038000000000	-.1977190000000000	-25.53235439999999
8.860000000000000	0.9012099999999998	-.5853949999999999	28.80254799999999
-4.280000000000000	-.3049799999999999	0.5755100000000000	-14.18602400000000
5.600000000000000	0.6231000000000000	-0.2484500000000000	18.09227999999999

Les valeurs propres de A sont 11, 0.001, 0.3, -0.2.

L'exponentielle de A est:

-44305.41903261383	-1.956706681670923	4429.698535801303	-150638.8603474294
47898.55216194489	1.856961696115510	-4789.601664750217	162853.8314734810
-23949.11674255609	-.3558354735574807	2396.083615326406	-81426.56992212243
29936.66149222237	0.5650362159750517	-2993.383631114249	101784.7897608499

L'exponentielle de A calculée par l'algorithme de WARD est:

-44305.41903261262	-1.956706681634610	4429.698535801191	-150638.8603474248
47898.55216194355	1.856961696076255	-4789.601664750093	162853.8314734758
-23949.11674255541	-.3558354735378685	2396.083615326343	-81426.56992211981
29936.66149222143	0.5650362159504739	-2993.383631114160	101784.7897608464

L'exponentielle de A calculée par l'algorithme de bloc diagonalisation:

-44305.41903261228	-1.956706681574356	4429.698535801172	-150638.8603474243
47898.55216194319	1.856961696018697	-4789.601664750069	162853.8314734752
-23949.11674255529	-.3558354735027677	2396.083615326340	-81426.56992211976
29936.66149222135	.5650362158932836	-2993.383631114171	101784.7897608465

Le produit $|T|\ |T^{-1}|$ a la forme suivante:

2.480000000000002	.2252053736425162	2.297094811153551	2.273671423494276
2.294794265523956	1.000000000000007	1.550400000000005	1.622523455445677
14.68289416537648	10.33333333333429	8.440000000000765	9.311172969299379
14.83415750028138	10.43978744538325	7.516646960676022	8.440000000000765

Soit A maintenant la matrice:

-21.40000000000000	-2.950000000000000	-.5649999999999996	5.419999999999999
-28.11999999999999	-4.930000000000000	-.8469999999999986	4.131999999999996
-11.60000000000000	-18.90000000000000	7.289999999999998	-18.03999999999999
18.59999999999999	-12.10000000000000	7.909999999999999	-20.95999999999999

Les valeurs propres de A sont -30, -10, -0.1, 1.
L'exponentielle de A est:

0,6671028164770958	-1,118778879892131	0,5893117774766309	-,5220630510777775
-2,030738804724715	3,201881742284651	-1,620080031029076	1,392627418480613
-,1471517764685395	-,7724514866075276	0,7393900068283522	-,8680810733416173
1,630969097075371	-2,990060071382210	1,658126945398642	-1,522177896029755

L'exponentielle de A calculée par l'algorithme de WARD est:

0,6671028164771114	-1,118778879892143	0,5893117774766316	-,5220630510777767
-2,030738804724758	3,201891742284685	-1,620080031029080	1,392627418480612
-,1471517764685291	-,7724514866075089	0,7393900068283283	-,8680810733415906
1,630969097075409	-2,990060071382230	1,658126945398635	-1,522177896029744

L'exponentielle de A calculée par l'algorithme de bloc diagonalisation:

0,6671028164770922	-1,118778879892115	0,5893117774766205	-,5220630510777649
-2,030738804724698	3,201891742284597	-1,620080031029042	1,392627418480573
-,1431517764685432	-,7724514866075096	,7393900068283384	-,8680810733416018
1,630969097075357	-2,990060071382163	1,658126945398611	-1,522177896029718

Le produit $|T|\ |T^{-1}|$ a la forme suivante:

1,000000000000001	0,4849717274859754	0,2446117753398687	0,6811393170495828
3,338751329677937	3,100000000000007	2,219287745029504	3,707861087763928
3,630242243106278	3,803923136559210	3,100000000000010	3,508561835545472
2,377193563005154	1,836962021710347	1,752512934988367	2,119999999999986

On observe dans les deux cas que chaque fois qu'un coefficient $(|T|\ |T^{-1}|)_{ij}$ est relativement petit alors le coefficient $(\mathrm{Exp}(A))_{ij}$ est calculé avec plus de précision par la méthode qu'utilise les approximants de Padé.

BIBLIOGRAPHIE :

(1) R. VARGA, On higher order stable implicit methods for solving parabolic partial differential equations.
J. Math. Phys. ,40 (1961), pp. 220 - 231

(2) C. UNDERHILL et A. WRAGG, Convergence properties of PADE Approximants to exp (2) and their derivatives. J. INST. Maths. Applics (1973) 11 , pp. 361 - 367

(3) H. PADE, Thèse. Ann. Ec. N. Sup. (1892) 9

(4) CLEVE MOLER et CHARLES VANLOAN, Nineteen dubious ways to compute the exponential of a matrix. Siam Review, Vol 20, n° 4, October 1978, pp. 801-836

(5) R.C. WARD, Numerical Computation of the matrix exponential with accuracy estimate. Sian. J. Numer. Anal. Vol 14, n° 4, Sept 1977, pp. 600-610

(6) C.A. BAVELY, G.W. STEWART, An algorithme for computing reducing subspaces by block diagonalisation. T.R. 489, October 1976.

A RELIABLE METHOD FOR RATIONAL INTERPOLATION

by H. WERNER, Institut für Numerische und Instrumentelle Mathematik
Universität Münster
Roxeler Strasse 64
D - 4400 MUNSTER (GERMANY)

1. Introduction

In this paper I will describe an algorithm for rational interpolation that avoids the usual shortcoming which many other algorithms have, namely the accidental stop due to a zero arising in a denominator of one of the many quotients formed to set up a rational interpolant. Of course the method cannot help circumvent the case of non existence of the interpolant due to unattainable points. (Werner-Schaback [6], p. 59-77) It will, however, even help to detect those points while in other algorithms one will have to check separately for points where the interpolants will lead to the undetermined form $\frac{0}{0}$.

Using the terminology introduced by Graves-Morris [3] in his talk at this conference, the algorithm can be called "reliable". The accuracy in reproducing the input data is in addition remarkably high. The algorithm is easily adapted to coalescence of points, i.e. the case that values of the function and its derivatives are given.

We use the notation of Werner-Schaback [6] and will not repeat the detailed motivation of the algorithm found in Werner [5] together with an account for the literature and we only mention the contributions of Meinguet [4], Wuytack [7], Claessens [2] and point out the relationship to the ideas of Bultheel [1] developed at this conference. Here we choose the more formal mathematical description of the algorithm.

2. Statement of Problem

Let $\mathcal{R}(\ell,m)$ denote the rational functions with degree of numerator and denominator not exceeding ℓ or m respectively. Given $\ell,m \in \mathbb{N}$, $n=\ell+m$ and $n+1$ interpolation points (x_j,f_j), $j \in J := \{0,\dots,n\}$ with $x_k \neq x_j$ for $k \neq j$.

Find a rational function $R(x) \in \mathcal{R}(\ell,m)$ such that

$$(2.1) \qquad R(x_j) = f_j \quad \text{for every} \quad j \in J \, .$$

If we write

$$(2.2) \qquad R(x) = \frac{a_0+a_1x+\dots+a_\ell x^\ell}{b_0+b_1x+\dots+b_mx^m} = \frac{p(x)}{q(x)}$$

then (2.1) may be transformed into a system of n+1 homogeneous linear equations for n+2 unknown polynomial coefficients

$$p(x_j) \quad - \quad q(x_j) \quad \cdot \quad f_j = 0 \quad \text{i.e.}$$

$$(2.3) \qquad a_0+a_1x_j+\dots+a_\ell x_j^\ell - (b_0+b_1x_j+\dots+b_mx_j^m)f_j = 0 \text{ for every } j \in J$$

(linearised interpolation problem) which has of course non trivial solutions.

One can prove (compare Werner-Schaback [6])

Lemma 2.1: There is a unique pair $(p^*(x),q^*(x))$ such that the degree of $q^*(x)$ is minimal, the highest coefficient of $q^*(x)$ is equal to 1 and

$$p^*(x_j) = f_j \cdot q^*(x_j) \quad \forall\, j \in J .$$

Furthermore every solution of (2.3) is of the form

$$(s(x)\cdot p^*(x) \;,\; s(x)\cdot q^*(x)),$$

$s(x)$ a polynomial of degree ∂s less or equal $d := \min(\ell-\partial p^*, m-\partial q^*)$.

Here the symbol ∂g is to denote the degree of a polynomial g. We also use the notation

$$w(x;i,k) := \begin{cases} (x-x_{i+1})\cdot\ldots\cdot(x-x_k) & \text{if } i<k \\ 1 & \text{if } i=k \end{cases} .$$

It is no restriction to complement the <u>assumption</u>

i) $x_k \neq x_j$ for every pair $(k,j) \in J\times J$ with $k \neq j$

by the <u>technical assumptions</u>

ii) $\ell \geq m$

and

iii) $f_j \neq 0$ for every $j \in J$.

We start out with iii). By renumbering we may achieve

$$f_0,\ldots,f_\nu = 0 \;,\; f_{\nu+1} \neq 0,\ldots,f_n \neq 0 .$$

If $\nu < \ell$, let

$$p^*(x) = w(x;-1,\nu)\cdot\tilde{p}(x) \;, \qquad \partial\tilde{p} \leq \ell-\nu-1 .$$

Then determine an interpolant $\tilde{R}(x)$ (if possible) of the class $\mathcal{R}(\ell-\nu-1,m)$ such that

$$\tilde{R}(x_j) = \frac{\tilde{p}(j)}{q(x_j)} = \tilde{f}_j = \frac{f_j}{w(x_j;-1,\nu)} \quad \text{for} \quad j=\nu+1,\ldots,n \ .$$

If $\nu \geq \ell$, let $p^*(x) \equiv 0$

$$q^*(x) = w(x;\nu,n) \quad .$$

It is clear that $(p^*(x),q^*(x))$ solves the linearised interpolation problem and that in case $\nu \geq \ell$

$$(x_j,f_j) \quad \text{for} \quad j > \nu \quad \text{are unattainable points}$$

for the function $R(x) \equiv 0$.

If condition iii) is met the condition ii) can be achieved in case $\ell < m$ by interchanging the role of numerator and denominator along with replacing the given data by $(x_j,1/f_j)$.

Above we found a special case in which the solution of the linearised problem could be found by inspection. A more general setting is found in the following lemma which will be used to terminate the algorithm as described in section 3.

Lemma 2.2: Let $r = \ell-m$, $\nu \geq \ell$.
Suppose there are $\nu+1$ points (without loss of generality $x_0,\ldots,x_\nu$) such that the divided differences satisfy

$$\Delta^{r+1}(x_0,\ldots,x_r,x_k)f = 0 \quad \text{for} \quad k=r+1,\ldots,\nu$$

while

$$\Delta^{r+1}(x_0,\ldots,x_r,x_k)f \neq 0 \quad \text{for} \quad k=\nu+1,\ldots,n \ .$$

Let $p_1(x)$ be an r-th degree polynomial to interpolate $(x_0,f_0),\ldots,(x_r,f_r)$.

Then

$$q^*(x) = \begin{cases} (x-x_{\nu+1})\ldots(x-x_n) & \text{if} \quad \nu < n \\ 1 & \nu = n \end{cases} \quad \text{and}$$

$$p^*(x) = q^*(x) \cdot p_1(x)$$

solve the linearised interpolation problem.

This is almost obvious: The highest coefficient of $q^*(x)$ is 1.

The degree

$$\partial q^* = n-\nu \leq m \ ,$$

$$\partial p^* = \partial q^* + \partial p_1 \leq m + r = \ell \ .$$

Thus the degrees do not exceed the prescribed values.

Furthermore

$$p^*(x) = q^*(x) = 0 \quad \text{for} \quad x = x_{\nu+1},\ldots,x_n \ ,$$

$$\frac{p^*(x)}{q^*(x)} = p_1(x) \quad \text{else, therefore} \quad p_1(x_j) = f_j \ \text{for} \ j \leq \nu .$$

This proves that (2.3) is solved.

Now we can formulate the new algorithm.

3. Generalized Thiele Continued Fraction.

At first we define the generalized form of the continued fraction that will be used and investigate its properties.

Let $\ell \geq m$ and

$$R_o(x) = p_o(x) + \cfrac{w_o(x)}{p_1(x) + \cfrac{w_1(x)}{p_2(x) + \cfrac{w_2(x)}{p_3(x) \; \cdots \; \cfrac{}{p_{last}(x)}}}} \tag{3.1}$$

with a monotonically increasing sequence $k_{-1} = -1,\ k_o, k_1, \ldots, k_{last}$, the $p_j(x)$ polynomials of degree less than $k_j - k_{j-1}$ and
$w_j(x) = w(x;k_{j-1},k_j)$.

The numbers $k_1,\ldots,k_{last} = n$ should be determined together with $p_o(x),\ldots$ to have $R_o(x) \in \mathcal{R}(\ell,m)$ interpolate the given data.

With

$$R_{last}(x) = p_{last}(x) \tag{3.2}$$

we may define

$$R_j(x) = p_j(x) + \frac{w_j(x)}{R_{j+1}(x)} \quad . \tag{3.3}$$

These rational functions can be transcribed in homogeneous form as in the case of ordinary continued fractions (Werner-Schaback [6], p. 68 ff.)

$$\vec{R}_j(x) = \begin{pmatrix} z_j(x) \\ n_j(x) \end{pmatrix} \quad \text{from} \quad R_j(x) = \frac{z_j(x)}{n_j(x)} \quad . \tag{3.4}$$

Then (3.3) can be transformed into the equations

$$\vec{R}_j(x) = T_j(x) \cdot \vec{R}_{j+1}(x), \text{ with}$$
$$T_j(x) := \begin{pmatrix} p_j(x) & w_j(x) \\ 1 & 0 \end{pmatrix} \quad . \tag{3.5}$$

Obviously, if $w_j(x)$ is different from zero,

$$(3.6) \qquad T_j^{-1}(x) = \frac{1}{w_j(x)} \begin{pmatrix} 0 & w_j(x) \\ 1 & -p_j(x) \end{pmatrix} .$$

Now we try to choose $p_0(x),\ldots$ to have $R_0(x)$ satisfy the interpolation condition.

Let $\partial z_0 = \ell_0 = \ell$, $\partial n_0 = m_0 = m$.

A) Let $r_0 \geq 0$ be a fixed integer with $0 \leq \ell_0 - m_0 - r_0 \leq 1$. Interpolate (x_j, f_j) for $j=0,\ldots,r_0$ to find $p_0(x)$. Renumbering the other data points, if necessary, we may choose $k_0 \geq r_0$ such that

$$(3.7) \qquad \begin{aligned} p_0(x_j) &= f_j \quad \text{for} \quad j=k_{-1}+1,\ldots,k_0, \text{ (here } k_{-1}+1 = 0) \\ &\neq f_j \quad \text{for} \quad j=k_0+1,\ldots,n \quad (\text{if } k_0 < n). \end{aligned}$$

From

$$(3.8) \qquad \vec{R}_1(x) = T_0^{-1}(x) \cdot \vec{R}_0(x)$$

interpolation conditions for $R_1(x)$, the quotient of the components of $\vec{R}_1(x)$ at x_j, $j=k_0+1,\ldots,n$ are obtained. The matrix $T_0^{-1}(x)$ is not singular for the values $x = x_j$, because $w(x_j;k_{-1},k_0) = (x_j-x_0)\ldots(x_j-x_{k_0}) \neq 0$. Again we request that the degree of the denominator of R_1 should not exceed that of the numerator:

$$\partial n_1 \leq \partial z_1$$

Due to (3.5) the degrees satisfy

(3.9) $$\partial z_0 = \max(\partial z_1 + r_0, \partial w_0 + \partial n_1) \geq \partial z_1 + r_0$$

$$\partial n_0 = \partial z_1 , \quad \text{hence} \quad r_0 \leq \partial z_0 - \partial n_0 .$$

By the choice of k_0 we have furthermore

$$\partial w_0 > r_0$$

and

(3.10) $$\partial n_1 \leq \partial z_0 - \partial w_0 \leq \partial z_0 - r_0 - 1 .$$

We request $\partial n_1 \leq \partial z_1 = \partial n_0$ showing that one may ask for

(3.11) $$r_0 \geq \partial z_0 - \partial n_0 - 1$$

in order to ensure that the degree of the numerator is not smaller than the degree of the denominator.

B) If $k_0 \geq \ell_0$, only $n-k_0 = \ell_0 + m_0 - k_0 \leq m_0$ data are left for interpolation and lemma 2.2 provides the solution of the linearised interpolation problem:

$$\vec{R}_1(x) = \begin{pmatrix} z_1(x) \\ \\ n_1(x) \end{pmatrix} , \quad z_1(x) = n_1(x) = w(x;k_0,n) .$$

If $k_0 < n$ there are unattainable points namely at $x_{k_0+1}, \ldots, x_n$. The determination of the continued fraction is finished.

If $k_0 < \ell_0$ we have to apply the above reduction scheme described in A) to $R_1(x)$. Its values are given at $x_{k_0+1}, \ldots, x_n$ by

$$f^1(x_j) = \frac{w(x_j;k_{-1},k_0)}{f(x_j) - p_0(x_j)} ,$$

as is seen from (3.8). The degrees should be replaced appropriately: $\ell_o \leftarrow \partial z_1$, $m_o \leftarrow \partial n_1$, r_1 selected as specified before. (It is obvious, how we should index the quantities that come up.)

The iteration defined by this procedure will come to a stop since the sum of the degree of numerator and denominator of $R_j(x)$ is strictly declining as we show next. (We consider the case j=0 again.)

Since $r_o \geq 0$, we have $\partial w(x;k_{-1},k_o) \geq r_o+1 \geq 1$. From (3.9) and (3.10)

$$\partial z_1 + \partial n_1 \leq \partial n_o + \partial z_o - \partial w_o < \partial n_o + \partial z_o ,$$

and this shows our assertion.

Hence there will be a j_o such that

$$k_{j_o} \geq \ell_{j_o} = \partial z_{j_o}$$

and by B) the iteration ends.

With

$$\text{(3.12)} \qquad \vec{R}_{j_o+1}(x) = \begin{pmatrix} w(x;k_{j_o},n) \\ w(x;k_{j_o},n) \end{pmatrix}$$

and the calculated $p_j(x)$, $w_j(x)$ we form $T_j(x)$ and

$$\text{(3.13)} \qquad \vec{R}_o(x) = T_o(x)\cdots T_{j_o}(x)\cdot \vec{R}_{j_o+1}(x) \quad .$$

The components of $\vec{R}_o$ provide a solution of the linearised interpolation problem, as may be deduced from the above construction.

The numbers $r_o, r_1, \ldots$ show that there is a certain option which we will control by an extra parameter in the implementation.

Whether or not the rational interpolation problem has a genuine solution depends on the zeros of $n_o(x)$ and $z_o(x)$. Let σ satisfy $k_{j-1} < \sigma \leq k_j$.

By construction

$$T_o(x_\sigma),\ldots,T_{j-1}(x_\sigma) \text{ are not singular.}$$

Hence

$$\vec{R}_o(x_\sigma) = \begin{pmatrix} 0 \\ 0 \end{pmatrix} \text{ is equivalent to } \vec{R}_j(x_\sigma) = \begin{pmatrix} 0 \\ 0 \end{pmatrix} .$$

Now we have

$$T_j(x_\sigma) = \begin{pmatrix} p_j(x_\sigma) & 0 \\ 1 & 0 \end{pmatrix}$$

due to the definition of $w_j(x)$. Therefore by (3.5) we find

$$\vec{R}_j(x_\sigma) = \begin{pmatrix} p_j(x_\sigma) \\ 1 \end{pmatrix} \cdot z_{j+1}(x_\sigma) . \tag{3.14}$$

We can state this result as

<u>Lemma 3.1:</u> Let σ satisfy $k_{j-1} < \sigma \leq k_j$. Then x_σ is the coordinate of an unattainable point iff

$$z_{j+1}(x_\sigma) = 0 . \tag{3.15}$$

This concludes the mathematical description of the algorithm.

If $\ell_o = m_o$ or $m_o + 1$ and if the classical Thiele algorithm goes through without any exceptional situation then it is reproduced by the above algorithm if $r_o = \ldots = r_{j_o} = 0$ is used which amounts to taking $p_o, p_1, \ldots$ as constant, namely the inverse divided differences.

4. The Evaluation of the Generalized Continued Fraction.

The evaluation of the partial fraction $R_o(x)$, whence the coefficients of the $p_j(x)$ and the numbers k_j have been spezified, can be done by a Horner-like scheme.

One may write

$$R_o(x) = (a_o + (x-x_o) \cdot (a_1 + \ldots + (x-x_{r_o-1}) (a_{r_o} + (x-x_{r_o}) \cdot (0 + \ldots$$

(4.1)

$$\ldots + (x-x_{k_o-1}) \cdot (0 + (x-x_{k_o}) / R_1(x)) \ldots)$$

and obviously r_o multiplications and 1 division are needed to procede from the value of $R_1(x)$ to $R_o(x)$, evaluating the inner most paranthesis first and procede up to the outer one.

Together with the coefficients a_j one may therefore store indicators $L1_j$ to show whether a multiplication ($L1_j = +1$) or a division ($L1_j = -1$) should be used in connecting the current value of the calculation with the next term $(x - x_j)$.

If the coefficients of the p_j are stored in one sequence and are supplemented by zeros as indicated by (4.1) an implementation could look like the following piece of program:

$$R = A(n+1)$$

```
For I = N to 1 Step -1
   If L1(I) > 0  then   R = A(I) + (x - x(I))*R
                 else   R = A(I) + (x - x(I))/R ;
```

A more sophisticated implementation will take precautions in case R vanishes during the calculation and check that no singularity arises, no unattainable points are met and that N is reduced to the degree actually needed in the specific case.

5. Implementation of the Algorithm.

In order to perform the calculations we assume that the degrees and data points are given

$$L\,,\,M \quad \text{and} \quad X(J)\,,\,F(J) \text{ for } J=1,\ldots,N+1 \quad \text{with} \quad N=L+M\ .$$

Furthermore let the technical assumptions made in section 2, i.e.

$$L \geq M \quad \text{and} \quad F(J) \neq 0 \quad \text{for every} \quad J$$

be satisfied.

Specify also a parameter $N8 \in \{0,1\}$ to define

$$r_j = L_j - M_j - N8 \quad .$$

Here $N8 = 1$ corresponds to the classical case of continued fractions. $N8 = 0$ splits R_j into a polynomial of as high a degree as possible and a rational function with denominator degree higher than numerator degree. The cases could also be mixed.

The algorithm is written in BASIC but without Input/Output statements, hence should be appropriately be supplemented. A complete program together with print out of all interesting quantities and also a plot program for the rational interpolant and the error curve in case the input data are derived from a function f(x) that is given in an x-interval and can be called as a subroutine is available from the author.

The following lines of code assume that the data F(I) are already stored in A(I), where the coefficients of R(x) are built up.

The integer J counts the data points that have already been processed, it corresponds to the numbers k_j of section 3. If it is compared to N after the program is terminated, one can see whether unattainable points have been detected. The structure of the program can be summarized in the following sentences.

1: After initialisation of J difference quotients (to furnish the coefficients of the polynomials p_j) are calculated, if the degree L_j exceeds M_j.

2: In this section the inverse difference quotients are formed. If a point is detected for which this inversion is impossible it will be interpolated automatically by the function constructed by the data handled before. This point is put aside in 3. These points are counted by the integer K1.

4: After the calculation of the inverse difference quotients, the points are rearranged if K1 > O. The coefficients are processed such that the data still unmatched are shifted to the end of the data points, while those put aside previously are stored in front.

5: At this point L1(J) is put to -1 referring to the inverse difference quotient A(J) calculated before.

From the current degrees (L_c, M_c) the degrees of the next iteration are calculated: L_{new}, M_{new}.

Only if $M_{new} \geq 0$ should we go back to 1.

If the numbers of points to be processed is less or equal $M_c = \partial n_c$ by Lemma 2.2 the iteration may be stopped. If there are indeed points left, they are unattainable: $J \leq N$. Otherwise the algorithm has terminated regularly.

If at this last step K1 was positive, it accounts for the K1 last coefficients being zero, therefore one may start the evaluation of the generalized continued fraction with

$$A(N+1-K1) \quad \text{instead of} \quad A(N+1)$$

in section 4.

It is left to the reader to apply Lemma 3.1 to check for unattainable points.

We may summarize: After completion of the program the x-coordinates (possibly reordered), the coefficients of the polynomials p_j and the indicators are found in X(J), A(J), and L1(J). We remark that from the indicators one can read off the values k_j.

```
        J      = O
1:   IF  L ≤ M+N8                     THEN  2
     FOR K     = 1  TO  L-M-N8  STEP  1
         J     = J + 1
         L1(J) = 1
     FOR I     = J+1  TO  N+1  STEP  1
         A(I)  = (A(I) - A(J))/(X(I) - X(J))
     NEXT I
     NEXT K
     IF  J ≥ N                        THEN  6

2:       J     = J + 1
         K1    = O
     FOR I     = N+1  TO  J+1  STEP  -1
         F     = A(I) - A(J)
         D     = X(I) - X(J)
     IF  ABS(F) ≤ ABS(D)*10^-10       THEN  3
         K     = I + K1
         A(K)  = D / F
         X(K)  = X(I)
                                      GOTO  4
3:       K1    = K1 + 1
         Z(K1) = X(I)

4:   NEXT I
     IF  K1 = O                       THEN  5
         K2    = J + K1
     FOR K     = 1  TO  K1  STEP  1
         L1(J) = 1
         J     = J + 1
         X(J)  = Z(K)
         A(J)  = O
     FOR I     = K2  TO  N+1  STEP  1
         A(I)  = A(I)*(X(I) - X(J))
     NEXT I
     NEXT K

5:       L1(J) = -1
         L     = M
         M     = N - L - J
     IF  M ≥ O                        THEN  1
6:   STOP
```

6. Examples.

The following examples were run on the Tektronix 4051 calculator. The manufacturer states that it uses 14 digits for storage and calculation.

1) Interpolation of e^x in 6 equidistant points X(1) = 0, X(2) = 0.2,... by $R(x) \in \mathcal{R}(3,2)$. For N8 = 1 the results are

I	X(I)	A(I)	L1(I)
1	0	1.	-1
2	0.2	0.903 331 113 226	-1
3	0.4	-2.221 402 758 16	-1
4	0.6	-2.382 501 038 43	-1
5	0.8	2.713 227 456 03	-1
6	1.0	3.295 732 786 34	0

hence $R(x)=1+\underline{(x-0)}/0.903\ldots+\underline{(x-0.2)}/-2.221\ldots \qquad + \underline{(x-0.8)}/3.295\ldots$

For the same data but N8 = 0 one obtains

I	X(I)	A(I)	L1(I)
1	0	1.	1
2	0.2	1.107 013 790 8	-1
3	0.4	1.632 014 200 24	1
4	0.6	-0.560 830 905 687	-1
5	0.8	18.289 820 250 1	1
6	1.0	-0.809 036 601 456	0 ,

hence

$$R(x) = 1 + (x-0)\cdot\left(1.107\ldots+\cfrac{x-0.2}{1.632\ldots+(x-0.4)\left(-0.560+\cfrac{x-0.6}{18.2\ldots+(x-0.8)\cdot(-0.809\ldots)}\right)}\right)$$

2) Interpolation of $|x|$ by $R \in \mathcal{R}(3,2)$ on ± 0.2, ± 0.6, ± 1. For N8 = 1 and input of X(I) in ascending order the output is

I	X(I)	A(I)	L1(I)
1	-1	1	1
2	1	0	-1
3	-0.6	1.6	1
4	0.6	0	-1
5	-0.2	0.8	1
6	0.2	0	0

The X(I) have been rearranged.

Since the last coefficient is zero the effective degree N is reduced to 4. This reflects the fact that $|x|$ is a symmetric function and the selected data are symmetric.

$$R(x) = 1 + (x+1)\left(0 + \frac{(x-1)}{\cdots}\right) = 1 + \cfrac{x^2 - 1}{1.6 + \cfrac{x^2 - 0.36}{0.8}} \quad .$$

It is apparent that the number of point operations is further reduced to 3, even less than $N = 4$.

For N8 = 0:

I	X(I)	A(I)	L1(I)
1	-1.0	1	1
2	-0.6	-1	1
3	-0.2	0	-1
4	0.2	0.96	1
5	0.6	0.8	1
6	1.0	1.0	0

There is one zero. It is due to the fact that the first 3 data points lie on a straight line. Here N remains 5.

$$R(x) = 1 + (x+1)\left(-1 + (x+0.6)\,(0 + \frac{x+0.2}{0.96+(x-0.2)\,(0.8+(x-0.6))})\right)$$

$$= -x + \frac{(x+1)\,(x+0.6)\,(x+0.2)}{0.96+(x-0.2)\,(0.2+x)}$$

This form shows that -x indeed gives the correct interpolation for the 3 data points with $x < 0$.

3) We construct an example with unattainable points by interpolation of

$$f(x) = \begin{cases} \frac{x^2+1}{1+x/4} + 1 & \text{for} \quad x = 0.6 \quad \text{and} \quad 1.4 \, , \\ \frac{x^2+1}{1+x/4} & \text{else} \end{cases}$$

with

$$X(I) = 0.2 \cdot (I-1) \; , \; I = 1,\ldots,11 \; , \quad f(I) := f(X(I)) \quad .$$

We ask for an interpolant $R \in \mathcal{R}(6,4)$.

N8 = 1:

I	X(I)	A(I)	L1(I)
1	0	1	1
2	0.2	-0.047 619 047 619	-1
3	0.4	1.087 058 823 53	-1
4	0.6	-0.224 995 927 676	-1
5	0.8	0.046 809 740 909 6	-1
6	1.0	4.272 614 975 29	1
7	2.0	0	1
8	1.8	0	1
9	1.6	0	1
10	1.2	0	-1
11	1.4	$6.385\ 255\ 363 \cdot 10^{-4}$	0

From these numbers it is apparent that again the points 1.2, 1.6, 1.8, 2.0 are detected to satisfy the rational function interpolating the first 6 points.

From Lemma 2.2 it is immediately discovered by the computer that the 11th point (x = 1.4) is unattainable.

To discover that there lies another unattainable point at x = 0.6 one has to use Lemma 3.1.

The effective N is given as 5 by the computer.

The same interpolation problem will lead to the following figures for N8 = 0:

I	X(I)	A(I)	L1(I)
1	0	1	1
2	0.2	-0.047 619 047 619 1	1
3	0.4	0.919 913 419 914	-1
4	0.6	0.048 465 222 348 9	1
5	0.8	-26.331 737 876 4	-1
6	1	0.015 844 923 168 2	1
7	1.2	0.039 612 307 920 6	1
8	2	0	1
9	1.8	0	1
10	1.6	0	-1
11	1.4	0.344 359 355 944	0

In this case the effective degree is equal to N = 6. Again the 11th point is immediately seen to be unattainable, while x = 0.6 is only found by a closer look.

We use the notation of section 3. There the points were indexed $0,\ldots,n$. With this notation we find $k_{-1} = -1$, $k_0 = 2$, $k_1 = 4$, $k_2 = 9$. Thus $x_3 = 0.6$ and we have to test $z_2(0.6)$. Since

$$z_2(x) = (x-1.4)(0.015\,844\,923\,1682 + (x-1)\cdot 0.039\,612\,307\,9206)$$

$$n_2(x) = (x-1.4)$$

up to round off errors $z_2(0.6) = 0$. By Lemma 3.1 we conclude that 0.6 is an unattainable point.

It is clear that theoretically we obtain the same interpolation for N8 = 0 and N8 = 1. There is no apparent difference in accuracy as far as the worked examples show so far. From the foregoing examples it seems, however, preferable to use N8 = 1, since the effective N is smaller.

LITERATURE

[1] BULTHEEL, A.: Recursive Algorithms for the Padé Table: two approaches (this volume)

[2] CLAESSENS, G.: The rational Hermite interpolation problem and some related recurrence formulas. Comp. and Maths. with Appls. 2 (1976), 117-123

[3] GRAVES-MORRIS, P.R.: The Numerical Calculation of Padé Approximants (this volume)

[4] MEINGUET, J.: On the solubility of the Cauchy interpolation problem. In Approximation theory, ed. by A. Talbot, London and New York, Academic Press 1970, p. 137-163

[5] WERNER, H.: Ein Algorithmus zur rationalen Interpolation in: Numerische Methoden der Approximationstheorie, Band 5, Birkhäuser-Verlag 1979

[6] WERNER, H. - SCHABACK, R.: Praktische Mathematik II, 2. Aufl. Springer Verlag, Berlin-Heidelberg-New York 1979

[7] WUYTACK, L.: On the osculatory rational interpolation problem. Math. Comp. 29 (1975), 837-843

RATIONAL PREDICTOR-CORRECTOR METHODS FOR NONLINEAR VOLTERRA INTEGRAL EQUATIONS OF THE SECOND KIND.

T.H. CLARYSSE

ABSTRACT : The Volterra integral equation of the second kind is approximated by rational predictor-corrector formulas, derived through osculatory rational interpolation. A fourth order method is treated explicitly. Convergence and A-stability are considered. For some nonlinear and singular equations, numerical results are included, and compared with results from a analougous linear method.

1. INTRODUCTION.

Consider the nonlinear Volterra integral equation of the second kind :

$$f(x) = g(x) + \int_{x_0}^{x} K(x,y,f(y))dy \ ; \quad x_0 \leq x \leq a \ . \tag{1}$$

It will be assumed that g(x) is continuous and bounded for all $x \in [x_0,a]$, that $K(x,y,z)$ is uniformly continuous in x and y for all $(x,y,z) \in R$, where $R = \{(x,y,z) : x_0 \leq y \leq x \leq a, -\infty < z < +\infty\}$ and that $K(x,y,z)$ satisfies a uniform Lipshitz condition with respect to z, namely,

$$|K(x,y,z_1) - K(x,y,z_2)| \leq L.|z_1 - z_2| \ ; \quad x_0 \leq x \leq y \leq a$$

for some constant L.

Then (1) has a unique solution $f(x) \in C[x_0,a]$ (See [1], p. 43 and [9]).

Denoting a partition of $J = [x_0,a]$ by

$$J(h) = \{x_k : x_k = x_0 + kh \ ; \ k = 0,1,\dots,N \ ; \ h = \frac{a-x_0}{N}\}$$

it follows from (1) that :

$$f(x_k) = g(x_k) + \sum_{j=1}^{k} \int_{x_{j-1}}^{x_j} K(x_k,y,f(y))dy \ ; \quad x_k \in J(h) \tag{2}$$

For approximating

$$I_j^k = \int_{x_{j-1}}^{x_j} K(x_k,y,f(y))dy \ ; \quad x_k \in J(h); \quad 1 \leq j \leq k \ .$$

several techniques can be used. The use of linear quadrature rules for instance is well known (See [2], p. 825). Linear predictor and predictor-corrector techniques also have been considered (See [3], [4]). Hoping to improve the results of such linear methods for nonlinear equations, we will approximate I_j^k by means of rational techniques.

2. APPROXIMATING METHODS USING OSCULATORY RATIONAL INTERPOLATION.

We begin with remarking that the calculation of

$$I_k = \sum_{j=1}^{k} I_j^k = \int_{x_0}^{x_k} K(x_k,y,f(y))dy \tag{3}$$

is equivalent with finding the value in x_k of the solution of the initial value problem :

$$\begin{cases} (q_k)'(z) = K(x_k,z,f(z)) \\ q_k(x_0) = 0 \end{cases} \tag{4}$$

For solving (4), we then can use the osculatory rational interpolation formulas as proposed by Lambert [5]; Thacher [6]; Luke, Fair, Wimp [7]. Among other rational techniques, we have first considered these formulas because they need no higher derivatives of K(x,y,z), few starting values and are based on a predictor-corrector approach.

In the following we will restrict ourselves to fourth order formulas. These were obtained by imposing osculatory interpolation conditions on a rational function of degree 2 in numerator and denominator.

We will denote the approximations to $f(x_j)$, $q_k(x_j)$ and $K(x_k,x_j,f(x_j))$ respectively by f_j, q_j^k and $K_j^k = K(x_k,x_j,f_j)$.
The predictor (See [7],p.6) then can be written with respect to problem (4) as:

$$q_j^k = q_{j-1}^k + \frac{(q_{j-1}^k - q_{j-3}^k)((\Delta_{j-1}^k)^2 - 4\bar{K}_{j-2}^{\,k}\,\bar{K}_{j-1}^{\,k}) + 6\bar{K}_{j-1}^{\,k}\,\Delta_{j-1}^k\Delta_{j-2}^k}{15\Delta_{j-1}^k\,\Delta_{j-2}^k - 6\bar{K}_{j-2}^k(q_{j-1}^k - q_{j-3}^k) - ((\Delta_{j-1}^k)^2 - 4\bar{K}_{j-2}^{\,k}\,\bar{K}_{j-1}^{\,k}) - 6\bar{K}_{j-1}^{\,k}\,\Delta_{j-2}^k} \qquad (5a)$$

where $\Delta_i^k = q_i^k - q_{i-1}^k$; $\bar{K}_i^{\,k} = h.K_i^k$; $x_k \in J(h)$, $3 \leqslant j \leqslant k$

The corresponding corrector becomes :

$$q_j^k = q_{j-1}^k + \frac{\Delta_{j-1}^k(\bar{K}_{j-1}^{\,k} - 2\Delta_{j-1}^k) + 2t_o\sqrt{((\Delta_{j-1}^k)^2 - \bar{K}_{j-2}^{\,k}\,\bar{K}_{j-1}^{\,k})((\Delta_{j-1}^k)^2 + \bar{K}_j^{\,k}\,N_{j-1}^k)}}{N_{j-1}^k} \qquad (5b)$$

where

$$\begin{cases} N_{j-1}^k = 4\Delta_{j-1}^k - \bar{K}_{j-1}^{\,k} - 4\bar{K}_{j-2}^{\,k} \ ; \quad x_k \in J(h),\ 2 \leqslant j \leqslant k \\ t_o = \mathrm{sgn}((\Delta_{j-1}^k)^2 - \bar{K}_{j-2}^{\,k}\,\bar{K}_{j-1}^{\,k}) \end{cases}$$

The choice of the factor t_o is ours, and was obtained through series expansion of (5b) and the requirement that it must be a fourth order method.
If in (5b) the denominator $N_{j-1}^k = 0$, then it has to be replaced by the following corrector, of the same order (See [6], p.366)

$$q_j^k = q_{j-1}^k + \frac{4\bar{K}_j^{\,k}(4K_{j-2}^k - K_{j-1}^k)^2 + \bar{K}_{j-1}^{\,k}(4K_{j-2}^k + K_{j-1}^k)^2}{4(16(K_{j-2}^k)^2 - (K_{j-1}^k)^2)} \qquad (5c)$$

In general one calculates at each step, using (5a), a first approximation and then uses (5b) or (5c) to correct the approximation until convergence.
For our special problem (4) this can be simplified.
At the k-th step ($k \geq 2$) of calculating (2), we know $f_i (i = 0,\dots,k-1)$. Therefore K_j^k $(i=0,\dots,k-1)$ are known. So we can applie (5b) or (5c) directly for $j = 2,\dots,$ $k-1$. Since only for $j = k$ we need to calculate K_k^k, for which we first must obtain a predicted value for f_k with (5a). For $j=1$ ($k \geq 1$) and $j=2$ ($k = 2$) starting values must be supplied for instance by using linear Runge-Kutta of order 4.

Since

$$q_k(x_j) - q_k(x_{j-1}) = I_j^k$$

the approximate values of I_j^k will be defined by

$$q_j^k - q_{j-1}^k = \tilde{I}_j^k \quad .$$

We will further denote P_j^k for the predicted value of I_j^k, C_j^k for the corrected value and R_j^k for approximations obtained through linear Runge-Kutta methods of order 4 (see [2], [3]).
If C_j^k can be calculated (see §6), we set :

$$\tilde{I}_j^k = C_j^k \qquad \text{for } 2 \leq j \leq k$$

Otherwise : $\qquad \tilde{I}_j^k = P_j^k \qquad \text{for } 3 \leq j \leq k$

and $\qquad \tilde{I}_j^k = R_j^k \qquad \text{for } j = 1,2$

Formulas (5a,b) then can be written as :

$$P_j^k = \frac{(\tilde{I}_{j-2}^k + \tilde{I}_{j-1}^k)((\tilde{I}_{j-1}^k)^2 - 4\bar{K}_{j-2}^{\,k}\,\bar{K}_{j-1}^{\,k}) + 6\bar{K}_{j-1}^{\,k}\,\tilde{I}_{j-2}^k\,\tilde{I}_{j-1}^k}{15\tilde{I}_{j-2}^k\,\tilde{I}_{j-1}^k - 6\bar{K}_{j-2}^{\,k}(\tilde{I}_{j-2}^k + \tilde{I}_{j-1}^k) - ((\tilde{I}_{j-1}^k)^2 - 4.\bar{K}_{j-2}^{\,k}\,\bar{K}_{j-1}^{\,k}) - 6\bar{K}_{j-1}^{\,k}\,\tilde{I}_{j-2}^k}, \quad 3 \leq j \leq k \quad (6a)$$

$$C_j^k = \frac{\tilde{I}_{j-1}^k(\bar{K}_{j-1}^{\;k} - 2\tilde{I}_{j-1}^k) + 2t_1 \sqrt{((\tilde{I}_{j-1}^k)^2 - \bar{K}_{j-2}^{\;k}\,\bar{K}_{j-1}^{\;k})((\tilde{I}_{j-1}^k)^2 + \bar{K}_j^{\;k}\,M_{j-1}^k)}}{M_{j-1}^k}, \quad 2 \leq j \leq k \quad (6b)$$

where
$$\begin{cases} M_{j-1}^k = 4\tilde{I}_{j-1}^k - \bar{K}_{j-1}^{\;k} - 4\bar{K}_{j-2}^{\;k} \\ t_1 = \mathrm{sgn}((\tilde{I}_{j-1}^k)^2 - \bar{K}_{j-1}^{\;k}\,\bar{K}_{j-2}^{\;k}) \end{cases}$$

If $M_{j-1}^k = 0$, we use instead of (6b), the following :

$$C_j^k = \frac{4\bar{K}_j^{\;k}(4K_{j-2}^k - K_{j-1}^k)^2 + \bar{K}_{j-1}^{\;k}(4K_{j-2}^k + K_{j-1}^k)^2}{4(16(K_{j-2}^k)^2 - (K_{j-1}^k)^2)}, \quad 2 \leq j \leq k, \quad (6c)$$

With these notations, the proposed algorithm for solving (2) now can be summarized as follows :

<u>step o</u> : $f_0 = f(x_0) = g(x_o)$

<u>step 1</u> : $f_1 = g(x_1) + R_1^1$

<u>step 2</u> :
$$f_2^{(0)} = g(x_2) + R_1^2 + R_2^2 \quad (7a)$$
$$f_2^{(i)} = g(x_2) + R_1^2 + C_2^2\,(f_2^{(i-1)}), \text{ for } i \geq 1 \quad (7b)$$

<u>step k</u> :
$$f_k^{(0)} = g(x_k) + R_1^k + \Sigma_{j=2}^{k-1}\,C_j^k + P_k^k \quad (8a)$$
$$f_k^{(i)} = g(x_k) + R_1^k + \Sigma_{j=2}^{k-1}\,C_j^k + C_k^k\,(f_k^{(i-1)}), \text{ for } i \geq 1 \quad (8b)$$

Here $C_k^k(f_k^{(i-1)})$ means that in calculating C_k^k the approximation $h.K(x_k, x_k, f_k^{(i-1)})$ is used in determining $\bar{K}_k^{\;k}$.

Normally one uses (6a) in step k ($k \geq 3$), only once to calculate P_k^k ($j=k$).

The final approximation to I_j^k is, apart from the starting values R_1^k, essentially always given by C_j^k. However it is possible (and it occurs) that for some j ($2 \leq j \leq k-1$), C_j^k cannot be calculated. This due to the fact that the factor under the square root is negative. In order to continue the calculations it is then necessary to approximate for those $j<k$, I_j^k by P_j^k. Since $f(x_k)$ is iterativly approximated by $f_k^{(i)}$, we should know how many iterations are required to obtain good results. In practice of course we only can use a finite number of iterations and we will return to this subject in §5 in connection with the problems treated there. For the theoretical part we suppose iteration until convergence.

If calculation of C_k^k in (8b) is impossible, then $f_k^{(o)}$ is considered as the final approximation to $f(x_k)$.

3. CONVERGENCE

We define :

$$\psi_k(f) = g(x_k) + R_1^k + \sum_{j=2}^{k-1} \tilde{I}_j^k + C_k^k(f) \ . \qquad (9)$$

In (7b) and (8b) we then are interested in the fix point f_k, for which

$$\psi_k(f_k) = f_k \ .$$

Let

$$F_k = \{f_i \in \mathbb{R} \mid ((\tilde{I}_{k-1}^k)^2 - \bar{K}_{k-2}^k \bar{K}_{k-1}^k)((\tilde{I}_{k-1}^k)^2 + h\, K(x_k, x_k, f_i) M_{k-1}^k) \geq o\}$$

then

LEMMA 1. If $f_1, f_2 \in F_k$, then there exists a constant W, so that :

$$|\psi_k(f_1) - \psi_k(f_2)| \leq hWL|f_1 - f_2| \qquad (10)$$

Proof. From (9) follows

$$|\psi_k(f_1) - \psi_k(f_2)| = |C_k^k(f_1) - C_k^k(f_2)| \tag{11}$$

Substituting (6b), yields :

$$|\psi_k(f_1)-\psi_k(f_2)| = \frac{|2t_1|}{|M_{k-1}^k|} \cdot |D_k^k(f_1)-D_k^k(f_2)| \tag{12}$$

where $D_k^k(f_i) = \sqrt{((\tilde{I}_{k-1}^k)^2 - \bar{K}_{k-1}^k\ \bar{K}_{k-2}^k)((\tilde{I}_{k-1}^k)^2 + hK(x_k,x_k,f_i)M_{k-1}^k)}$.

Setting

$$y_i = hK(x_k,x_k,f_i) \tag{12*}$$

(12) becomes :

$$|\psi_k(f_1)-\psi_k(f_2)| = \frac{2}{|M_{k-1}^k|} \cdot |D_k^k(y_1) - D_k^k(y_2)| \tag{13}$$

where $D_k^k(y_i) = \sqrt{((\tilde{I}_{k-1}^k)^2 - \bar{K}_{k-1}^k\ \bar{K}_{k-2}^k)((\tilde{I}_{k-1}^k)^2 + y_i\ M_{k-1}^k)}$

If $y_1 = y_2$, then (13) equals zero and (10) is valid. Hence, without loss of generality, we can suppose $y_2 < y_1$. Then using Lagrange's formula (see [10], p.149) we obtain from (13) :

$$|\psi_k(f_1) - \psi_k(f_2)| = \frac{2}{|M_{k-1}^k|}\ |y_1 - y_2||E_k^k(\xi)|,\ y_2 < \xi < y_1 \tag{14}$$

where $E_k^k(\xi) = \frac{1}{2}\ \dfrac{M_{k-1}^k((\tilde{I}_{k-1}^k)^2 - \bar{K}_{k-2}^k\ \bar{K}_{k-1}^k)}{\sqrt{((\tilde{I}_{k-1}^k)^2 - \bar{K}_{k-1}^k\ \bar{K}_{k-2}^k)((\tilde{I}_{k-1}^k)^2 + \xi\ M_{k-1}^k)}}$

If

$$U_k = \min_{f_i \in F_k} ((\tilde{I}^k_{k-1})^2 - \bar{K}^{\,k}_{k-1}\,\bar{K}^{\,k}_{k-2})((\tilde{I}^k_{k-1})^2 + y_i \cdot M^k_{k-1})$$

there exists a constant $T_k \neq o$, such that

$$((\tilde{I}^k_{k-1})^2 - \bar{K}^{\,k}_{k-2}\,\bar{K}^{\,k}_{k-1})((\tilde{I}^k_{k-1})^2 + \xi M^k_{k-1}) \geq T_k > U_k .$$

Hence

$$|E^k_k(\xi)| \leq \frac{1}{2} \frac{|M^k_{k-1}| \cdot |(\tilde{I}^k_{k-1})^2 - \bar{K}^{\,k}_{k-2}\,\bar{K}^{\,k}_{k-1}|}{\sqrt{T_k}} .$$

Substituting this in (14) yields :

$$|\psi_k(f_1) - \psi_k(f_2)| \leq |y_1 - y_2| \frac{|(\tilde{I}^k_{k-1})^2 - \bar{K}^{\,k}_{k-1}\,\bar{K}^{\,k}_{k-2}|}{\sqrt{T_k}} \tag{15}$$

Using (12^*) and setting $W_k = \dfrac{(\tilde{I}^k_{k-1})^2 - \bar{K}^{\,k}_{k-1}\,\bar{K}^{\,k}_{k-2}}{\sqrt{T_k}}$ we find from (15) :

$$|\psi_k(f_1) - \psi_k(f_2)| \leq h|K(x_k,x_k,f_1) - K(x_k,x_k,f_2)||W_k| \tag{16}$$

Finally denoting $W = \sup_{2 \leq k \leq N} |W_k|$ and using the Lipshitz condition for $K(x,y,z)$ with repect to z, it follows from (16) that

$$|\psi_k(f_1) - \psi_k(f_2)| \leq hWL|f_1 - f_2| .$$

Restarting from (11), it is easy to see that substitution of (6c) also yields result (10), due to the linearity of (6c). q.e.d.

LEMMA 2.	Given that $hWL<1$ and $F_k \neq \phi$ $(k \geq 2)$, then the sequence $\{f_k^{(i)}\}$ defined by eqn. (7b) and (8b) has a unique limit f_k in terms of $f_0, f_1, \ldots, f_{k-1}$, and this solution may be found by the proposed rational predictor-corrector technique.

The proof is analogous to the one in Baker [2], pag. 923 and is based on the fix point theorem.

Suppose now for a moment that f_k^* is an approximate solution in x_k of the special problem :

$$f(x) = \int_{x_0}^{x} \Phi(x,y)dy \qquad (16^*)$$

obtained with the proposed algorithm setting $g(x) \equiv o$ and $K(x,y,z) = \phi(x,y)$ Here $\Phi(x,y)$ is a continuous function for $x_0 \leq y \leq x \leq a$. In solving (16^*) the algorithm can be reduced, in step k, to a single calculation of (8b). Indeed, to find f_k^* here, there is no information required about the f_i^* for any index $i \leq k$. If we denote

$$\varepsilon(h) = |\varepsilon_0(h)| + |\varepsilon_1(h)|$$

with $\varepsilon_i(h)$ the error on the starting values f_i^*, then :

Definition 1. :	The proposed method is said to be consistent if $\varepsilon(h) \to o$ as $h \to o$ and $$\lim_{\substack{h\to o \\ Nh=a}} \sup_{2\leq k\leq N} \left|f_k^* - \int_{x_0}^{x_k} \Phi(x_k,y)dy\right| = o$$ for all functions $\Phi(x,y)$ which are continuous for $x_0 \leq y \leq x \leq a$.

It follows from the error analysis of the used formulas, that :

LEMMA 3. The proposed algorithm is consistent.

Finally we obtain the following result :

THEOREM : (a) If $F_k \neq \phi$ $(k \geq 2)$, then the proposed algorithm is convergent, or

$$\max_{2 \leq k \leq N} |f_k - f(x_k)| \to o$$

as $h \to o$ such that $Nh = a$.

(b) For the absolute error then holds : $|f_i - f(x_i)| = O(h^4)$ $(i=o,1,\dots,N)$.

Assumption (a) follows from the consistency and (b) follows from the fact that the starting algorithm and the rational interpolation formulas are of fourth order.

4. STABILITY.

DEFINITION 2. A method is said to be A-stable if when applied to the problem

$$f(x) = 1 + \lambda \int_0^x f(y)dy \ , \quad \lambda < 0 \qquad (17)$$

with arbitrary step h, then $\lim_{k \to \infty} f_k = o$, for h fixed.

Analytic techniques as described in [2], to investigate A stability are rather difficult applicable on the above method, and this is subject of further research. Therefore we have chosen here for a numerical verification. For $\lambda = -1$ and $\lambda = -12$ we have solved (17) for several stepsizes and some of the results are shown in tables 1 and 2.

Table 1 : $\lambda = -1$; $f(x) = e^{-x}$; $h = 0.01$

x_k	f_k	$\varepsilon_k = \lvert f(x_k) - f_k \rvert$
0.53	0.5886	2.83 E-12
1.03	0.3570	3.37 E-12
1.53	0.2165	3.05 E-12
1.98	0.1380	2.53 E-12

Table 2 : $\lambda = -12$; $f(x) = e^{-12x}$

	h = 0.01		h = 0.05	h = 0.1
x_k	f_k	ε_k	ε_k	ε_k
0.53	1.729 E-3	2.E-9	5.E-6	1.E-5
1.03	4.286 E-6	1.E-11	1.E-9	5.E-7
1.53	1.062 E-7	3.E-14	5.E-12	1.E-10
1.98	4.799 E-11	2.E-16	5.E-13	5.E-12
2.98	2.94 E-15	1.E-21	3.E-18	—

5. NUMERICAL EXAMPLES.

To get an idea of the practical reliability of the method, it was applied to several linear, non linear and even singular equations. Some representative problems are given below. In applying the algorithm we used a different number of iterations on identical problems. We found for instance no significant difference in accurracy, between the use of three or ten iterations.

For comparing results a program of P. Pouzet [8] was chosen. This program uses also Runge-Kutta fourth order starting values, and was adapted to calculate linear Adams-Bashforth fifth order predictor values in approximating (4). This program calculates no corrected values. Theoretically the approximations obtained by this program are of $O(h^5)$.

All numerical results were obtained from a PDP 11/45 digital computer of the University of Antwerp (U.I.A.) in double precision (16 digits of accuracy). In tables 3 upto 7, the first two columns contain the discretization point x_k, and the exact value of f(x) in x_k. The following two columns give the results obtained with the linear algorithm of P. Pouzet and the rational algorithm from §2. The last column gives the absolute error for the rational approximations.

Problem 1. (linear)
$$\begin{cases} f(x) = 1 - \int_0^x f(y)dy \\ f(x) = e^{-x} \end{cases}$$

Table 3 : h = 0.01

x_k	$f(x_k)$	f_k(linear)	f_k(rat.)	(rat.)ε_k
0.53	5.88604969678 E-1	5.8860496967 0 E-1	5.8860496967 5 E-1	2.8 E-12
1.03	3.57006960569 E-1	3.570069605 58 E-1	3.5700696056 5 E-1	3.4 E-12
1.53	2.16535667316 E-1	2.165356673 05 E-1	2.1653566731 2 E-1	3.0 E-12
1.98	1.38069237311 E-1	1.380692373 03 E-1	1.380692373 08 E-1	2.5 E-12

Problem 2.: (non-linear) $\begin{cases} f(x) = e^{-x^2} + x.e^{-\frac{x^2}{2}} - x + \int_0^x xy\sqrt{f(y)}dy \\ f(x) = e^{-x^2} \end{cases}$

Table 4 : h = 0.01

x_k	$f(x_k)$	f_k(linear)	f_k(rat)	ε_k(rat)
0.53	7.5510384208 E-1	7.5510384 199 E-1	7.551038420 6 E-1	2.9 E-11
1.03	3.4614414037 E-1	3.46144140 21 E-1	3.46144140 18 E-1	1.9 E-10
1.53	0.9624098236 E-1	0.96240982 20 E-1	0.9624098 147 E-1	8.8 E-10
1.98	0.1983315989 E-1	0.1983315 520 E-1	0.198331 23 E-1	3.6 E-8

Problem 3.: (non-linear) $\begin{cases} f(x) = e^{-x} + \int_0^x e^{-(x-y)} (f(y) + e^{-f(y)})dy \\ f(x) = \ln(x+e) \end{cases}$

Table 5 : h = 0.01

x_k	$f(x_k)$	f_k(lin)	f_k(rat)	ε_k(rat)
0.53	1.17812618838 E0	1.1781261883 5 E0	1.1781261883 9 E0	1.3 E-11
1.03	1.32129755591 E0	1.321297555 85 E0	1.3212975559 3 E0	2.3 E-11
1.53	1.44651462554 E0	1.446514625 45 E0	1.4465146255 7 E0	3.2 E-11
1.98	1.54719687346 E0	1.547196873 35 E0	1.547196873 51 E0	3.9 E-11

Although in the following problems, our initial conditions upon the given functions are not fulfilled, we give them here because they illustrate the possibilities of this technique for singular equations.

Problem 4.: (singular) $\begin{cases} f(x) = tg(x_0) + \int_{x_0}^{x} \frac{f(y)}{\sin y \cdot \cos y} dy \\ f(x) = tg(x) \end{cases}$

Table 6 : $x_0 = 1.E-6$; $h = 0.01$

x_k	$f(x_k)$	f_k(linear)	f_k(rat.)	ε_k(rat)
0.53	5.8593044566 E-1	5.85930 190 E-1	5.8593044 29 E-1	2.7 E-9
1.03	1.6652817242 E 0	1.66528 09 E 0	1.6652817 17 E 0	7.3 E-9
1.53	2.4504423636 E 1	2.4 438 E 1	2.4504423 53 E 1	1.0 E-7
1.98	-2.3057590714 E 0	0.6479 E 3	-2.3057 6066 E 0	1.6 E-6

Problem 5.: (singular) $\begin{cases} f(x) = -\frac{x}{x-1} + \frac{x_0+1}{x_0-1} - \int_{x_0}^{x} \frac{2f(y)}{y-1} dy \\ f(x) = \frac{1}{x-1} \end{cases}$

Table 7 : $x_0 = 0.005$; $h = 0.01$

x_k	$f(x_k)$	f_k(linear)	f_k(rat) (*)	ε_k(rat)
0.305	-1.43884892 E 0	-1.43884 796 E 0	-1.438848 85 E 0	7.0 E-8
0.605	-2.53164557 E 0	-2.5316 2299 E 0	-2.531645 21 E 0	3.6 E-7
0.995	-2.00000000 E 2	-2.168 E 1	-1.9999 64 E 2	3.5 E-3
1.285	3.50877 E 0	2.335 E 0	3.5087 5 E 1	1.6 E-5

(*) : In calculating the values in this column, it occured several times that I_j^k could not be approximated by the corrector C_j^k for $j<k$, and

the predictor P_j^k had to be used uncorrected. This seems to cause a loss of accuracy, although the results remain usefull.

These numerical results indicate that for the first three problems there is little difference in accurracy between the linear and rational algorithm. For the singular problems, the rational algorithm gives better approximations in the neighbourhood of the singularity, as could be expected. In comparing these results it must also be taken in account that the linear algorithm is of $O(h^5)$ while the rational algorithm is of $O(h^4)$.

6. COMPUTABILITY OF THE CORRECTOR.

Trying to indicate when C_j^k cannot be calculated, we can start from the fact that there will be no problem if

$$((\tilde{I}_{j-1})^2 - \overline{K}_{j-2}\, \overline{K}_{j-1})((\tilde{I}_{j-1})^2 + \overline{K}_j\, M_{j-1}) \geq 0 \;. \tag{18}$$

where we have left away all upper right indices k.

Using series expansion of $K(x_k, x_i, f(x_i))$, $(i=j,j-1,j-2)$ in $z = x_{j-2}$, with respect to the second variable, and denoting :

$$K_{j-2}^{(n)} = [D_y^{(n)}\; K(x,y,f(y))]\,(x_k, x_{j-2})$$

condition (18) can be written :

$$\frac{1}{16}\, A_{j-2}^2 + \frac{h}{12}\, B_{j-2}\, A_{j-2} + O(h^2) \geq 0 \tag{19}$$

where : $A_{j-2} = 3K_{j-2}^{(1)} - 2K_{j-2}\, K_{j-2}^{(2)}$

and $\quad B_{j-2} = [D_y\; A(x,y)]\,(x_k, x_{j-2}) = A'_{j-2}$.

Condition (19) is equivalent to :

$$\frac{1}{16} A_{j-2}^2 (1+\frac{16}{12} h \frac{A'_{j-2}}{A_{j-2}}) + O(h^2) \geq o$$

So taking only in account terms upto O(h), we can deduce :

$$(20) \quad \begin{cases} \text{If } \dfrac{A'_{j-2}}{A_{j-2}} \geq o \Rightarrow \forall h, \text{ (17) is fulfilled.} \\ \\ \text{If } \dfrac{A'_{j-2}}{A_{j-2}} < o \Rightarrow h < -\dfrac{12}{16} \dfrac{A_{j-2}}{A'_{j-2}} . \end{cases}$$

From (20) follows that if A(x,y) becomes zero, or A'(x,y) becomes infinit at some discretization point (x_k, x_{j-2}), problems will occur in calculating C_j^k. Although (20) gives some insight into restriction (18) it is in practice of no use, since A(x,y) only can be calculated if f(y) is known. This is only the case for special test-equations, in which (20) seems to give indeed correct indications about arrays where difficulties are to be expected.

7. CONCLUSIONS.

The here proposed method based on rational osculatory interpolation gives in general good results of at least $O(h^4)$, and in practice even frequently of $O(h^5)$. If the problem is less suited for these methods, this will be indicated by the fact that the corrector gives difficulties. In that case the methods stay applicable, but with less accuracy.

REFERENCES.

1. F.G. TRICOMI : Integral equations, Interscience, New York (1957).

2. C.T.H. BAKER : The numerical treatment of integral equations, Oxford (1977).

3. P. POUZET : Etude en vue de leur traitement numérique des équations intégrales de type Volterra. Rev. Franç. Trait. de l'inform. (1963), pp. 79-112.

4. L. GAREY : Predictor-corrector methods for nonlinear Volterra integral equations of the second kind, BIT 12 (1972), pp. 325-333

5. J.D. LAMBERT and B. SHAW : On the numerical solution of $y' = f(x,y)$ by a class of formulae based on rational approximation. Mathematics of Computation 19 (1965), pp. 456-462.

6. H.C. THACHER, Jr. : Closed rational integration formulas. The Computer Journal, Vol.8, (1966) pp. 362-367.

7. Y.L. LUKE, W. FAIR, J. WIMP : Predictor-corrector formulas based on rational interpolants. Computers & Maths. with Applic., Vol 1, (1975), pp. 3-12.

8. P. POUZET : Algorithme de résolution des équations intégrales de type Volterra par des méthodes parpas. Rev. Franç. Trait. de l'inform. (1964), pp. 169-173.

9. T. SATO : Sur l'équation intégrale non linéaire de Volterra. Compositio Mathematica 11 (1953), pp. 271-290.

10. V.I. SMIRNOV : A course of higher mathematics, part I. Elementary Calculus, Pergamon Press 1964.

T.H. CLARYSSE
Department of Mathematics
University of Antwerp
Universiteitsplein 1

B - 2610 Wilrijk (Belgium)

On a Summability Method

Yudell L. Luke*

ABSTRACT

In previous studies we applied Lanczos' τ-method to get polynomial and rational approximations to series of hypergeometric type. It was shown that the approximations could be viewed as a weighted sum of the partial sums of the given series. This we call a summability method. The τ-method starts from the differential equation satisfied by the given function. But in view of the summability feature, the approximations can be found directly from the hypergeometric series. In the present paper, we explore the applicability of the direct summability process for functions of hypergeometric type to functions which are not of hypergeometric type. Our treatment is expository. We treat two examples. The first converges in $|z|<2$. The second is the divergent but asymptotic series related to the logarithm of the gamma function. Main diagonal Padé approximations are also constructed. The numerics indicate that all of the approximations converge at least for all $z>0$. These approximations are quite remarkable in that they appear to converge in a region where the given series upon which they are based are divergent.

---

* Department of Mathematics, University of Missouri, Kansas City, Missouri, 64110. This research was sponsored by the National Science Foundation under Grant MCS-78-01351.

I. Introduction

In previous studies, see [1, 2, 3], we applied Lanczos' τ-method to get polynomial and rational approximations for functions defined by a differential equation possessing an at least formal series solution beginning with a constant. Usually the differential equation is linear, but this is not a prerequisite. In the event that the function is a generalized hypergeometric series, it was shown that the approximations could be viewed as a weighted sum of the partial sums of the given series. This we call a summability method.

The problem of getting polynomial and rational approximations for a function defined at least formally by a series can be dealt with in yet another fashion. Here it is not necessary to suppose that the function is defined by a differential equation or indeed by any functional equation. Thus the series need not be hypergeometric. The idea is to weight the partial sums so that a summability process is achieved directly. This is in contrast to the τ-method for hypergeometric series where the summability process is achieved indirectly. If the given series is hypergeometric, we can make the results of the τ-method and the above direct summability method coalesce. In the error analyses for the hypergeometric series, each technique has certain advantages, but we shall not further discuss this here.

In the present paper, we explore the applicability of the direct summability process described above for functions of hypergeometric type to functions which are not of hypergeometric type. Our choice of weights for non-hypergeometric series is motivated by our experience with the τ-method and by the success enjoyed in achieving approximations for functions of the hypergeometric family. The present work is by far and large expository. The aim is to see and observe what happens before pursuing precise theorems and proofs. Further, for the sake of clarity and to avoid complexities, our findings are not presented in the most general form possible.

In Section II, we explain the direct summability process and indicate its relationship to the τ-method when applied to series of hypergeometric type. We then show how to adapt this summability technique for series not necessarily of the hypergeometric class. Two examples are treated in Section III. Rational approximations of the Padé type are also constructed and numerical comparisons are made with the rational approximations derived from the summability process.

II. The Direct Summability Process

Let

$$H(z) = S_k(z) + M_k(z), \tag{1}$$

where

$$S_k(z) = \sum_{r=0}^{k} h_r z^r . \tag{2}$$

In practice $H(z)$, for example, might be defined by an integral and (1) arises from this integral by repeated partial integrations. Here $S_k(z)$, a partial sum of the infinite series representation for $H(z)$, is a polynomial approximation to $H(z)$ and $M_k(z)$ is the remainder. Multiply (1) by the weight function $\gamma^{-k}U_{n,k}$ and sum from $k = 0$ to $k = n$. Then

$$H(z) = F_n(z,\gamma) + R_n(z), \tag{3}$$

$$F_n(z,\gamma) = \frac{A_n(z,\gamma)}{B_n(\gamma)}, \quad R_n(z) = \frac{T_n(z,\gamma)}{B_n(\gamma)}, \tag{4}$$

$$\begin{aligned} A_n(z,\gamma) &= \sum_{k=0}^{n} \gamma^{-k} U_{n,k}\, S_k(z) \\ &= \sum_{k=0}^{n} h_k (z/\gamma)^k \sum_{r=0}^{n} U_{n,r+k}\, \gamma^{-r} \\ &= \sum_{k=0}^{n} \gamma^{-k} \sum_{r=0}^{n-k} h_r U_{n,r+k}\, (z/\gamma)^r, \end{aligned} \tag{5}$$

$$B_n(\gamma) = \sum_{k=0}^{n} U_{n,k}\, \gamma^{-k}, \tag{6}$$

$$T_n(z,\gamma) = \sum_{k=0}^{n} \gamma^{-k} U_{n,k}\, M_k(z) \tag{7}$$

So, $H_n(z,\gamma)$ is an approximation to $H(z)$ and $R_n(z)$ is the remainder. If γ is fixed, $H_n(z,\gamma)$ is a polynomial. But if we set $\gamma = z$, then with $F_n(z,z) \equiv F_n(z)$ and $A_n(z,z) \equiv A_n(z)$, we have the approximation

$$F_n(z) = \frac{A_n(z)}{B_n(z)} = \frac{z^n A_n(z)}{z^n B_n(z)} \tag{8}$$

which is the ratio of two polynomials in z, each of degree n. In this event

$$z^n B_n(z) H(z) - z^n A_n(z) = O(z^{n+1}). \tag{9}$$

Presently, we discuss our choice for $U_{n,k}$, but first it is necessary to consider the generalized hypergeometric series defined formally at least by

$$ {}_pF_q(\alpha_1,\ldots,\alpha_p;\rho_1,\ldots,\rho_q;z) = \sum_{k=0}^{\infty} \frac{\prod_{j=1}^{p}(\alpha_j)_k z^k}{\prod_{j=1}^{q}(\rho_j)_k k!} . \qquad (10) $$

We often use the shorthand notation

$$ {}_pF_q(\alpha_p;\rho_q;z) = {}_pF_q\left(\begin{matrix}\alpha_p\\ \rho_q\end{matrix}\middle| z\right) = \sum_{k=0}^{\infty} \frac{(\alpha_p)_k z^k}{(\rho_q)_k k!} . \qquad (11) $$

Thus $(\alpha_p)_k$ stands for $\prod_{j=1}^{p}(\alpha_j)_k$, etc. Where no confusion can result, we often refer to (10) as ${}_pF_q(z)$ or more simply as ${}_pF_q$. The ${}_pF_q$ series converges for all z if $p \le q$. It converges for $|z|<1$ if $p = q + 1$. It diverges if $p>q + 1$ unless $z = 0$ or the series terminates. Nonetheless, when it diverges, the ${}_pF_q$ is the asymptotic expansion of a well defined function, call it $F(z)$, in an appropriate region of the complex plane as $|z|\to 0$.

When the τ-method was applied to the differential equation satisfied by ${}_pF_q(z)$ and this was related to the direct summability method enunciated above, it was found that

$$ U_{n,k} = \left\{\frac{(-n)_k\,(n+\lambda)_k}{(\beta+1)_k\,k!}\right\} \left\{\frac{k!(\rho_q)_k(c_f)_k}{(\alpha_p+1)_k(d_g)_k}\right\} , $$

$$ \lambda = \alpha + \beta + 1 . \qquad (12) $$

Here

$$ (c_f)_k = \prod_{j=1}^{f} (c_j)_k , \quad (d_g)_k = \prod_{j=1}^{g} (d_j)_k . \qquad (13) $$

We shall not enter into a discussion of the τ-method and its philosophy. For a thorough treatment of the subject, see

[1, 2]. Under certain circumstances the rational approximations achieved by use of the τ-method are of the Padé type. We do not discuss Padé approximations here, suffice it to say that with the notation of (8), the right hand side of (9) is $0(z^{2n+1})$ for Padé approximations. For convience, we speak of the approximations in (3) - (7), with $\gamma = z$, as rational approximations when they are not of the Padé type. This serves to distinguish these approximations from the Padé ones even though the latter, of course, are also rational.

In the hypergeometric case, we always took $f = g$, and then almost always $f = g = 0$, though in a few important situations we had $f = g = 1$. Also, in most instances, we put $\alpha = \beta = 0$, for then both $A_n(z)$ and $B_n(z)$ satisfy the same recurrence formula. Under rather liberal restrictions, the polynomial and rational approximations for ${}_pF_q(z)$ converge uniformly on compact subsets of the z - plane which exclude $z = 0$ if $p \le q$; and exclude $z = 0$, $z = 1$ and $|\arg(1-z)| = \pi$ if $p = q+1$. For the polynomial approximations we always require $0 < z/\gamma \le 1$. Recall that when $p > q+1$, the ${}_pF_q$ series, though divergent, is the asymptotic expansion of $F(z)$ in some region of the complex plane as $|z| \to 0$. In this case, under rather liberal conditions, the rational approximations converge uniformly to $F(z)$ on compact subsets of the z-plane, $|\arg(-z)| < \pi$, z fixed, $z \ne \infty$.

We are now in a position to see how to adapt the summability process for functions of hypergeometric type to those functions not of this class. Notice that in the hypergeometric case, $U_{n,k}$ depends on the α_j's and ρ_j's, the parameters in the generalized hypergeometric series (9) or (11). Comparing (11) and (1), we have $h_k = (\alpha_p)_k/(\rho_q)_k\, k!$, and by a proper choice of the c_j's and d_j's, we could make the second curly bracket in the expression for $U_{n,k}$ equal to the reciprocal of h_k. For the general problem at hand where the series representation for for $H(z)$ is not of hypergeometric type, all of this suggests that we explore the possibility of choosing

$$U_{n,k} = \left\{\frac{(-n)_k\ (n+\lambda)_k}{(\beta+1)_k k!}\right\} \frac{1}{h_k} \tag{14}$$

Clearly we must hypothesize that, in effect, no h_k vanishes.

III. Examples

A. Solution of a Convolution Integral Equation

Consider

$$\int_0^z k(z-t)H(t)dt = H(z) + q(z) \tag{15}$$

$$q(z) = 2k(z) = -\frac{4}{(z+2)^2}. \tag{16}$$

This equation was studied by Friedlander [4]. Numerical solutions were explored by Fox and Goodwin [5]. Exponential approximations based on rational approximations for the Laplace transform of $H(z)$ were developed by Luke [1], Vol. 2, pp. 265-269.

Assume the solution

$$H(z) = \sum_{r=0}^{\infty} h_r z^r, \quad |z|<2 . \tag{17}$$

Expand $q(z)$ and $k(z)$ by the binomial expansion. Substitute these together with the series for $h(z)$ in (15), integrate and equate like powers of z. It is readily deduced that the h_r's can be generated by the recurrence formula

$$h_r = \frac{(-)^r}{2^r}\left\{(r+1) + \sum_{m=0}^{r-1} \frac{(-2)^m m!(r-m)!h_m}{r!}\right\},$$

$$r = 1, 2, \ldots, \quad h_0 = 1 . \tag{18}$$

In the numerics for the summability process we take $\alpha = \beta = 0$ whence $\lambda = 1$. For the rational approximations, we write

$$H(z) = F_n(z) + R_n(z), \quad F_n(z) = A_n(z)/B_n(z), \tag{19}$$

$$A_n(z) = \sum_{k=0}^{n} v_k z^{n-k}, \quad B_n(z) = \sum_{k=0}^{n} u_k z^{n-k}, \tag{20}$$

$$u_k = \frac{(-n)_k(n+1)_k}{(k!)^2 h_k}, \quad v_k = \sum_{m=0}^{n-k} u_{m+k} h_m . \tag{21}$$

For the polynomial approximations, we put

$$H(z) = F_n(z) + R_n(z), \quad F_n(z) = \sum_{r=0}^{n} t_r z^r \tag{22}$$

$$t_r = \frac{q_r h_r}{\sum_{k=0}^{n} u_k \gamma^{-k}} \qquad q_r = \sum_{k=0}^{n-r} u_{r+k}\gamma^{-r-k} \tag{23}$$

For the main diagonal Padé approximations, we have

$$H(z) = F_n(z) + R_n(z), \quad F_n(z) = A_n(z)/B_n(z), \tag{24}$$

$$A_n(z) = \sum_{k=0}^{n} a_k z^k, \quad B_n(z) = \sum_{k=0}^{n} b_k z^k . \tag{25}$$

The coefficients in the polynomials for the Padé approximations were found by solving systems of linear equations. The efficacy of this procedure will be discused in a future paper.

All the numerics were done on an Amdahl 470/V7 operating under OS MVS Release 3.7 and using the FORTRAN IV H-Extended Compiler. The precision used is quadruple.

Before describing the numerical data it is helpful to correlate the notation used in the paper with that of the computer. We have the following table.

Paper Notation	Computer Notation	
h_r	H(I)	
n	N	
u_k, v_k	U(I), V(I)	(26)
$A_n(z)$, $B_n(z)$, $F_n(z)$	AN(Z), BN(Z), FN(Z)	
a_k, b_k	A(I), B(I)	
t_k	TERM (I)	

Table 1 gives the coefficients h_r for $r = 1, 2, \ldots, 31$. Calculations for the rational, polynomial and Padé approximations were carried out for $n = 1, 2, \ldots, 15$ and for $z = 0.1\ (0.1)\ 1.0\ (0.5)\ 5.0$ except that for the polynomial approximations $\gamma = 2$ and so $0 < z \leq 2$. We present only a small sample of this material. Thus in Table 2, we give the coefficients v_k and u_k which define the numerator and denominator polynomials respectively in the rational approximations for $n = 4, 5, 6$ and 7.

Using the coefficients in Table 2, the polynomials $A_n(z)$ and $B_n(z)$ for the rational approximations are computed. These as well as $F_n(z)$, the rational approximation to $H(z)$ for $z = 0.5, 1.0, 2.0,$ and 4.0 are posted in Table 3. If Δ is the forward difference operator, then in all situations $\Delta F_n(z) = F_{n+1}(z) - F_n(z)$, that is, the first differences as well

TABLE 1. COEFFICIENTS IN THE TAYLOR SERIES FOR H(Z)

I	H(I)
0	0.1000000000000000000000000000000000Q+01
1	-0.1500000000000000000000000000000000Q+01
2	0.1375000000000000000000000000000000Q+01
3	-0.9791666666666666666666666666666667Q+00
4	0.6015625000000000000000000000000000Q+00
5	-0.3393229166666666666666666666666667Q+00
6	0.1829644097222222222222222222222222Q+00
7	-0.9671223958333333333333333333333333Q-01
8	0.5081370520213293650793650793650794Q-01
9	-0.2670032427214230599647266313933300Q-01
10	0.1405417225348255621693121693121217Q-01
11	-0.7407929477199293280022446689113300Q-02
12	0.3906531876484565761366976644754400Q-02
13	-0.2059267366428379713698724115390800Q-02
14	0.1084400050489234811289235272370200Q-02
15	-0.5702353160777117138514372103988500Q-03
16	0.2993763879711137109750025273493800Q-03
17	-0.1569075357362617813696822248620800Q-03
18	0.8209750031627285418086487604230400Q-04
19	-0.4288372538053038522846061083402000Q-04
20	0.2236448294809399628609049335049000Q-04
21	-0.1164559535670131804121398618812300Q-04
22	0.6055301439691561276376197893660400Q-05
23	-0.3144242013021411788745334196459200Q-05
24	0.1630557830791987235999788545628500Q-05
25	-0.8445567455317473192832580593660800Q-06
26	0.4369425584185869096250725278693700Q-06
27	-0.2258142199470977714050985525952300Q-06
28	0.1165832756125880811964469489367100Q-06
29	-0.6013176786746781994833897895598000Q-07
30	0.3098684750399979008906329230657800Q-07
31	-0.1384802013536639279103448277099500Q-07

TABLE 2. COEFFICIENTS IN THE POLYNOMIALS FOR THE RATIONAL APPROXIMATIONS

N-4

I	V(I)	U(I)
0	9.99999999999999999999999999999995Q-01	1.00000000000000000000000000000000Q+00
1	-2.19213410702772404900064474532550Q+00	1.33333333333333333333333333333333Q+01
2	1.09864603481624758220502901353960Q+01	6.54545454545454545454545454545450Q+01
3	-3.15667311411992263056092843326880Q+01	1.42978723404255319148936170212770Q+02
4	1.16363636363636363636363636363640Q+02	1.16363636363636363636363636363640Q+02

N-5

I	V(I)	U(I)
0	-1.00000000000000000000000000000002Q+00	1.00000000000000000000000000000000Q+00
1	-1.40883038843555490899590440747410Q+00	2.00000000000000000000000000000000Q+01
2	7.67150943144150309284778023041480Q+00	1.52727272727272727272727272727270Q+02
3	2.21569922704783337366084218682980Q+01	5.71914893617021276595744680851060Q+02
4	-6.67103886136886904346612711923520Q+01	1.04727272727272727272727272727270Q+03
5	7.42655410590943975441289332310050Q+02	7.42655410590943975441289332310050Q+02

N-6

I	V(I)	U(I)
0	1.00000000000000000000000000000000Q+00	1.00000000000000000000000000000000Q+00
1	2.34862285406894344041348710835630Q+00	2.80000000000000000000000000000000Q+01
2	-2.91917511647650621099271604625770Q+01	3.05454545454545454545454545454550Q+02
3	1.48911534598843887143843161092690Q+02	1.71574468085106382978723404255320Q+03
4	-7.34772070027617907463811094138770Q+01	5.23636363636363636363636363636360Q+03
5	5.93965773172190158510332578314180Q+02	8.16920951650038372985418265541060Q+03
6	5.05016249555212904756256671806420Q+03	5.05016249555212904756256671806420Q+03

N-7

I	V(I)	U(I)
0	-1.00000000000000000000000000000205Q+00	1.00000000000000000000000000000000Q+00
1	1.17591829130277617134703659643280Q+01	3.73333333333333333333333333333330Q+01
2	-2.59381383850579812456175341996820Q+01	5.49818181818181818181818181818180Q+02
3	-5.15409230644678326570419097831820Q+01	4.28936170212765957446808510638300Q+03
4	1.20135731358038530055247550006620Q+03	1.92000000000000000000000000000000Q+04
5	-6.68673933836748548513248573660660Q+02	4.90152570990023023791250959324630Q+04
6	1.24220350271658970730424147936040Q+04	6.56521124421776776183133673348350Q+04
7	3.54867182766745203635139683608210Q+04	3.54867182766745203635139683608210Q+04

TABLE 3. RATIONAL APPROXIMATIONS FOR Z = 0.5

N	AN(Z)	BN(Z)
1	8.3333333333333333333333333333333Q-01	1.8333333333333333333333333333333Q+00
2	3.3409090909090909090909090909091Q+00	6.6136363636363636363636363636364Q+00
3	1.6729932301740812379110251450677Q+01	3.3459622823984526112185686653772Q+01
4	1.0311536911669890393294648613798Q+02	2.0594580109606705351386202450032Q+02
5	7.1567910113137217936290453068428Q+02	1.4296426567224720218629527752501Q+03
6	5.3456545572287810031247605989477Q+03	1.0678307782090522073031244119270Q+04
7	4.1676881021212783682797705358879Q+04	8.3252446842912065006007696482019Q+04
8	3.3279409332232978920796012326882Q+05	6.6477923091752772657464631341543Q+05
9	2.6918186533918466491286262402990Q+06	5.3770940327717447506361697637759Q+06
10	2.1926239999667206008430340355420Q+07	4.3799181592223230529344712929321Q+07
11	1.7936572417378055641677702937053Q+08	3.5829544530071106616899257215947Q+08
12	1.4719933693702724063081034515809Q+09	2.9404086104108511947855995250843Q+09
13	1.2115387668632435590281511299585Q+10	2.4201325196622844802516383527965Q+10
14	1.0000822342938011968279813445560Q+11	1.9977334640450201735840531082800Q+11
15	8.2799735514576911406249250026911Q+11	1.6539820104727977920447384595910Q+12

N	FN(Z)	1ST DIFF'S	F15(Z)-FN(Z)
1	4.5454545454545454545454545454545Q-01	5.061Q-02	4.606Q-02
2	5.0515463917525773195876288659794Q-01	-5.151Q-03	-4.546Q-03
3	5.0000361300392372226116237562234Q-01	6.882Q-04	6.048Q-04
4	5.0069177700107087979829187530365Q-01	-9.179Q-05	-8.334Q-05
5	5.0059999102999812262098995726844Q-01	8.782Q-06	8.450Q-06
6	5.0060877306743515807091658892891Q-01	-2.913Q-07	-3.324Q-07
7	5.0060848181258067202773699052443Q-01	-4.243Q-08	-4.115Q-08
8	5.0060843938070638265714010583777Q-01	2.311Q-10	1.281Q-09
9	5.0060843961181163025159583519281Q-01	1.062Q-09	1.050Q-09
10	5.0060844067370250695766284550935Q-01	3.633Q-12	-1.171Q-11
11	5.0060844067733559517481047650764Q-01	-1.419Q-11	-1.535Q-11
12	5.0060844066314879629667961813803Q-01	-1.269Q-12	-1.160Q-12
13	5.0060844066188028362872248446863Q-01	8.368Q-14	1.089Q-13
14	5.0060844066196396154063602184716Q-01	2.519Q-14	2.519Q-14
15	5.0060844066198915225114483576720Q-01		

TABLE 3. RATIONAL APPROXIMATIONS FOR Z = 1.0

N	AN(Z)	BN(Z)
1	3.33333333333333333333333333334Q-01	2.333333333333333333333333333333Q+00
2	2.81818181818181818181818181818Q+00	9.36363636363636363636363636364Q+00
3	1.39632495164410058027079303675 05Q+01	5.12437137330754352030947775628 63Q+01
4	9.45912314635718891038039974210 18Q+01	3.39130238555770470664087685364 28Q+02
5	7.03364693290739566927088358808 93Q+02	2.53557030420796525203703401316 11Q+03
6	5.69371946800970518380084767470 30Q+03	2.05059348747217584253858015978 46Q+04
7	4.83747168050475561364064210028 12Q+04	1.74231601035133675086935668249 65Q+05
8	4.23555818739079333895754844293 64Q+05	1.52554509349035819566486856187 76Q+06
9	3.77494621245541885651426169101 97Q+06	1.35964421858318059239808434465 74Q+07
10	3.39931103440566771734976913263 88Q+07	1.22434807221525739083213616137 11Q+08
11	3.07988872117954920259413702600 19Q+08	1.10930000185269759330025583799 41Q+09
12	2.80157581452478612064571363631 88Q+09	1.00905860179128807814627912811 00Q+10
13	2.55596927688315662187834153671 10Q+10	9.20597184407979555654358974274 31Q+10
14	2.33788588619559693083576900467 18Q+11	8.42048917414877376179898982449 67Q+11
15	2.14366979746105437452115712440 55Q+12	7.72097065127031218136400601565 81Q+12

N	FN(Z)	1ST DIFF'S	F15(Z)-FN(Z)
1	1.42857142857142857142857142857 14Q-01	1.581Q-01	1.348Q-01
2	3.00970873786407766990291262135 92Q-01	-2.848Q-02	-2.333Q-02
3	2.72487072056769712754312459895 07Q-01	6.436Q-03	5.155Q-03
4	2.78923023397738744579279873610 01Q-01	-1.524Q-03	-1.280Q-03
5	2.77399010440946626501393614643 97Q-01	2.630Q-04	2.435Q-04
6	2.77662028227179879002973289599 60Q-01	-1.597Q-05	-1.950Q-05
7	2.77646055696250129885479987249 12Q-01	-3.775Q-06	-3.529Q-06
8	2.77642280484812366144160186928 78Q-01	-7.373Q-08	2.459Q-07
9	2.77642206752374347274181607948 23Q-01	3.290Q-07	3.196Q-07
10	2.77642535774583360392731583290 74Q-01	6.012Q-09	-9.429Q-09
11	2.77642541786322231401901009747 25Q-01	-1.344Q-08	-1.544Q-08
12	2.77642528347849035773223717566 58Q-01	-2.369Q-09	-2.002Q-09
13	2.77642525979140066333410348136 68Q-01	2.262Q-10	3.667Q-10
14	2.77642526205365442320895521901 02Q-01	1.405Q-10	1.405Q-10
15	2.77642526345876174421063710574 83Q-01		

TABLE 3. RATIONAL APPROXIMATIONS FOR Z = 2.0

N	AN(Z)	BN(Z)
1	-6.6666666666666666666666666666664Q-01	3.3333333333333333333333333333333Q+00
2	3.2727272727272727272727272727272Q+00	1.6363636363636363636363636363636Q+01
3	8.2166344294003868471953578336559Q+00	1.0406189555125725338491295938104Q+02
4	9.5638942617666021921341070277240Q+01	7.8680593165699548678272082527400Q+02
5	7.0469339168204307571724824872226Q+02	6.6986786214226654454606316920778Q+03
6	6.8075654033701784950828497458448Q+03	6.1907266248088679872841531642037Q+04
7	6.6236863613382838747255279424708Q+04	6.0519327394259678982328196361397Q+05
8	6.7144078330294335473599156778335Q+05	6.1439527381052925438234343579522Q+06
9	6.9867092688531907979990853054972Q+06	6.3943389500014992398791099178203Q+07
10	7.3923199178360631881581109751979Q+07	6.7626517709258560711439020063472Q+08
11	7.8990098870298761244358402917079Q+08	7.2259944690212257896942955742518Q+09
12	8.4957319902183239684584864397585Q+09	7.7722263985412269528966928899049Q+10
13	9.1779110015289873544296919777517Q+10	8.3964410642773835164001391220006Q+11
14	9.9458066701545887551196149038438Q+11	9.0989380851016467794908838767611Q+12
15	1.0803909638853145727362566839191Q+13	9.8839523991726512189461274010397Q+13

N	FN(Z)	1ST DIFF'S	F15(Z)-FN(Z)
1	-1.9999999999999999999999999999999Q-01	4.000Q-01	3.093Q-01
2	2.0000000000000000000000000000000Q-01	-1.210Q-01	-9.069Q-02
3	7.8959107806691449814126394052047Q-02	4.259Q-02	3.035Q-02
4	1.2155340824166459073566624273971Q-01	-1.635Q-02	-1.225Q-02
5	1.0519886555363366346905517167077Q-01	4.765Q-03	4.109Q-03
6	1.0996391564262223806905299375372Q-01	-5.165Q-04	-6.563Q-04
7	1.0944745499545236851007389106167Q-01	-1.626Q-04	-1.399Q-04
8	1.0928482231619609144737975908081Q-01	-2.084Q-05	2.276Q-05
9	1.0926398058475696958211224806949Q-01	4.698Q-05	4.360Q-05
10	1.0931096511012574260141180209901Q-01	2.845Q-06	-3.380Q-06
11	1.0931381031211627135487320705106Q-01	-4.951Q-06	-6.226Q-06
12	1.0930885893664770419866407619313Q-01	-1.702Q-06	-1.274Q-06
13	1.0930715682119611099031759575801Q-01	1.775Q-07	4.279Q-07
14	1.0930733429695033879395202532636Q-01	2.504Q-07	2.504Q-07
15	1.0930758468401264960407109036974Q-01		

TABLE 3. RATIONAL APPROXIMATIONS FOR Z = 4.0

N	AN(Z)	BN(Z)
1	-2.6666666666666666666666666666666Q+00	5.3333333333333333333333333333333Q+00
2	1.0181818181818181818181818181818Q+01	3.6363636363636363636363636363636Q+01
3	-2.5129593810444874274661508704061Q+01	2.9969825918762088974854932301740Q+02
4	2.8158349451966473243068987749837Q+02	2.8448845905867182462927143778207Q+03
5	-6.3358243363403305272314596133596Q+01	3.0000930072099648037336840589563Q+04
6	1.4808629994909451989709819311168Q+04	3.4228084195420356725518045788129Q+05
7	1.4338942671405336513748961742548Q+05	4.1415310288892544351181139113687Q+06
8	1.6807807957104645577567083715078Q+06	5.2343149948114597858835833329168Q+07
9	2.1002480943207267160790126813303Q+07	6.8299509639535734769721404626069Q+08
10	3.0311091208026334525812079737042Q+08	9.1200348705909637792564877129468Q+09
11	4.1604636594412576053080493196596Q+09	1.2380397400139663848840604735302Q+11
12	5.6208554376304819048299619705518Q+10	1.7003494389282892573991208405077Q+12
13	7.6982470514915619463218624111805Q+11	2.3545435214099816370766838273437Q+13
14	1.0728751112388471844800891244557Q+13	3.2793468322315669712456481442619Q+14
15	1.5053098408788368058964759801934Q+14	4.5862499329341865523473723488551Q+15

N	FN(Z)	1ST DIFF'S	F1S(Z)-FN(Z)
1	-5.0000000000000000000000000000000Q-01	7.800Q-01	5.328Q-01
2	2.8000000000000000000000000000000Q-01	-3.638Q-01	-2.472Q-01
3	-8.3849648905410987195373812474183Q-02	1.828Q-01	1.167Q-01
4	9.8978881411000232072406590856343Q-02	-1.011Q-01	-6.616Q-02
5	-2.1118759722161209994831398768602Q-03	4.538Q-02	3.493Q-02
6	4.3264559916242155610182291747683Q-02	-8.642Q-03	-1.044Q-02
7	3.4622323414660002515115646293458Q-02	-2.512Q-03	-1.800Q-03
8	3.2110807190177639326186157644070Q-02	-1.360Q-03	7.114Q-04
9	3.0750558904525147208509556735618Q-02	2.485Q-03	2.072Q-03
10	3.3235718545078571549410778671495Q-02	3.695Q-04	-4.135Q-04
11	3.3605251309576889817095342262630Q-02	-5.482Q-04	-7.830Q-04
12	3.3057060560288375764436108729949Q-02	-3.618Q-04	-2.348Q-04
13	3.2695284591221260828765821827201Q-02	2.084Q-05	1.270Q-04
14	3.2716122024481473071554846510365Q-02	1.061Q-04	1.061Q-04
15	3.2822237402918229749431031341892Q-02		

TABLE 4. COEFFICIENTS IN THE POLYNOMIALS FOR THE PADÉ APPROXIMATIONS

N=4

I	A(I)	B(I)
0	1.0	1.0
1	-2.82977310921498023634304563256140-01	1.21702268907850197636569543674390+00
2	1.00529722577859690972741804202000-01	5.51063756195612655521284959317800-01
3	-1.48152183340221064957613641976950-02	1.17540885143123325950001515922850-01
4	1.18095435094491154145391196253530-03	9.88516701936235098276614859654110-03

N=5

I	A(I)	B(I)
0	1.0	1.0
1	-5.57243906422094825828353131079380-01	9.42756093577905174171646868920620-01
2	1.87456821773324842369290410968930-01	2.26590962140182603626760714349870-01
3	-4.55830728273808358230757978166340-02	-2.28195916200598782022825043910240-02
4	6.20443540916431584956401390641040-03	-1.80346833353110983975848324264130-02
5	-4.44246772716589043453037527645960-04	-2.05281207879735273145329608042740-03

N=6

I	A(I)	B(I)
0	1.0	1.0
1	4.39423672871959312404875795295090-01	1.93942367287195931240487579529510+00
2	-7.03079982641555111664084442056130-03	1.52710470948152341749067284852210+00
3	3.91227466500916513448537983210010-02	6.42238927340099389690825519240020-01
4	-2.94562755865695381202672266461950-03	1.58100300934857591737643939037350-01
5	3.92710718500532478300315567378290-04	2.23930285167705816842735370723590-02
6	-1.65607408170094457833335866020160-05	1.52157874768161755706173818742600-03

N=7

I	A(I)	B(I)
0	1.0	1.0
1	9.67012972824591776490956792882750-01	2.46701297282459177649095679288280+00
2	2.33659941591203427379755465151630-01	2.55917940082809109211619065447580+00
3	3.04786909752415679539588709412870-02	1.45627162125023118011984592910780+00
4	1.93551413679127887732337270020910-02	4.98945266328713755107302330457330-01
5	-1.56503237826805196168059101378110-03	1.05602976408447078519580238937960-01
6	2.45572693545978734207422232083510-04	1.31761946055603306099442766245840-02
7	-1.26713029316898244923007637063250-05	7.84589961308295483919088323463210-04

TABLE 5. PADÉ APPROXIMATIONS FOR Z = 0.5

N	AN(Z)	BN(Z)
1	-5.55555555555555555555555555555554Q-02	6.94444444444444444444444444444444Q-01
2	8.03305041152263374485596707818930Q-01	1.60423096707818930041152263374490Q+00
3	8.52621658764466663608983473222234Q-01	1.70317262699896038771422892874070Q+00
4	8.81865682538897204585403868395380Q-01	1.76158771716975471474334203197900Q+00
5	7.62918262630286569533251405921880Q-01	1.52398202028557139229452133246970Q+00
6	1.22267239153659904109548926144020Q+00	2.44237270534138313972202468356560Q+00
7	1.54689583533445811585015485593810Q+00	3.09003146908369074091169857651900Q+00
8	1.42154925851333383181479403203370Q+00	2.83964300848286879099247933924440Q+00
9	1.48471361341117450735191064658170Q+00	2.96581817807112794301116005323660Q+00
10	1.81162759290493054963562869626130Q+00	3.61885147303805634507196700165860Q+00
11	1.75842231250049782017618058934880Q+00	3.51257024387205284666443147723004Q+00
12	2.05364005576174932558810611778800Q+00	4.10228811373230447080351342152420Q+00
13	2.18006122004038425925383617333820Q+00	4.35482313713594309348503265623930Q+00
14	7.06200581881514951322181480886640Q+00	1.41068452810674644478228437542630Q+01
15	2.71964745053573446970450168510180Q+00	5.43268396940999752461414200201550Q+00

N	FN(Z)	1ST DIFF'S	F15(Z)-FN(Z)
1	-7.99999999999999999999999999999997Q-02	5.807Q-01	5.806Q-01
2	5.00741512685879193554851897871660Q-01	-1.336Q-04	-1.331Q-04
3	5.00607892147028441952533495816130Q-01	5.499Q-07	5.485Q-07
4	5.00608442000119010432203633099740Q-01	-1.431Q-09	-1.338Q-09
5	5.00608440568955741463134570238500Q-01	9.313Q-11	9.304Q-11
6	5.00608440662089596948275028145400Q-01	-1.017Q-13	-9.860Q-14
7	5.00608440661987906940795946758110Q-01	3.095Q-15	3.092Q-15
8	5.00608440661991001691332357665060Q-01	-2.622Q-18	-2.628Q-18
9	5.00608440661990999069801659191560Q-01	-6.808Q-21	-6.749Q-21
10	5.00608440661990999062993777068620Q-01	5.892Q-23	5.881Q-23
11	5.00608440661990999063052701814030Q-01	-1.187Q-25	-1.194Q-25
12	5.00608440661990999063052583082720Q-01	-6.840Q-28	-6.840Q-28
13	5.00608440661990999063052582398680Q-01	-2.022Q-32	5.258Q-32
14	5.00608440661990999063052582398660Q-01	7.280Q-32	7.280Q-32
15	5.00608440661990999063052582398730Q-01		

TABLE 5. PADÉ APPROXIMATIONS FOR Z = 1.0

N	AN(Z)	BN(Z)
1	-1.111111111111111111111111111111Q+00	3.888888888888888888888888888889Q-01
2	6.6075102880658436213991769547326Q-01	2.3644547325102880658436213991770Q+00
3	7.4708779563974376614838936537785Q-01	2.6910864235044107080065434420340Q+00
4	8.0391814767328447238412978871070Q-01	2.8955124974366003088197480605811Q+00
5	5.9039003116029690752397245845169Q-01	2.1264399686839194484670869503726Q+00
6	1.4689461421146619818535790085117Q+00	5.2907822178928919105653533773543Q+00
7	2.2491746157712957975459393864324Q+00	8.1009730222069435084477393108097Q+00
8	1.9667677466274551518437551885921Q+00	7.0838130324631198670043492796812Q+00
9	2.1453828939008266884488649608916Q+00	7.7271408022127379229365369273813Q+00
10	3.0777118863523645482732004681859Q+00	1.1085160211773311286689000154434Q+01
11	2.9331911275699225339979217105911Q+00	1.0564632032337674389653896548794Q+01
12	3.9324000680722933745901921387780Q+00	1.4163536543062535742873057578063Q+01
13	4.38085045261748520381841611141093Q+00	1.5778744380338847747388824878600Q+01
14	2.4000690419272171566126658493178Q+01	8.6444575813145094357762725228829Q+01
15	6.6080725436916782433408687353981Q+00	2.3800649814774812570800668561973Q+01

N	FN(Z)	1ST DIFF'S	F15(Z)-FN(Z)
1	-2.8571428571428571428571428571429Q+00	3.137Q+00	3.135Q+00
2	2.7945175677145654302186446209072Q-01	-1.836Q-03	-1.809Q-03
3	2.7761568306188561336345587634676Q-01	2.710Q-05	2.684Q-05
4	2.7764278288731058708808457416587Q-01	-3.195Q-07	-2.565Q-07
5	2.7764246339185241795171777619120Q-01	6.310Q-08	6.297Q-08
6	2.7764252649576734901099059040820Q-01	-1.467Q-10	-1.312Q-10
7	2.7764252634908225214047212009033Q-01	1.551Q-11	1.545Q-11
8	2.7764252636458818694002983800774Q-01	-5.287Q-14	-5.330Q-14
9	2.7764252636453531962846469481822Q-01	-4.472Q-16	-4.339Q-16
10	2.7764252636453487239280256512311Q-01	1.341Q-17	1.332Q-17
11	2.7764252636453488580583736056138Q-01	-9.591Q-20	-9.775Q-20
12	2.7764252636453488570992374609643Q-01	-1.835Q-21	-1.833Q-21
13	2.7764252636453488570808896301902Q-01	-1.244Q-25	1.354Q-24
14	2.7764252636453488570808883861367Q-01	1.478Q-24	1.478Q-24
15	2.7764252636453488570809031699243Q-01		

TABLE 5. PADÉ APPROXIMATIONS FOR Z = 2.0

N	AN(Z)	BN(Z)
1	-3.2222222222222222222222222222222Q+00	-2.2222222222222222222222222222221Q-01
2	5.3806584362139917695473251028808Q-01	4.3528806584362139917695473251029Q+00
3	6.2279693925912336222843951138219Q-01	5.7331494901313274931584338546682Q+00
4	7.3653779141138444931952976811473Q-01	6.7367901563942387981408012156863Q+00
5	3.5572996144976123543837602080180Q-01	3.2550743828695688754642137667162Q+00
6	2.1280829342802068048850677854556Q+00	1.9468740369736811017650054593506Q+01
7	4.3861911912096431051848276585937Q+00	4.0127039995096446937726925579238Q+01
8	3.6786234875145564995449608927975Q+00	3.3653855068744142989287989808735Q+01
9	4.3764793126101447521341601374482Q+00	4.0038183134842006278974492378556Q+01
10	8.0917270118672227142695268666960 2Q+00	7.4027094579170097770854031004476Q+01
11	7.5870431624324996589998087915522Q+00	6.9409998747441468548195118977276Q+01
12	1.2906438065185810769732703339010Q+01	1.1807443700523016138883925193001Q+02
13	1.5533163156309791924379051083499Q+01	1.4210500878154345731302927116075Q+02
14	1.5465530833158606713268915605519Q+02	1.4148627505818001758971359342755Q+03
15	3.2216361376665935276482972764258Q+01	2.9473110339929688453681137258825Q+02

N	FN(Z)	1ST DIFF'S	F15(Z)-FN(Z)
1	1.4500000000000000000000000000001Q+01	-1.438Q+01	-1.439Q+01
2	1.2361143937603403450720869770740Q-01	-1.498Q-02	-1.430Q-02
3	1.0863085644830397474450471043867Q-01	6.998Q-04	6.768Q-04
4	1.0933067147895322639996751509096Q-01	-4.593Q-05	-2.303Q-05
5	1.0928474117883633233241969465358Q-01	2.294Q-05	2.290Q-05
6	1.0930768472254147088357427973617Q-01	-6.593Q-08	-4.462Q-08
7	1.0930761879634378249837060202686Q-01	2.159Q-08	2.130Q-08
8	1.0930764038771476369973740873454Q-01	-2.815Q-10	-2.876Q-10
9	1.0930764010621768871458703574934Q-01	-6.776Q-12	-6.151Q-12
10	1.0930764009944124121331601315327Q-01	6.411Q-13	6.256Q-13
11	1.0930764010008236304886027606013Q-01	-1.469Q-14	-1.551Q-14
12	1.0930764010006767316641978636103Q-01	-8.200Q-16	-8.162Q-16
13	1.0930764010006685316709236465018Q-01	-1.133Q-19	3.803Q-18
14	1.0930764010006685305381700918493Q-01	3.916Q-18	3.916Q-18
15	1.0930764010006685696980967858028Q-01		

TABLE 5. PADÉ APPROXIMATIONS FOR Z = 4.0

N	AN(Z)	BN(Z)
1	-7.4444444444444444444444444444444Q+00	-1.4444444444444444444444444444444Q+00
2	9.4238683127572016460905349794243Q-01	1.0201646090534979423868312757202Q+01
3	4.5002675472794263786329098442067Q-01	1.7566301273525050069562292650858Q+01
4	8.3071665802424549991012476796398Q-01	2.4738330261560465106191572155837Q+01
5	-1.3556368783277641089139877347566Q-02	2.1706740234257976297878633673540Q-01
6	4.7292800062129485097960405311263Q+00	1.4393118618395909751294093196025Q+02
7	1.4707829339543793821111937018672Q+01	4.4770815711174133772058746785214 9Q+02
8	1.2592438097423051171978474549119Q+01	3.8325042309003698586620902715782Q+02
9	1.7744829608465862636821563144164Q+01	5.4006723067650471592467920513809Q+02
10	4.6784850161180324614375614559356Q+01	1.4239063181716891353552110025769Q+03
11	4.4290102261751742474755974381469Q+01	1.3479780167558437317192011768863Q+03
12	1.0978170138358326211437933164217Q+02	3.3412278039565295281629297814318Q+03
13	1.4897146058468584756351966691960Q+02	4.5339758838686252881729338435833Q+03
14	2.8171680337362982434124542339224Q+03	8.5741066615292879733554790465269Q+04
15	5.0670374671506783864366014718699Q+02	1.5421628806010060946412102796904Q+04

N	FN(Z)	1ST DIFF'S	F15(Z)-FN(Z)
1	5.1538461538461538461538461538462Q+00	-5.061Q+00	-5.121Q+00
2	9.2375958047599838644461476401775 2Q-02	-6.676Q-02	-5.952Q-02
3	2.5618754211291928192500884686774Q-02	7.961Q-03	7.238Q-03
4	3.3580142606271635210682200834510Q-02	-9.603Q-02	-7.234Q-04
5	-6.2452347229376849398285481424901Q-02	9.531Q-02	9.531Q-02
6	3.2857924203920854422016240655069Q-02	-6.547Q-06	-1.230Q-06
7	3.2851376741135861708395574427997Q-02	5.568Q-06	5.318Q-06
8	3.2856945064519109150722479611163Q-02	-2.402Q-07	-2.505Q-07
9	3.2856704870314288171694053574043Q-02	-1.369Q-08	-1.035Q-08
10	3.2856691177023900048148526489938Q-02	3.599Q-09	3.345Q-09
11	3.2856694776331734836730906950518Q-02	-2.242Q-10	-2.541Q-10
12	3.2856694552099913666816111913776 7Q-02	-3.047Q-11	-2.982Q-11
13	3.2856694521624938210167173107511Q-02	-7.863Q-15	6.552Q-13
14	3.2856694521617074981094482404263Q-02	6.630Q-13	6.630Q-13
15	3.2856694522280104554657159704207Q-02		

TABLE 6. COEFFICIENTS IN THE POLYNOMIAL APPROXIMATIONS , $\gamma = 2.0$

N=4

I	TERM(I)
0	1.0000000000000000000000000000000Q+00
1	-1.4694969254369288458260675748318Q+00
2	1.1606311704317499442776415675683Q+00
3	-5.0068177944434975285495142321459Q-01
4	8.8967300808957533007302906740439Q-02

N=5

I	TERM(I)
0	1.0000000000000000000000000000000Q+00
1	-1.4928344076925120854017208500986Q+00
2	1.3027469442328301944673519051604Q+00
3	-7.4911677322917272160297481749492Q-01
4	2.5478959984366803010861156091426Q-01
5	-3.7619359614311551640965536982581Q-02

N=6

I	TERM(I)
0	1.0000000000000000000000000000000Q+00
1	-1.4984492935027160257372926030512Q+00
2	1.3536777856623453538877732919537Q+00
3	-8.8668229621312055012754558727574Q-01
4	4.1136638910390191336038780142497Q-01
5	-1.1723399057142879639315786290708Q-01
6	1.49255500363582522278558695632413Q-02

N=7

I	TERM(I)
0	1.0000000000000000000000000000000Q+00
1	-1.4996827459784058812932528125905Q+00
2	1.3692806149995949155366965380906Q+00
3	-9.4662743316819075831618020381007Q-01
4	5.1335358887225416949608384315330Q-01
5	-2.0344573313725115067611336483363Q-01
6	5.0424893627230066347110718303414Q-02
7	-5.6709156359948718831059749440189Q-03

TABLE 7. POLYNOMIAL APPROXIMATIONS

Z=0.5

N	FN(Z)	1ST DIFF'S	F15(Z)-FN(Z)
1	7.0000000000000000000000000000000Q-01	-1.750Q-01	-1.994Q-01
2	5.2500000000000000000000000000000Q-01	-2.975Q-02	-2.439Q-02
3	4.9525092936802973977695167286245Q-01	3.134Q-03	5.358Q-03
4	4.9838456375948918986246410824560Q-01	1.994Q-03	2.224Q-03
5	5.0037868056058693160861374858040Q-01	2.609Q-04	2.298Q-04
6	5.0063956147054307422694435704584Q-01	-1.870Q-05	-3.112Q-05
7	5.0062085669316799043298873984140Q-01	-1.111Q-05	-1.242Q-05
8	5.0060974231139963807221732153733Q-01	-1.407Q-06	-1.302Q-06
9	5.0060833509180456710510147442557Q-01	5.173Q-08	1.056Q-07
10	5.0060838682155486626712632158512Q-01	4.629Q-08	5.386Q-08
11	5.0060843310762528479935418128394Q-01	7.563Q-09	7.573Q-09
12	5.0060844067094781857706104779454Q-01	3.007Q-10	9.984Q-12
13	5.0060844097164785315262101996418Q-01	-2.035Q-10	-2.907Q-10
14	5.0060844076815500393415953097441Q-01	-8.722Q-11	-8.722Q-11
15	5.0060844068093168025218059154683Q-01		

Z=1.0

N	FN(Z)	1ST DIFF'S	F15(Z)-FN(Z)
1	4.0000000000000000000000000000000Q-01	-1.667Q-01	-1.224Q-01
2	2.3333333333333333333333333333334Q-01	3.626Q-02	4.431Q-02
3	2.6959107806691449814126394052045Q-01	9.829Q-03	8.051Q-03
4	2.7941976635942887860392547626228Q-01	-1.454Q-03	-1.777Q-03
5	2.7796600354050186593030226149863Q-01	-3.619Q-04	-3.235Q-04
6	2.7760414451534014726872373577708Q-01	2.813Q-05	3.838Q-05
7	2.7763226957923648921123874336910Q-01	9.498Q-06	1.026Q-05
8	2.7764176732580155986587141049041Q-01	1.136Q-06	7.604Q-07
9	2.7764290333343645179683957888910Q-01	-8.409Q-08	-3.756Q-07
10	2.7764281924431232090775446079559Q-01	-2.652Q-07	-2.915Q-07
11	2.7764255408852089706540625488624Q-01	-6.237Q-08	-2.634Q-08
12	2.7764249171441861591395949362711Q-01	2.072Q-08	3.603Q-08
13	2.7764251243090553148496008484517Q-01	1.349Q-08	1.532Q-08
14	2.7764252591820552963570753350711Q-01	1.831Q-09	1.831Q-09
15	2.7764252774885203454274849349211Q-01		

as the quantities $F_g(z) - F_n(z)$ are also posted. Here g is the largest value of the index for which calculations are made. In the absence of information concerning convergence, we accept the number of common decimals in $F_n(z)$ and $F_{n-1}(z)$ as correct and approximate the error in $F_n(z)$ by $F_g(z) - F_n(z)$. Thus in Table 3, the machine prints the latter quantities and first differences. We recognize that acceptance of common decimals does not guarantee such accuracy, but it may be viewed as a measure of the accuracy. It seems to work very well when convergence is rapid, though it can be somewhat misleading if convergence is slow. For the rational approximation data at hand, the technique is quite accurate for $n \ll g$. The process weakens as n nears g , but even here it is sufficient to indicate the rate of convergence. In illustration, suppose $z = 1$. Since the Padé approximations given in Table 5 converge much more rapidly than do the rational approximations of Table 3, let us accept the Padé approximation for $n = 15$ as correct. In the display below, we post the error in the rational approximation, call it R_n, and the value of $F_{15}(1) - F_n(1)$ for $n = 10(1)14$.

n	R_n	$F_{15}(1) - F_n(1)$	
10	-9.41 (-9)	-9.43 (-9)	
11	-1.54 (-8)	-1.54 (-8)	
12	-1.98 (-9)	-2.00 (-9)	(27)
13	3.85 (-10)	3.67 (-10)	
14	1.59 (-10)	1.41 (-10)	

The following display is like that above except that $z = 4$.

n	R_n	$F_{15}(4) - F_n(4)$
10	-3.79 (-4)	-4.13 (-4)
11	-7.49 (-4)	-7.83 (-4)
12	-2.00 (-4)	-2.35 (-4)
13	1.61 (-4)	1.27 (-4)
14	1.41 (-4)	1.06 (-4)

In the previous section we noted that for ${}_{p+1}F_p(z)$, the Taylor series converges in the unit circle, but the rational approximations under rather liberal conditions converge in a much larger domain, namely for z fixed, $z \neq 1$, $|\arg(1-z)| < \pi$. For the above example, the Taylor series converges for $|z| < 2$. However, the rational approximations appear to converge for all $z > 0$. A similar statement applies to the Padé approximations given in Table 5. The advantage of rational and Padé approximations is manifest.

It is of interest to compare the rational and Padé approximations with the truncated Taylor series expansion of about the same number of terms. Thus if we speak of n as the order of the approximation given by (8) , then we compare such approximations with the truncated Taylor series of 2n terms. As already noted the the Padé approximations converge much more rapidly than the rational approximations. If $z = 0.5$ and $n = 10$, the rational approximation error is -1.17 (-9) while the truncated Taylor series of 20 terms give an error of 1.69 (-9) . For this value of z, the advantage of the rational approximation is not decisive. However the advantage improves significantly as z increases. Thus for $z = 1$ and $n = 10$, the error in the rational approximation is -9.43 (-9) while the error in the truncated Taylor series expansion of 20 terms is about 1,5 (-3).

The polynomials in the Padé approximations for $n = 4, 5, 6$ and 7 are presented in Table 4 and the Padé approximations for the same z and n used in the rational approximations are listed in Table 5. In Table 6, we give the coefficients in the polynomial approximations for $\gamma = 2$, and in Table 7, polynomial approximations for $z = 0.5$ and $z = 1.0$ are presented for the same n as in the previous tables.

It is of interest to compare the polynomial approximations with the truncated Taylor series expansion of the same number of terms. With $n = 10$ and $z = 0.5$, the error in the polynomial approximation (assume $F_{15}(0.5)$ for the Padé approximation is correct) is 5.38 (-8). The error using 11 terms of the Taylor series is -1.62 (-4). Again if $z = 1.0$, the corresponding errors are -2.93 (-7) and -4.85 (-3), respectively. Some care must be exercised in discussing the polynomial approximations since for n fixed, the error oscillates as z goes from 0 to γ and the maximum error in magnitude occurs in the vicinity of $z = \gamma$. This is illustrated in the next display where the true vales are $F_{15}(z)$ derived from the Padé approximations.

z	$R_{10}(z)$	z	$R_{10}(z)$	
0	0	1.0	-2.93 (-7)	
0.1	3.82 (-8)	1.1	-4.07 (-7)	
0.2	-1.36 (-8)	1.2	-2.62 (-7)	
0.3	-4.93 (-8)	1.3	4.08 (-7)	
0.4	5.55 (-9)	1.4	1.71 (-6)	
0.5	5.38 (-8)	1.5	3.38 (-6)	(29)
0.6	4.33 (-8)	1.6	4.81 (-6)	
0.7	-1.65 (-10)	1.7	5.32 (-6)	
0.8	-5.52 (-7)	1.8	4.61 (-6)	
0.9	-1.47 (-7)	1.9	2.55 (-6)	
		2.0	-3.38 (-6)	

B. Logarithm of the Gamma Function

Consider

$$\ln \Gamma(z) = (z-\tfrac{1}{2})\ln z - z + \tfrac{1}{2}\ln(2\pi) + (12z)^{-1} H(z), \qquad (30)$$

$$H(z) \sim \sum_{n=0}^{\infty} h_n z^{-2n} , \quad |z|\to\infty , \quad |\arg z|<\pi , \qquad (31)$$

$$h_n = \frac{6\, B_{2n+2}}{2(n+1)(n+1)} \qquad (32)$$

where the B_{2m} are the Bernoulli numbers, $B_0 = 1$, $B_2 = \frac{1}{6}$, $B_4 = -1/30$, etc. Notice that the series in (31) is divergent though asymptotic. It is well known that the Bernoulli numbers can be generated by

$$B_{2n} = -(2n+1)^{-1}\left[-\frac{(2n-1)}{2} + \sum_{k=1}^{n-1} \binom{2n+1}{2k} B_{2k}\right] , \quad n\geq 2 . \qquad (33)$$

Here $\binom{m}{k}$ is the usual symbol for the binomial coefficient $m!/k!(m-k)!$. It follows that the coefficients h_n can be generated by

$$h_n = \frac{3}{(2n+3)(n+1)} - \sum_{k=1}^{n} \frac{k(2k-1)\binom{2n+3}{2k} h_{k-1}}{(n+1)(2n+1)(2n+3)} , \quad n\geq 1 \qquad (34)$$

The rational and polynomial approximations as well as the Padé approximations are described by (19) - (25) with z replaced by $1/z^2$. Also in (23), replace γ by $1/\gamma^2$.

The format of the tables is like that of Example A with the added feature of giving approximations for $\ln \Gamma(z)$ which in the computer notation is LNGAMA (Z) . These are the values of the right hand side of (30) with $H(z)$ replaced by $F_n(z)$.

The coefficients in the expansion (31) are presented in Table 8. We have computed coefficients in the polynomials which define the rational and Padé approximations as well as the approximations themselves with $\alpha = \beta = 0$ for $n = 1, 2, \ldots, 15$ and $z = 1.0\ (0.5)\ 15.0$. Similar computations were made for the polynomial approximations with $\gamma = 2$ whence $z \geq 2$. As in Example A, only a small fraction of these data are given. Thus in Tables 9 and 11 we present the coefficients in the polynomials which define the rational approximations and the Padé approximations, respectively, for $n = 4, 5, 6$ and 7 . In Tables 10 and 12, we give values of the rational and Padé approximations, respectively, for $h(z)$, see (30), for $z = 1.0$, 2.0, 5.0 and 10.0 each for $n = 1\ (1)\ 15$. Also tabulated are the corresponding approximations for $\ln \Gamma(z)$. The coefficients in the polynomial approximations are given in Table 13 for $n = 4, 5, 6$, and 7 with $\gamma = 2$. The corresponding approximations for $H(z)$ and $\ln \Gamma(z)$ are noted in Table 14 for z and n as in the rational and Padé approximations. Since $\gamma = 2$, we use the polynomial approximation for $z \geq 2$ only. Also all polynomials spoken of above are polynomials in $1/z^2$. For any z , there is an optimum number of terms of the asymptotic expansion that can be used to produce the maximum accuracy. Since the series diverges, use of more terms causes the accuracy to deteriorate. On the other hand, for fixed z the accuracy in the rational, Padé and polynomial approximations improves as n increases. These approximations appear to converge in the region where the series for $H(z)$, upon which the approximations are based, diverges. In this respect the approximations are quite remarkable. Further if n is fixed the error in the Padé and rational approximation also improves as z increases which is expected. The errors in the polynomial approximations for n fixed and z increasing are discussed later.

Notice that the rational and Padé approximations are good even
for z = 1 . For this value of z , the optimum asymptotic
approximation uses 3 terms for which the approximation for H(z)
is 0.969 whereas the true value is 0.973 . Notice that for n≥5
the rational approximations are better than the Padé. Again, if
z=2 , eight terms of the asymptotic expansion produces the optimum
approximation for H(z) as 0.99216 whereas the true value is
0.99218.

TABLE 8.   COEFFICIENTS IN THE ASYMPTOTIC SERIES FOR H(Z)

N	H(N)
0	1.0000000000000000000000000000000000000Q+00
1	-3.3333333333333333333333333333333333333Q-02
2	9.5238095238095238095238095238095238097Q-03
3	-7.1428571428571428571428571428571428577Q-03
4	1.0101010101010101010101010101010101014Q-02
5	-2.3010323010323010323010323010323010356Q-02
6	7.6923076923076923076923076923076923527Q-02
7	-3.5460784313725490196078431373733381Q-01
8	2.1557324684259668779792618803933Q+00
9	-1.6709186602870813397129186609154Q+01
10	1.6083436853002070393374741225020Q+02
11	-1.8821794155120242076763816007493Q+03
12	2.6317240000000000000000006333359Q+04
13	-4.3330525504469987228607922434315Q+05
14	8.2976672262157568053007462044166Q+06
15	-1.8285865847288899430740066115363Q+08
16	4.5948090166969696969697252782080Q+09
17	-1.3058719242941269306818500816486Q+11
18	4.1678434051800270270274378747992Q+12
19	-1.4843522570723129345107905257233Q+14
20	5.8654567775169520209190330073782Q+15
21	-2.5584400753103248676531478670769Q+17
22	1.2261303558308400931092869948609Q+19
23	-6.4290566607960244339482833456774Q+20
24	3.6738939164458600995351738434067Q+22
25	-2.2799900911679044897175734739549Q+24
26	1.5316048840594601078027480090667Q+26
27	-1.1103416611344499855599777818841Q+28
28	8.6625871142227332528032216102433Q+29
29	-7.2542200871950313609362631081470Q+31
30	6.5048045658841242120377671360815Q+33
31	-6.2315493783769395595980402521794Q+35
32	6.3643906261438091253549380397620Q+37
33	-6.9159904177986888381041840429548Q+39
34	7.9813868578215238773175947983246Q+41
35	-9.7648540297805535864774190708980Q+43
36	1.2644360344114223888744952373205Q+46
37	-1.7301816720423462067958512686524Q+48
38	2.4980827829215026748350825191380Q+50
39	-3.8004271978340373375847144139775Q+52
40	6.0840077640040122546060425516900Q+54
41	-1.0235967455363913723267083974759Q+57
42	1.8077007872783704581845320932536Q+59
43	-3.3472197359552486003102980817861Q+61
44	6.4912234040891296493916512481805Q+63
45	-1.3170428773560694392320084289058Q+66
46	2.7930049832296372484792154018954Q+68

TABLE 9. COEFFICIENTS IN THE POLYNOMIALS FOR THE RATIONAL APPROXIMATIONS

N=4

I	V(I)	U(I)
0	9.9999999999999999999999999999995Q-01	1.0000000000000000000000000000000Q+00
1	4.2216666666666666666666666666667Q+02	6.0000000000000000000000000000000Q+02
2	8.8626666666666666666666666666665Q+03	9.4499999999999999999999999999998Q+03
3	1.9368999999999999999999999999998Q+04	1.9599999999999999999999999999998Q+04
4	6.9299999999999999999999999999976Q+03	6.9299999999999999999999999999976Q+03

N=5

I	V(I)	U(I)
0	-9.9999999999999999999999999999857Q-01	1.0000000000000000000000000000000Q+00
1	5.7678895320791123974915581283160Q+02	9.0000000000000000000000000000000Q+02
2	1.9952440906898215147129763627593Q+04	2.2050000000000000000000000000000Q+04
3	7.6425301013024602026049204052092Q+04	7.8399999999999999999999999999994Q+04
4	6.2004946454413892908827785817634Q+04	6.2369999999999999999999999999978Q+04
5	1.0951606367583212735166425470317Q+04	1.0951606367583212735166425470317Q+04

N=6

I	V(I)	U(I)
0	1.0000000000000000000000000000063Q+00	1.0000000000000000000000000000000Q+00
1	7.4294515195369030390738060781512Q+02	1.2600000000000000000000000000000Q+03
2	3.8490849975880366618427399903521Q+04	4.4099999999999999999999999999999Q+04
3	2.2586651114327062228654124457307Q+05	2.3519999999999999999999999999998Q+05
4	3.0794881099855282199710564399411Q+05	3.1184999999999999999999999999989Q+05
5	1.2006727004341534008683068017349Q+05	1.2046767004341534008683068017349Q+05
6	1.2011999999999999999999999999930Q+04	1.2011999999999999999999999999930Q+04

N=7

I	V(I)	U(I)
0	-1.0000000000000000000000000000103Q+00	1.0000000000000000000000000000000Q+00
1	9.1885529104202017566988497395176Q+02	1.6800000000000000000000000000000Q+03
2	6.6861732450039759909097780149698Q+04	7.9379999999999999999999999999998Q+04
3	5.5575122743468593996751951609756Q+05	5.8799999999999999999999999999995Q+05
4	1.1207745352989515147070700581274Q+06	1.1434499999999999999999999999996Q+06
5	7.1769299451697926355506915549097Q+05	7.2280602026049204052098408104092Q+05
6	1.5583339010229471938070223942403Q+05	1.5615599999999999999999999999909Q+05
7	9.6782969311584185789328172516389Q+03	9.6782969311584185789328172516389Q+03

TABLE 10. RATIONAL APPROXIMATIONS FOR Z = 1.0

N	AN(Z)	BN(Z)
1	5.900000000000000000000000000000Q+01	6.100000000000000000000000000000Q+01
2	7.899999999999999999999999999999Q+02	8.109999999999999999999999999999Q+02
3	6.137333333333333333333333333331Q+03	6.310999999999999999999999999997Q+03
4	3.558483333333333333333333333329Q+04	3.658099999999999999999999999996Q+04
5	1.699100836951278340569223347804 7Q+05	1.746726063675832127351664254702 9Q+05
6	7.051293873130728412928123492519 3Q+05	7.248906700434153400868306801732 9Q+05
7	2.627510032025151636274061451515 2Q+06	2.701151317191650459099916898291 2Q+06
8	8.983044813869704425597950972127 8Q+06	9.234807003666626713254718262970 6Q+06
9	2.860872433718620108634417298897 8Q+07	2.941052656797658702595460466060 4Q+07
10	8.582596389110192224434171641344 2Q+07	8.823136099608731017550895059746 6Q+07
11	2.446105104221678308403936109114 5Q+08	2.514660797590269127342303398460 4Q+08
12	6.667382946755631881376259832442 4Q+08	6.854246128706502529029739323987 8Q+08
13	1.747308664343362429517683607729 8Q+09	1.796279554758918662921374616140 7Q+09
14	4.421815092093550867876232730514 7Q+09	4.545742946840473812447131841396 7Q+09
15	1.084443727263393932091080675317 9Q+10	1.114836853871960361526402460673 4Q+10

N	FN(Z)	1ST DIFF'S	F15(Z)-FN(Z)
1	9.672131147540983606557377049180 3Q-01	6.893Q-03	5.524Q-03
2	9.741060419235511713933415536374 8Q-01	-1.624Q-03	-1.368Q-03
3	9.724819098927798024613109385728 6Q-01	2.863Q-04	2.557Q-04
4	9.727681947823551388243441495129 5Q-01	-3.362Q-05	-3.059Q-05
5	9.727345760076822433704261440519 0Q-01	4.371Q-06	3.025Q-06
6	9.727389473378668938840572727344 0Q-01	-1.872Q-06	-1.346Q-06
7	9.727370752249960416503688084388 9Q-01	6.128Q-07	5.262Q-07
8	9.727376879996559141721942871358 6Q-01	-1.115Q-07	-8.653Q-08
9	9.727375764953687777160522079408 2Q-01	4.108Q-08	2.497Q-08
10	9.727376175791728161434958668657 3Q-01	-1.950Q-08	-1.611Q-08
11	9.727375980751416371864145257519 6Q-01	4.508Q-09	3.393Q-09
12	9.727376025835631190319938561005 2Q-01	-1.850Q-09	-1.116Q-09
13	9.727376007337963109696704590706 9Q-01	9.748Q-10	7.341Q-10
14	9.727376017086361634443144832096 4Q-01	-2.408Q-10	-2.408Q-10
15	9.727376014678672462622955238498 4Q-01		

N	LNGAMA(Z)	1ST DIFF'S	
1	-4.603738991523948316921215178815 5Q-04	5.744Q-04	4.604Q-04
2	1.140366983020060631081992087393 9Q-04	-1.353Q-04	-1.140Q-04
3	-2.130763759560801456101871331231 2Q-05	2.386Q-05	2.131Q-05
4	2.549436535670015691748865028415 3Q-06	-2.802Q-06	-2.549Q-06
5	-2.521280204046054680849233926659 5Q-07	3.643Q-07	2.521Q-07
6	1.121494949829373345091334825023 4Q-07	-1.560Q-07	-1.122Q-07
7	-4.385991092141547286289114355998 0Q-08	5.106Q-08	4.385Q-08
8	7.204644067961345927000271164153 9Q-09	-9.292Q-09	-7.211Q-09
9	-2.087379860076665912932649105899 6Q-09	3.424Q-09	2.081Q-09
10	1.336270476458954391977759681584 8Q-09	-1.625Q-09	-1.343Q-09
11	-2.890654551208023864483877788624 6Q-10	3.757Q-10	2.827Q-10
12	8.663633503299589108065924995837 3Q-11	-1.541Q-10	-9.297Q-11
13	-6.751089897219772533849365061151 5Q-11	8.124Q-11	6.117Q-11
14	1.372575540068927667308610992674 1Q-11	-2.006Q-11	-2.006Q-11
15	-6.338321031145636606897449662448 0Q-12		

TABLE 10. RATIONAL APPROXIMATIONS FOR Z = 2.0

N	AN(Z)	BN(Z)
1	5.97500000000000000000000000000000Q+01	6.02500000000000000000000000000000Q+01
2	6.69812499999999999999999999999999Q+02	6.75062499999999999999999999999999Q+02
3	3.58175520833333333333333333333331Q+03	3.61001562499999999999999999999998Q+03
4	1.23327669270833333333333333333331Q+04	1.24300039062499999999999999999997Q+04
5	3.15434332889569766039556198745560Q+04	3.17921542191457127351664254703111Q+04
6	6.49558630941108523275446213216360Q+04	6.54680388487444600217076700432940Q+04
7	1.13240980889721787025339471050270Q+05	1.14133884195974327361494322316460Q+05
8	1.72854545367014398131884542245710Q+05	1.74217500610816932956901348206780Q+05
9	2.36579175730353708981494675707260Q+05	2.38444598666361023617800341298920Q+05
10	2.95455246540629614417992772078010Q+05	2.97784906325103282699222248257490Q+05
11	3.41197528878918352617133279627950Q+05	3.43887865812140276796961174665280Q+05
12	3.68155292597688816743548363724520Q+05	3.71058191056568405506844748739390Q+05
13	3.74254888383342086936736214983320Q+05	3.77205882055296883552494786280690Q+05
14	3.60861980569688380961690720685580Q+05	3.63707371382625356644992863683360Q+05
15	3.31869538174984170670080937618210Q+05	3.34486324053970726904107646848500Q+05

N	FN(Z)	1ST DIFF'S	F15(Z)-FN(Z)
1	9.91701244813278008298755186721990Q-01	5.217Q-04	4.755Q-04
2	9.92222942320155541153596889176930Q-01	-5.128Q-05	-4.624Q-05
3	9.92171663615260206341443005065480Q-01	5.573Q-06	5.039Q-06
4	9.92177236636444305890809609602440Q-01	-5.797Q-07	-5.337Q-07
5	9.92176656904899113389323072945180Q-01	4.850Q-08	4.602Q-08
6	9.92176705402510615249747090201490Q-01	-2.687Q-09	-2.473Q-09
7	9.92176702715914102133270851699210Q-01	3.189Q-10	2.139Q-10
8	9.92176703034861982480748674903290Q-01	-1.296Q-10	-1.050Q-10
9	9.92176702905241852549008206495800Q-01	2.701Q-11	2.458Q-11
10	9.92176702932249061935421110660290Q-01	-2.846Q-12	-2.429Q-12
11	9.92176702929403136887518542937570Q-01	6.313Q-13	4.173Q-13
12	9.92176702930034396350943077890880Q-01	-2.593Q-13	-2.139Q-13
13	9.92176702929775117693764456318930Q-01	5.474Q-14	4.533Q-14
14	9.92176702929829860798480859700550Q-01	-9.409Q-15	-9.409Q-15
15	9.92176702929820451480215545848120Q-01		

N	LNGAMA(Z)	1ST DIFF'S	
1	-1.98107548560437480406152937053550Q-05	2.174Q-05	1.981Q-05
2	1.92664126385345424445564191699920Q-06	-2.137Q-06	-1.927Q-06
3	-2.09971440118829595289529393383930Q-07	2.322Q-07	2.100Q-07
4	2.22377758853182949856596465406940Q-08	-2.416Q-08	-2.224Q-08
5	-1.91770516436926695336773905984700Q-09	2.021Q-09	1.918Q-09
6	1.03028648208250714017940617892440Q-10	-1.119Q-10	-1.030Q-10
7	-8.91287317160246258632106055051370Q-12	1.329Q-11	8.913Q-12
8	4.37662184287578004718201890926740Q-12	-5.401Q-12	-4.377Q-12
9	-1.02421690428007280312979840163620Q-12	1.125Q-12	1.024Q-12
10	1.01083486820464870390427143949850Q-13	-1.186Q-13	-1.012Q-13
11	-1.74967235088087847230864189094840Q-14	2.630Q-14	1.739Q-14
12	8.80575413388017166509131306453670Q-15	-1.080Q-14	-8.914Q-15
13	-1.99752324856239383290223306131560Q-15	2.281Q-15	1.889Q-15
14	2.83439447954413734517747764633160Q-16	-3.921Q-16	-3.921Q-16
15	-1.08615479766996783181520282108360Q-16		

TABLE 10. RATIONAL APPROXIMATIONS FOR Z = 5.0

N	AN(Z)	BN(Z)
1	5.99600000000000000000000000000Q+01	6.00400000000000000000000000000Q+01
2	6.36361599999999999999999999999Q+02	6.37201599999999999999999999999Q+02
3	2.92271726933333333333333333331Q+03	2.92657606399999999999999999998Q+03
4	7.71896728789333333333333333309Q+03	7.72915840255999999999999999975Q+03
5	1.35553631400759695127834056922170Q+04	1.35732598716856127351664254703160Q+04
6	1.73219629687050852522913651711172Q+04	1.73448326267647096034732272068690Q+04
7	1.71331004702836101849761161422224Q+04	1.71557207789686496837663917812680Q+04
8	1.37173162817310901879171830846450Q+04	1.37354268349935875399924962958400Q+04
9	9.19433554375182445232775163579860Q+03	9.20647454384182341774624625255870Q+03
10	5.29528058076927883258450281072160Q+03	5.30227177778101796622631302278348Q+03
11	2.67481035773831572063902062084830Q+03	2.67834182806793408521656553423370Q+03
12	1.20474740267389424673213494862360Q+03	1.20633799383297180488496804144680Q+03
13	4.90388244404732844670270375373510Q+02	4.91035689009580744317551679137860Q+02
14	1.82411816415932105596909962053500Q+02	1.82652649159674485878178446685630Q+02
15	6.25851716981730378810196826567660Q+01	6.26678009867454996791748857084950Q+01

N	FN(Z)	1ST DIFF'S	F15(Z)-FN(Z)
1	9.98667554963357761492338441039310Q-01	1.418Q-05	1.392Q-05
2	9.98681735890179811224579473748970Q-01	-2.715Q-07	-2.645Q-07
3	9.98681464420442049147208952684630Q-01	7.205Q-09	6.969Q-09
4	9.98681471625255961369956924705570Q-01	-2.454Q-10	-2.360Q-10
5	9.98681471379843205363369936671130Q-01	9.835Q-12	9.428Q-12
6	9.98681471389678657618019317481410Q-01	-4.246Q-13	-4.071Q-13
7	9.98681471389254018641708648336140Q-01	1.819Q-14	1.751Q-14
8	9.98681471389272206254863119978790Q-01	-6.960Q-16	-6.765Q-16
9	9.98681471389271510294663837679340Q-01	1.995Q-17	1.950Q-17
10	9.98681471389271530240894486843220Q-01	-5.363Q-19	-4.492Q-19
11	9.98681471389271529704630664596600Q-01	1.060Q-19	8.705Q-20
12	9.98681471389271529810665484384560Q-01	-2.128Q-20	-1.898Q-20
13	9.98681471389271529789382503186190Q-01	2.475Q-21	2.302Q-21
14	9.98681471389271529791857837936270Q-01	-1.730Q-22	-1.730Q-22
15	9.98681471389271529791684877336960Q-01		

N	LNGAMA(Z)	1ST DIFF'S	
1	3.17805359840751372350861904327410Q+00	2.363Q-07	2.319Q-07
2	3.17805383475629409100415639381930Q+00	-4.524Q-09	-4.408Q-09
3	3.17805383023179846163620021846820Q+00	1.201Q-10	1.161Q-10
4	3.17805383035187869350657935133520Q+00	-4.090Q-12	-3.933Q-12
5	3.17805383034778848090646956820130Q+00	1.639Q-13	1.571Q-13
6	3.17805383034795240511071372454810Q+00	-7.077Q-15	-6.785Q-15
7	3.17805383034794532779444188006240Q+00	3.031Q-16	2.919Q-16
8	3.17805383034794563092132778792310Q+00	-1.160Q-17	-1.127Q-17
9	3.17805383034794561932199113321810Q+00	3.324Q-19	3.250Q-19
10	3.17805383034794561965442831070420Q+00	-8.938Q-21	-7.487Q-21
11	3.17805383034794561964549058033340Q+00	1.767Q-21	1.451Q-21
12	3.17805383034794561964725782732990Q+00	-3.547Q-22	-3.163Q-22
13	3.17805383034794561964690311097660Q+00	4.126Q-23	3.837Q-23
14	3.17805383034794561964694436655570Q+00	-2.883Q-24	-2.883Q-24
15	3.17805383034794561964694148387910Q+00		

TABLE 10. RATIONAL APPROXIMATIONS FOR Z = 10.0

N	AN(Z)	BN(Z)
1	5.999000000000000000000000000000Q+01	6.001000000000000000000000000000Q+01
2	6.315900999999999999999999999999Q+02	6.318000999999999999999999999999Q+02
3	2.830594832333333333333333333331Q+03	2.831536000999999999999999999998Q+03
4	7.124576688843333333333333333309Q+03	7.126945600009999999999999999976Q+03
5	1.157931832043735055475156777615 4Q+04	1.158316842658331273516642547031 7Q+04
6	1.324369383302794722768933912197 3Q+04	1.324809734156015440086830680166 5Q+04
7	1.130952647036772908776447131776 7Q+04	1.131328687112414779298491565973 4Q+04
8	7.531844575402494312838116748472 8Q+03	7.534348902538805498910499878487 9Q+03
9	4.045069009875732213906071796274 4Q+03	4.046413989314469752032747044471 8Q+03
10	1.799140052733586679805738409921 1Q+03	1.799738264129344840817379840648 7Q+03
11	6.771788873357054826436916465787 0Q+02	6.774040483101140918011186618659 6Q+02
12	2.195886738787661662606973310504 0Q+02	2.196616867867352336090988949326 0Q+02
13	6.227443906175181802405660896500 3Q+01	6.229514522026048165968317600512 0Q+01
14	1.564406655036740847294146193088 3Q+01	1.564926817926350788612709515990 0Q+01
15	3.519583265461249735789270819492 7Q+00	3.520753521670331204209113731474 8Q+00

N	FN(Z)	1ST DIFF'S	F15(Z)-FN(Z)
1	9.996667222129645059156807198800 2Q-01	8.942Q-07	8.898Q-07
2	9.996676163868919932111438412244 6Q-01	-4.415Q-09	-4.383Q-09
3	9.996676119723237569153313171430 6Q-01	3.152Q-11	3.122Q-11
4	9.996676120038486974298351381213 1Q-01	-3.080Q-13	-3.041Q-13
5	9.996676120035407405148693332685 5Q-01	3.868Q-15	3.810Q-15
6	9.996676120035446088756242485408 3Q-01	-5.914Q-17	-5.810Q-17
7	9.996676120035445497381068116072 9Q-01	1.059Q-18	1.038Q-18
8	9.996676120035445507973210277218 7Q-01	-2.146Q-20	-2.100Q-20
9	9.996676120035445507758570470194 3Q-01	4.717Q-22	4.612Q-22
10	9.996676120035445507763287224119 1Q-01	-1.072Q-23	-1.048Q-23
11	9.996676120035445507763180059970 4Q-01	2.378Q-25	2.332Q-25
12	9.996676120035445507763182437661 7Q-01	-4.678Q-27	-4.611Q-27
13	9.996676120035445507763182390879 7Q-01	6.837Q-29	6.734Q-29
14	9.996676120035445507763182391563 5Q-01	-1.031Q-30	-1.031Q-30
15	9.996676120035445507763182391553 2Q-01		

N	LNGAMA(Z)	1ST DIFF'S	
1	1.280182747266654811083387922857 3Q+01	7.451Q-09	7.415Q-09
2	1.280182748011799750656134142125 1Q+01	-3.679Q-11	-3.653Q-11
3	1.280182748008120943792554298355 0Q+01	2.627Q-13	2.602Q-13
4	1.280182748008147214576316384872 5Q+01	-2.566Q-15	-2.535Q-15
5	1.280182748008146957945553913368 4Q+01	3.224Q-17	3.175Q-17
6	1.280182748008146961169187875797 8Q+01	-4.928Q-19	-4.842Q-19
7	1.280182748008146961119906611267 0Q+01	8.827Q-21	8.652Q-21
8	1.280182748008146961120789289780 5Q+01	-1.789Q-22	-1.750Q-22
9	1.280182748008146961120771403129 9Q+01	3.931Q-24	3.843Q-24
10	1.280182748008146961120771796192 7Q+01	-8.930Q-26	-8.736Q-26
11	1.280182748008146961120771787262 4Q+01	1.981Q-27	1.943Q-27
12	1.280182748008146961120771787460 5Q+01	-3.898Q-29	-3.842Q-29
13	1.280182748008146961120771787456 6Q+01	5.701Q-31	5.608Q-31
14	1.280182748008146961120771787456 7Q+01	-9.244Q-33	-9.244Q-33
15	1.280182748008146961120771787456 7Q+01		

TABLE 11. COEFFICIENTS IN THE POLYNOMIALS FOR THE PADÉ APPROXIMATIONS

N=4

I	A(I)	B(I)
0	1.0	1.0
1	1.26132776884720362660691693706130+01	1.26466110218053695994025027039460+01
2	3.79067160935681851766075626129610+01	3.83187459847712213061114555602350+01
3	2.70375678400565295861235725439340+01	2.82015583155317096334758353225710+01
4	2.33308148453684756608918703893170+00	2.98842491620673069946580836308520+00

N=5

I	A(I)	B(I)
0	1.0	1.0
1	2.38999571400418484978303350328320+01	2.39332904733751818311636683661660+01
2	1.60123155486334213003731439361260+02	1.60911408025922909540960704497270+02
3	3.42537221120780794765924327353030+02	3.47680141479041318590326220661360+02
4	1.96754376629019153374538572455250+02	2.06972076142782458434467079246030+02
5	1.46608772697208619042473039208560+01	1.91793341461962559341224733822740+01

N=6

I	A(I)	B(I)
0	1.0	1.0
1	4.04365364967936091271138333860450+01	4.04698698301269424604471667193780+01
2	5.08243001947737603107535249779820+02	5.09582473799218024999073964860940+02
3	2.39629454879455844125440315482940+03	2.41290234677995973787379659968940+03
4	4.07560205917706973118028465717010+03	4.15145794047461859026407978957990+03
5	1.99473598822649605990399493150260+03	2.11339199579728588622118009412050+03
6	1.32544010133311084508559725099570+02	1.76394736069658022083970549205130+02

N=7

I	A(I)	B(I)
0	1.0	1.0
1	6.32230337616892972682918265619800+01	6.32563670950226306016251598953130+01
2	1.33481979979824790025313156613590+03	1.33691882155855817841604288098960+03
3	1.17109338795895496370620069252960+04	1.17549025415739299322415738293300+04
4	4.34942455083351376221183348782330+04	4.38737847630323838839923789142860+04
5	6.23805475508376262162615768611350+04	6.37399890674358311059910594699980+04
6	2.69273272751160331737563608778160+04	2.87059859936164275940150101526950+04
7	1.62991935450631759051101200877140+03	2.20063749430940196541167636333880+03

TABLE 12. PADÉ APPROXIMATIONS FOR Z = 1.0

N	AN(Z)	BN(Z)
1	-7.60476190476190476190476190476200Q+00	-7.57142857142857142857142857142870Q+00
2	3.04479964479964479964479964480030Q+00	3.12937062937062937062937062937130Q+00
3	1.26529879278113048480799743822990Q+01	1.30066607499267760840341370216140Q+01
4	8.08906431066335985948894915664390Q+01	8.31553402383150312384556019498380Q+01
5	7.38975587645896871546271978123230Q+02	7.59676250267318124331040146153100Q+02
6	9.14885614477596652908189155176750Q+03	9.40519936275086720390254816417520Q+03
7	1.47542016401944601437230714943950Q+05	1.51676475048621555290669366770530Q+05
8	3.00603008182790565241364807094980Q+06	3.09027065793234494321767488204860Q+06
9	7.55170951576819096429510412865980Q+07	7.76334447819972335848660597204930Q+07
10	2.29346001628720129904333639006000Q+09	2.35773496909282553681389533375500Q+09
11	8.28365081258859278411032890026330Q+10	8.51580562046916340303065400111530Q+10
12	3.50955840885344317707232162871600Q+12	3.60791689557231239426129710089680Q+12
13	1.72376078698151540615073815420870Q+14	1.77207097197378831123541427100380Q+14
14	9.71600999941136118464627223260850Q+15	9.98831228260058460093272837662490Q+15
15	6.22935251843446043328065729472870Q+17	6.40393770700669956299949880122200Q+17

N	FN(Z)	1ST DIFF'S	F15(Z)-FN(Z)
1	1.00440251572327044025157232704400Q+00		
2	9.72975082025361354970293517779550Q-01	-3.143Q-02	-3.166Q-02
3	9.72808330368925604791667996783450Q-01	-1.668Q-04	-2.372Q-04
4	9.72765463706948546225381312289500Q-01	-4.287Q-05	-7.049Q-05
5	9.72750678181479261478888781535640Q-01	-1.479Q-05	-2.763Q-05
6	9.72744520547842542460569717054100Q-01	-6.158Q-06	-1.284Q-05
7	9.72741595917550115453247399265250Q-01	-2.925Q-06	-6.683Q-06
8	9.72740065376408345903251178930360Q-01	-1.531Q-06	-3.758Q-06
9	9.72739202411302844657556920821470Q-01	-8.630Q-07	-2.228Q-06
10	9.72738686218682583536573863469280Q-01	-5.162Q-07	-1.365Q-06
11	9.72738362261046952339134182224260Q-01	-3.240Q-07	-8.484Q-07
12	9.72738150693111537720117177628760Q-01	-2.116Q-07	-5.245Q-07
13	9.72738007813274259796984508243000Q-01	-1.429Q-07	-3.129Q-07
14	9.72737908518982944634996547067680Q-01	-9.929Q-08	-1.700Q-07
15	9.72737837786707177108774616075250Q-01	-7.073Q-08	-7.073Q-08

N	LNGAMA(Z)	1ST DIFF'S	
0	-8.10614667953272582196702635943840Q-02		
1	2.63874284827861180129409699261760Q-03	8.370Q-02	8.106Q-02
2	1.97900401195213611875295539117690Q-05	-2.619Q-03	-2.639Q-03
3	5.89406874987551296873613756978830Q-06	-1.390Q-05	-1.977Q-05
4	2.32184691845396577817909640720560Q-06	-3.572Q-06	-5.874Q-06
5	1.08971979601357023713486691896650Q-06	-1.232Q-06	-2.302Q-06
6	5.76583659620318710546160124328880Q-07	-5.131Q-07	-1.070Q-06
7	3.32864468584734767019677720212840Q-07	-2.437Q-07	-5.569Q-07
8	2.05319373437272267334649812032180Q-07	-1.275Q-07	-3.132Q-07
9	1.33405614645501792813140738266550Q-07	-7.191Q-08	-1.856Q-07
10	9.03895629570750442250280553836920Q-08	-4.302Q-08	-1.137Q-07
11	6.33930933211419242515909709753330Q-08	-2.700Q-08	-7.070Q-08
12	4.57624320365903395012080122683270Q-08	-1.763Q-08	-4.371Q-08
13	3.38557789300967451120925321631280Q-08	-1.191Q-08	-2.608Q-08
14	2.55812546538332461153279224124620Q-08	-8.275Q-09	-1.417Q-08
15	1.96868983398727276210785531248790Q-08	-5.894Q-09	-5.894Q-09

TABLE 12. PADÉ APPROXIMATIONS FOR Z = 2.0

N	AN(Z)	BN(Z)
1	-1.15119047619047619047619047619050Q+00	-1.14285714285714285714285714285720Q+00
2	1.46333319458319458319458319458330Q+00	1.47486888111888111888111888111900Q+00
3	2.74674220775283678957570018872130Q+00	2.76839973959915319090135081078700Q+00
4	6.95406477501587597564348171383300Q+00	7.00889726300865923630043988933470Q+00
5	2.31177196295843682024810241395660Q+01	2.33000023215088866665201118266720Q+01
6	9.82170883554837972180137903334610Q+01	9.89915283104052553013617635898750Q+01
7	5.20706774673952099372515804181030Q+02	5.24812538752558795831846214867030Q+02
8	3.37522204724979709743392259142370Q+03	3.40183561707654881225543013437060Q+03
9	2.62896009481911530400871554569860Q+04	2.64968940189309288848361688150610Q+04
10	2.42445443942924844091611844497570Q+05	2.44357122290746388888758395309730Q+05
11	2.61380759283190614311369251306990Q+06	2.63441742287908973982528737232050Q+06
12	3.25824805402820746426895743371960Q+07	3.28393928658932624922240275710000Q+07
13	4.65131048484167598447484622838620Q+08	4.68798599188707611050627934446950Q+08
14	7.54015970626534182801923190449670Q+09	7.59961374216110960465232440044250Q+09
15	1.37769656028986635487989938266270Q+11	1.38855967512503163620333099554080Q+11

N	FN(Z)	1ST DIFF'S	F15(Z)-FN(Z)
1	1.00729166666666666666666666666670Q+00	-1.511Q-02	-1.511Q-02
2	9.92178500283404713690681982280740Q-01	-1.623Q-06	-1.797Q-06
3	9.92176876938497229001314649989000Q-01	-1.458Q-07	-1.740Q-07
4	9.92176731098317496179441912321700Q-01	-2.182Q-08	-2.817Q-08
5	9.92176709280571693737744152180980Q-01	-4.548Q-09	-6.349Q-09
6	9.92176704732822528826488918044140Q-01	-1.197Q-09	-1.801Q-09
7	9.92176703536150643769375040067510Q-01	-3.742Q-10	-6.042Q-10
8	9.92176703161917405857817891661850Q-01	-1.337Q-10	-2.299Q-10
9	9.92176703028223852722837440424340Q-01	-5.309Q-11	-9.624Q-11
10	9.92176702975135918330372413021250Q-01	-2.297Q-11	-4.315Q-11
11	9.92176702952161763928224277616210Q-01	-1.068Q-11	-2.017Q-11
12	9.92176702941484189594911547188790Q-01	-5.270Q-12	-9.496Q-12
13	9.92176702936213986203280311085930Q-01	-2.738Q-12	-4.226Q-12
14	9.92176702933475567114470453393360Q-01	-1.488Q-12	-1.488Q-12
15	9.92176702931988053239041845914350Q-01		

N	LNGAMA(Z)	1ST DIFF'S	
0	-4.13406959554092940938220814071220Q-02	4.197Q-02	4.134Q-02
1	6.29790155701817017289029703989340Q-04	-6.297Q-04	-6.298Q-04
2	7.48897325689766230011879090848210Q-08	-6.764Q-08	-7.489Q-08
3	7.25036142378123269567575315252150Q-09	-6.077Q-09	-7.250Q-09
4	1.17368726824698799827294909806600Q-09	-9.091Q-10	-1.174Q-09
5	2.64614526478583924933752407126630Q-10	-1.895Q-10	-2.645Q-10
6	7.51249779406149568447173259135680Q-11	-4.986Q-11	-7.503Q-11
7	2.52636493965685452623577445753080Q-11	-1.559Q-11	-2.517Q-11
8	9.67059781692033074545553161806380Q-12	-5.571Q-12	-9.580Q-12
9	4.10003310296281194389244548949390Q-12	-2.212Q-12	-4.010Q-12
10	1.88803583661010246876361325622740Q-12	-9.573Q-13	-1.798Q-12
11	9.30779403187263493553670226076890Q-13	-4.449Q-13	-8.406Q-13
12	4.85880472632566392411239286970460Q-13	-2.196Q-13	-3.957Q-13
13	2.66288664647931554792021706925210Q-13	-1.141Q-13	-1.761Q-13
14	1.52187869280854150934947680265350Q-13	-6.198Q-14	-6.198Q-14
15	9.02081244713288393096924243629470Q-14		

TABLE 12. PADÉ APPROXIMATIONS FOR Z = 5.0

N	AN(Z)	BN(Z)
1	6.558095238095238095238095238095Q-01	6.571428571428571428571428571428Q-01
2	1.071988882228882228882228882228Q+00	1.073404195804195804195804195804Q+00
3	1.231891426800679467822021352147Q+00	1.233517855379911832546952806868Q+00
4	1.566918230318954579148619971966Q+00	1.568986984547828256713011522984Q+00
5	2.234622909009541360033100674911Q+00	2.237573213309884666492064629828Q+00
6	3.594651459247929468047449381862Q+00	3.599397367658567568625411543979Q+00
7	6.531976396387549350732773869652Q+00	6.540600365100275077582252803513Q+00
8	1.339090469815979870912176549409Q+01	1.340858430018896242552014444073Q+01
9	3.088238557248931869314321877465Q+01	3.092315864189274962405827916673Q+01
10	7.982585183374962985365616963080Q+01	7.993124346514953257835666638165Q+01
11	2.303233207686797466491188320908Q+02	2.306274096066643276080775261322Q+02
12	7.387095214616057830409105893075Q+02	7.396848170558153298724793203801Q+02
13	2.622698787222546571059776534035Q+03	2.626161456238990022784988123107Q+03
14	1.026633525116391615278173177392Q+04	1.027988957968986671026940199281Q+04
15	4.413714608878661294944351429669Q+04	4.419541901321867478643004769619Q+04

N	FN(Z)	1ST DIFF'S	F15(Z)-FN(Z)
1	9.979710144927536231884057971014Q-01	7.105Q-04	7.105Q-04
2	9.986814719181778299623101599429Q-01	-5.249Q-10	-5.289Q-10
3	9.986814713932686121469376276617Q-01	-3.925Q-12	-3.997Q-12
4	9.986814713893436923921674311911Q-01	-6.974Q-14	-7.216Q-14
5	9.986814713892739496876659885603Q-01	-2.291Q-15	-2.420Q-15
6	9.986814713892716584863157853938Q-01	-1.189Q-16	-1.287Q-16
7	9.986814713892715395567905611545Q-01	-8.783Q-18	-9.765Q-18
8	9.986814713892715307734629383094Q-01	-8.578Q-19	-9.818Q-19
9	9.986814713892715299156517862364Q-01	-1.051Q-19	-1.239Q-19
10	9.986814713892715298105778381135Q-01	-1.552Q-20	-1.887Q-20
11	9.986814713892715297950596406820Q-01	-2.681Q-21	-3.357Q-21
12	9.986814713892715297923782858556Q-01	-5.295Q-22	-6.754Q-22
13	9.986814713892715297918488172186Q-01	-1.173Q-22	-1.460Q-22
14	9.986814713892715297917315416906Q-01	-2.870Q-23	-2.870Q-23
15	9.986814713892715297917028384530Q-01		

N	LNGAMA(Z)	1ST DIFF'S	
0	3.161409139158124427483746735923Q+00	1.663Q-02	1.664Q-02
1	3.178041989399670321203553499208Q+00	1.184Q-05	1.184Q-05
2	3.178053830356760724649785238589Q+00	-8.748Q-12	-8.815Q-12
3	3.178053830348012237686195696384Q+00	-6.542Q-14	-6.662Q-14
4	3.178053830347946822356949526443Q+00	-1.162Q-15	-1.203Q-15
5	3.178053830347945659978541169066Q+00	-3.819Q-17	-4.033Q-17
6	3.178053830347945621791851999013Q+00	-1.982Q-18	-2.145Q-18
7	3.178053830347945619809693245276Q+00	-1.464Q-19	-1.628Q-19
8	3.178053830347945619663304451561Q+00	-1.430Q-20	-1.636Q-20
9	3.178053830347945619649007599027Q+00	-1.751Q-21	-2.066Q-21
10	3.178053830347945619647256366558Q+00	-2.586Q-22	-3.146Q-22
11	3.178053830347945619646997729934Q+00	-4.469Q-23	-5.595Q-23
12	3.178053830347945619646953040687Q+00	-8.824Q-24	-1.126Q-23
13	3.178053830347945619646944216210Q+00	-1.955Q-24	-2.433Q-24
14	3.178053830347945619646942261618Q+00	-4.784Q-25	-4.784Q-25
15	3.178053830347945619646941783231Q+00		

TABLE 12. PADÉ APPROXIMATION FOR Z = 10.0

N	AN(Z)	BN(Z)
1	9.1395238095238095238095238095238Q-01	9.1428571428571428571428571428571Q-01
2	1.0179206338106338106338106338106Q+00	1.0182590909090909090909090909091Q+00
3	1.0563150243937885047552825787010Q+00	1.0566662475707402299153318196103Q+00
4	1.1299505093927320830764142344318Q+00	1.1303262162590955119015766430889Q+00
5	1.2553563931800267042370623544023Q+00	1.2557737973165179930285330184151Q+00
6	1.4576269153382390135033751397479Q+00	1.4581115741229702488264021452794Q+00
7	1.7778644589297722161452325883266Q+00	1.7784555962221854897715695974456Q+00
8	2.2868075277000158818860876113801Q+00	2.2875678878069998744277419168322Q+00
9	3.1110360014508716854778338050591Q+00	3.1120704163013743750935643422543Q+00
10	4.4855449545594990440317181782888Q+00	4.4870363915957242352936401458516Q+00
11	6.8634043521138544509130892701799Q+00	6.8656864238685755048443125070482Q+00
12	1.1153371498033468652948793159164Q+01	1.1157079977493480919589024315741Q+01
13	1.9254992293081812858888368039226Q+01	1.9261394549425034156138322138740Q+01
14	3.5312094491236915013136326809123Q+01	3.5323835710215755078246091677611Q+01
15	6.8769942219616285391458496461930Q+01	6.8792808123278926222575034117955 9Q+01

N	FN(Z)	1ST DIFF'S	F15(Z)-FN(Z)
1	9.9963541666666666666666666666667Q-01	3.220Q-05	3.220Q-05
2	9.9966761200417574710597115146770Q-01	-6.308Q-13	-6.312Q-13
3	9.9966761200354493961750310053311Q-01	-3.882Q-16	-3.888Q-16
4	9.9966761200354455143675899244225Q-01	-6.580Q-19	-6.604Q-19
5	9.9966761200354455077873725527450Q-01	-2.403Q-21	-2.419Q-21
6	9.9966761200354455077633450726163Q-01	-1.609Q-23	-1.627Q-23
7	9.9966761200354455077631841847014Q-01	-1.763Q-25	-1.793Q-25
8	9.9966761200354455077631824213343Q-01	-2.908Q-27	-2.978Q-27
9	9.9966761200354455077631823922529Q-01	-6.766Q-29	-6.986Q-29
10	9.9966761200354455077631823915763Q-01	-2.112Q-30	-2.200Q-30
11	9.9966761200354455077631823915551Q-01	-8.628Q-32	-8.744Q-32
12	9.9966761200354455077631823915543Q-01	-1.926Q-34	-1.156Q-33
13	9.9966761200354455077631823915543Q-01	-2.889Q-33	-9.630Q-34
14	9.9966761200354455077631823915542Q-01	1.926Q-33	1.926Q-33
15	9.9966761200354455077631823915543Q-01		

N	LNGAMA(Z)	1ST DIFF'S	
0	1.2793496916648106739951248555907Q+01	8.330Q-03	8.331Q-03
1	1.2801827211786995628840137444796Q+01	2.683Q-07	2.683Q-07
2	1.2801827480081474871177131648836Q+01	-5.257Q-15	-5.260Q-15
3	1.2801827480081469614448061081745Q+01	-3.235Q-18	-3.240Q-18
4	1.2801827480081469611213221547511Q+01	-5.484Q-21	-5.504Q-21
5	1.2801827480081469611207738033034Q+01	-2.002Q-23	-2.016Q-23
6	1.2801827480081469611207718010134Q+01	-1.341Q-25	-1.356Q-25
7	1.2801827480081469611207717876061Q+01	-1.469Q-27	-1.494Q-27
8	1.2801827480081469611207717874591Q+01	-2.424Q-29	-2.482Q-29
9	1.2801827480081469611207717874567Q+01	-5.639Q-31	-5.824Q-31
10	1.2801827480081469611207717874567Q+01	-1.849Q-32	-1.849Q-32
11	1.2801827480081469611207717874567Q+01	0.0	0.0
12	1.2801827480081469611207717874567Q+01	0.0	0.0
13	1.2801827480081469611207717874567Q+01	0.0	0.0
14	1.2801827480081469611207717874567Q+01	0.0	0.0
15	1.2801827480081469611207717874567Q+01		

TABLE 13. COEFFICIENTS IN THE POLYNOMIAL APPROXIMATIONS, $\gamma = 2.0$

N=4

I	TERM(I)
0	1.0000000000000000000000000000000Q+00
1	-3.3333322858008119005560616883522Q-02
2	9.5166234507124949574403249529686Q-03
3	-6.7980670510901513820672698149423Q-03
4	5.6315348352226106123634187816099Q-03

N=5

I	TERM(I)
0	1.0000000000000000000000000000000Q+00
1	-3.3333332309430293956090382217022Q-02
2	9.5227560738538673947046752842821Q-03
3	-7.0646599856134809789241133004792Q-03
4	8.4335992047930273747852494752354Q-03
5	-7.9264839451565636714898598638349Q-03

N=6

I	TERM(I)
0	1.0000000000000000000000000000000Q+00
1	-3.3333333209028084903100211385324Q-02
2	9.5236304887359990080473124054839Q-03
3	-7.1239279129893480079969453201776Q-03
4	9.5072287526403063331091169643468Q-03
5	-1.4807225281937570416607240961864Q-02
6	1.4113757128646910527073541003486Q-02

N=7

I	TERM(I)
0	1.0000000000000000000000000000000Q+00
1	-3.3333333315507731211180257848486Q-02
2	9.5237752935604200978608570506832Q-03
3	-7.1379800542967498923716868586351Q-03
4	9.8908365235325156447743779406453Q-03
5	-1.8929529594086812162891862500468Q-02
6	3.2834108868889767670842200867795Q-02
7	-3.0069948325836892601054588386720Q-02

TABLE 14. POLYNOMIAL APPROXIMATIONS Z = 5.0

N	FN(Z)	1ST DIFF'S	F15(Z)-FN(Z)
1	9.986721991701244813278008298755 2Q-01	8.812Q-06	9.272Q-06
2	9.9868101101749837977965003240441Q-01	4.499Q-07	4.604Q-07
3	9.9868146092396299069573510040786Q-01	1.210Q-08	1.047Q-08
4	9.9868147302363872363190419518967Q-01	-1.266Q-09	-1.634Q-09
5	9.9868147175744388345292322053820Q-01	-3.284Q-10	-3.682Q-10
6	9.9868147142909010997173110305759Q-01	-4.024Q-11	-3.982Q-11
7	9.9868147138884543183373411230768Q-01	-9.694Q-13	4.260Q-13
8	9.9868147138787601922551221968132Q-01	1.028Q-12	1.395Q-12
9	9.9868147138890363833345675995794Q-01	3.242Q-13	3.678Q-13
10	9.9868147138922787327987856087970Q-01	5.011Q-14	4.358Q-14
11	9.9868147138927798247941886622108Q-01	-1.430Q-15	-6.525Q-15
12	9.9868147138927655245839765381553Q-01	-3.651Q-15	-5.095Q-15
13	9.9868147138927290140890361320744Q-01	-1.247Q-15	-1.444Q-15
14	9.9868147138927165419051706999990Q-01	-1.965Q-16	-1.965Q-16
15	9.9868147138927145770606755445623Q-01		

N	LNGAMA(Z)	1ST DIFF'S	
1	3.1780536758109598355058767497547Q+00	1.469Q-07	1.545Q-07
2	3.1780538226750827338134075697969Q+00	7.498Q-09	7.673Q-09
3	3.1780538301735238106620089875969Q+00	2.017Q-10	1.744Q-10
4	3.1780538303751850728776118058433Q+00	-2.110Q-11	-2.724Q-11
5	3.1780538303540818255412954562658Q+00	-5.473Q-12	-6.136Q-12
6	3.1780538303486092626499422543077Q+00	-6.707Q-13	-6.636Q-13
7	3.1780538303479385180143089711286Q+00	-1.616Q-14	7.100Q-15
8	3.1780538303479223611375052729181Q+00	1.713Q-14	2.326Q-14
9	3.1780538303479394881226376819227Q+00	5.404Q-15	6.130Q-15
10	3.1780538303479448920384113786048Q+00	8.352Q-16	7.264Q-16
11	3.1780538303479457271917370503605Q+00	-2.383Q-17	-1.087Q-16
12	3.1780538303479457033580533634870Q+00	-6.085Q-17	-8.491Q-17
13	3.1780538303479456425072284628102Q+00	-2.079Q-17	-2.406Q-17
14	3.1780538303479456217202553537568Q+00	-3.275Q-18	-3.275Q-18
15	3.1780538303479456184455145284977Q+00		

TABLE 14. POLYNOMIAL APPROXIMATIONS Z = 10.0

N	FN(Z)	1ST DIFF'S	F15(Z)-FN(Z)
1	9.99668049792531120331950207468880Q-01	-4.635Q-07	-4.378Q-07
2	9.99667586334598648273308027034530Q-01	2.268Q-08	2.567Q-08
3	9.99667609010233392918717168525640Q-01	2.682Q-09	2.993Q-09
4	9.99667611692013287321514861920020Q-01	2.794Q-10	3.115Q-10
5	9.99667611971396440487112205101940Q-01	2.924Q-11	3.215Q-11
6	9.99667612000636358516859747820400Q-01	2.806Q-12	2.908Q-12
7	9.99667612003442170118450197765730Q-01	1.577Q-13	1.024Q-13
8	9.99667612003599897790452610008940Q-01	-3.089Q-14	-5.535Q-14
9	9.99667612003569007330041053268850Q-01	-1.711Q-14	-2.446Q-14
10	9.99667612003551893030730108789120Q-01	-5.459Q-15	-7.344Q-15
11	9.99667612003546433881019314219990Q-01	-1.459Q-15	-1.885Q-15
12	9.99667612003544974589593950336110Q-01	-3.453Q-16	-4.259Q-16
13	9.99667612003544629289626024396320Q-01	-7.024Q-17	-8.056Q-17
14	9.99667612003544559048100550561040Q-01	-1.032Q-17	-1.032Q-17
15	9.99667612003544548725659436897150Q-01		

N	LNGAMA(Z)	1ST DIFF'S	
1	1.28018274837297111659540148076360Q+01	-3.862Q-09	-3.648Q-09
2	1.28018274798675617286868594561320Q+01	1.890Q-10	2.139Q-10
3	1.28018274800565253515589045323110Q+01	2.235Q-11	2.494Q-11
4	1.28018274800788735173455945130900Q+01	2.328Q-12	2.596Q-12
5	1.28018274800812017102886411576160Q+01	2.437Q-13	2.679Q-13
6	1.28018274800814453762722223871390Q+01	2.338Q-14	2.423Q-14
7	1.28018274800814687580355689742220Q+01	1.314Q-15	8.532Q-16
8	1.28018274800814700724328356609900Q+01	-2.574Q-16	-4.612Q-16
9	1.28018274800814698150123322313510Q+01	-1.426Q-16	-2.038Q-16
10	1.28018274800814696723931713068140Q+01	-4.549Q-17	-6.120Q-17
11	1.28018274800814696269002570501920Q+01	-1.216Q-17	-1.571Q-17
12	1.28018274800814696147394951721600Q+01	-2.877Q-18	-3.549Q-18
13	1.28018274800814696118619954394440Q+01	-5.853Q-19	-6.714Q-19
14	1.28018274800814696112766493938280Q+01	-8.602Q-20	-8.602Q-20
15	1.28018274800814696111906290512140Q+01		

Again for $n \geq 7$ and $z = 2$ the rational approximations are better than the Padé. This trend continues as z increases to about 3 or 4. For about $z \geq 4$, the Padé approximations are better than the rational approximations for almost all n. This is illustrated by the following display which presents the error in the various procedures for $n = 10$ as z varies from 1 to 8.

$R_n(z)$, $n = 10$

z	Padé	Rational	Polynomial, $\gamma=2$	Polynomial, $\gamma=4$
1.0	-5.25 (-5)	-8.48 (-6)		
2.0	-4.32 (-9)	-2.43 (-10)	-2.43 (-10)	
3.0	-1.12 (-12)	-1.40 (-12)	6.56 (-11)	
4.0	-9.37 (-16)	-5.40 (-15)	-2.14 (-12)	-5.40 (-15)
5.0	-1.89 (-18)	-4.49 (-17)	4.36 (-12)	-1.05 (-15)
6.0	-7.58 (-21)	-4.23 (-18)	3.85 (-12)	4.22 (-17)
7.0	-5.31 (-23)	-4.88 (-19)	1.01 (-14)	1.61 (-16)
8.0	-5.91 (-25)	-5.68 (-20)	-1.17 (-13)	-1.31 (-16)

Notice that for the polynomial approximations, the error oscillates and the largest error in magnitude occurs when $z = \gamma$.

IV. Acknowledgment

This is to acknowledge the programming assistance of Mr. Larry Horner.

REFERENCES

1. Y.L. Luke, "The Special Functions and Their Approximations," Vols. 1, 2, Academic Press, Inc., New York, 1969.

2. Y.L. Luke, "Mathematical Functions and Their Approximations," Academic Press, Inc., New York, 1975.

3. Y.L. Luke, "Algorithms for the Computation of Mathematical Functions," Academic Press, Inc., New York, 1977.

4. F.G. Friedlander, "The reflexion of sound pulses by convex parabolic reflectors," Proc. Cambridge Philos. Soc. 37 (1941), 134-139.

5. L. Fox and E.T. Goodwin, "The numerical solution of non-singular linear integral equations," Philos. Trans. Roy. Soc. London, A 245 (1953), 501-534.

PADE APPROXIMANTS AND RATIONAL FUNCTIONS AS TOOLS FOR FINDING POLES AND ZEROS OF ANALYTICAL FUNCTIONS MEASURED EXPERIMENTALLY

Maciej Pindor

Institute of Theoretical Physics, Warsaw University,
00-681 Warszawa, ul. Hoża 69, Poland

1. Motivation for the study and statement of the problem
2. Padé Approximants and rational functions as possible tools for solving the problem
3. Numerical experiment
4. Conclusions

1.

The fast convergence of Padé Approximants (P.A.) which takes place in many cases (see e.g. numerical examples in [1]) makes it possible to use them as a tool for guessing a function if we know only limited number of coefficients of its power expansion. Such a situation is the case in all practical applications, where, however, an additional difficulty arises - these coefficients are known with a limited accuracy only. Fortunately, it appeared [2] that P.A. act like a sort of a "noise filter" - singularities and even values of the function can be recovered also in the case when coefficients of the power expansion of the function are randomly disturbed.

On the other hand, it often happens in practice that one knows values of a function in a limited number of points and one needs to guess its values at other points. Here, rational functions, called P.A. of the type II in this context, appeared to be very useful (I always have in mind functions defined in the complex plane). In practice it is always the case, again, that the known values of the function are biased by random errors and, therefore, our problem will be as follows:

> Let f(z) be a function of z analytical in a circle centred on the origin; let f_i, i=1,...,N, be values of f on some set z_i, i=1,...,N, contained in the interval of the real axis symmetric with respect to the origin, biased by errors distributed normally around $f(z_i)$ with variances σ_i^2. How to find anything about positions of poles and zeros of f?

Problems of this type, or ones which can easily be reduced to it, often arise in physics. To distract you let me present an example from particle physics.

If we consider scattering of a point electrical charge (electron) on an extended charge (proton), some function related to a charge distribution in the scatterer, and called formfactor, can be proved to be analytical function of the momentum transfer considered as a complex variable. So called physical values of the momentum transfer i.e. the values actually met in Nature are real and negative, and the formfactor is measured experimentally there.

It can be proved that the formfactor f(t) has no singularities in the whole complex plane of t outside of a cut starting from some real t_o and going along the real axis to $+\infty$. The branch point at t_o is of a square root type, but there are infinitely many other, more complicated, branch points lying on the semiaxis $t > t_o$.

It appears that poles of the formfactor lying on the Remanian sheet reached by crossing the real semiaxis between t_o and the next branch point (it is so called unphysical sheet - the one on which values of the formfactor are measured is, of course, the physical sheet) are of high interest because their positions correspond to masses and life-times of some other particles.

One can easily find a transformation which maps the physical sheet onto the interior of the unit circle and the unphysical sheet onto its exterior. Infinity goes then into -1 and one can also place points, at which the formfactor was measured till now, symmetrically around the origin.

2.

Before I present two possible methods of solution of the problem, let us recall what can be done at all with randomly disturbed information on values of the function.

If we know that f(z) can be well approximated by some $g(z;y_1,\ldots,y_n)$ for $y_i=\overset{o}{y}_i$ (unknown) (or it is just this function) we can look for the "best" y_i's by minimizing :

$$\chi^2 = \sum_{i=1}^{N} \frac{\left[f_i - g(z_i;y_1,\ldots,y_n)\right]^2}{\sigma_i^2} \tag{1}$$

with respect to y_i's (it is the least-square approximation in numerical analysis and the minmal χ^2 fitting in physics).

If we do not know anything on a shape of f we can apply,

first, the most naive approach and look for the "most probable" optimal rational function, from a given class, approximating f. Such a function has poles and zeros in the complex plane, and we could conjecture that they approximate, somehow, poles and zeros of f.

Serious objections arise, because even if was known exactly at all z_i's, and the problem was that of the standard approximation by rational functions, we do not know anything about a possible convergence of complex plane sigularites of the optimal rational function to singularities of the function approximated. Additional difficulties arise in our case, because coefficients of the rational approximation found that way are known with randomly distributed errors. Nevertheless, it is interesting to see what will appear in numerical experiments and I shall present an example ~~below~~ in the next section.

We can note, however, that if P.A. approximates the function well, the optimal rational approximation of the same type will not differ drastically from it and its singularities should not lie far from those of the P.A..

A more sophisticated approach has been proposed by O. Dumbrajs [3]. It consists in observation that as ~~that~~ z_i's lie within the circle of convergence of the Taylor series, then the truncated series should approximate well the function f. If, therefore, we fit a polynomial to the data for f, we can expect that its coefficients would be close to the first coefficients of the Taylor series. Indeed, it is obvious that if a remainder of the n-th order Taylor formula is small on on the whole interval considered and it goes to zero, then coefficients of the best approximating polynomial of the n-th degree approach the first n+1 coefficients of the Taylor series.

Having found the approximate coefficients of the Taylor series, we can built P.A.s from them and see what are their poles and zeros. If errors of the coefficients found are not large(and the method of χ^2 fitting gives us estimates of these errors), we can expect that poles and zeros found in this way will reasonably approximate poles and zeros of actual P.A. of f. Such an expectation is based, as it was mentioned earlier, on results reported in [2] .

One can raise two objections here.

Both of them stem from the fact that that because $f(z_i)$ are known with a finite accuracy only, there is, generally, no sense to fit to f a polynomial of a too high degree. It may, therefore, happen that the remainder of the Taylor formula, of the order of the optimal polynomial fitted, is not small. Then coefficients of this polynomial may differ significantly from those of the Taylor series.

On the other hand, if, because of the same reason as before, we can determine, with a reasonable accuracy, only few terms of the Taylor expansion, we may not be able to calculate many enough P.A.s as to make sure which of their singularities and zeros approximate actual singularities and zeros of f.

3.

Let me present now a numerical experiment. It was done as follows: I took some function analytic in the circle of the radius $\sqrt{5}$ and calculated its values in 300 aquidistant points on the interval (-1.5,1.5). Then values of f at all z_i's were biased by errors produced with the use of a generator of normally distributed random numbers with the standard deviation $.001 \times f(z_i)$.

To make this part of my talk a suspense story let me not reveal a form of f, untill the very end.

I fitted polynomials of 5-th, 6-th, 7-th and 8-th degrees to the "experimental" data produced that way and the results are shown in the Table I. We see that the value of χ^2 stabilizes at 7-th degree, though the fit of the 6-th order polynomial was also not very unreasonable.

Unfortunately, only coefficients a_0 through a_5 remain re-relatively stable in orders 6,7 and 8. To make sure that they are really close to coefficients of the Taylor series, we can think of the following check: if we consider approximating f on a smaller interval (still symmetric with respect to the origin) then coefficients of the best approximating polynomial should be closer to coefficients of the Taylor ~~series~~ formula, as before, simply because a remainder would be smaller. We can, thus, neglect some data and fit a polynomial on an interval, say, (-1.25,1.25) (we have now only 250 points). It must be noted, however, that now, probably, lower order polynomials will provide a satisfactory fit, and therefore, a smaller number of coefficients can be ~~approximated~~ estimated. Results are shown in Table II. We can see that in fact χ^2 stabilizes now at 6-th order, but we also see that a_0 through a_5 in the 7-th order polynomial fitted on (-1.5,1.5) and in the 6-th order polynomial fitted on (-1.25,1.25) differ less than 1% (except for a_4 which differs by 1.5%).

Therefore, we guess theat $a_0 \div a_5$ are reasonable approximations of the corresponding coefficients of the Taylor series construct [2/2], [3/2], and [2/3] and calculate their zeros and poles. Results are shown in Table III. They are, unfortunately, rather unsatisfactory - although positions of poles and zeros from the two compared fits do not differ much, they differ significantly when we change a type of a P.A. . We can guess that a stability is not yet achieved and we do not know what is a relation of zeros and poles of [2/2], [3/2] and [2/3] (approximate positions of which we have probabely established) to zeros and poles of f.

We can easily get some feeling what is an influence of errors of $f(z_i)$ on zeros and poles of our P.A. using a very simple method. We can namely randomly disturb f_i's again using the same technique, as we used before to disturb $f(z_i)$. It will result in producing values distributed again normally around $f(z_i)$ but with a variance twice as large. Effect of this check can be seen in the same Table III - numbers in parentheses correspond to new positions of zeros and poles. They do not differ much from the previous ones - so we can conclude that random errors are, certainly, not the source of the discrepancy of results obtained from various P.A. .

Now let us see what can be achieved by fitting rational functions.

I have fitted ((m/n) means here a polynomial of mth degree divided by a polynomial of nth degree) (2/3), (3/2), (3/3), (2/4), and (4/2) on (-1.5,1.5) and (-1.25,1.25) (where I also fitted (2/2)). Values of χ^2 obtained in these fits are shown in Table IV.

We see that only functions (3/3) and (4/2) can approximate f well on (-1.5,1.5) and their poles and zeros, shown in Table V, quite resemble those of [3/2] constructed in the first approach. Also zeros and poles of (3/2), which is not quite unreasonable, as an approximation of f, lie in the neighbourhood.

We could conjecture therefore, that just these zeros and poles approximate those of f. Results of fitting of rational functions strongly suggest that [2/2] and [2/3] do not approximate f well and therefore their poles and zeros differ from those of [3/2].

Let us now see what happens when we fit rational functions on the interval (-1.25,1.25). Table IV shows that here only (3/2) and (2/4) provide reasonable fits ((3/3) and (4/2) possess abundant parameters with respect to (3/2) which is already good enough). Corresponding poles and zeros are in Table V. Now, if we had only data on this interval we would not be able to say whether poles and zeros of any of these two functions have anything in common with poles and zeros of f. When, however, we compare them with those obtained from [3/2] constructed from the fitted polynomial of sixth order, we see that poles and zeros of (3/2) and [3/2] are alike and we can suspect that they represent the corresponding features of f.

Eventually we can compare our conjectures and suspicions with reality - f was:

$$f(z)=\exp(.3\,\frac{z-\sqrt{5}}{z+\sqrt{5}})\times\frac{(\sqrt{z^2+4z+5}\,)^3}{z^2-6z+13} \qquad (2)$$

It has poles at $3\pm2i$ and vanishes at $2\pm i$ (where it has branch points) and at $-\sqrt{5}\approx-2.234$. Its Taylor series is:

$$f(z)\cong.6371+1.230z+1.014z^2+.4374z^3+.1417z^4+.02862z^5+ \\ .002416z^6-.001116z^7+\dots$$

4.

The example presented here was not the one to illustrate the most favorable situation - on the contrary it has shown that both methods alone can give no clear indications on poles and zeros of the interesting function. However, if we merge the evidence of both of them we can be able to reach definite conclusions. If both methods give conflicting evidence, then probably none of them can be trusted. The last statement can be illustrated by the same egzample as above if we select 100 points on the interval (-.5,.5). Then, the lowest order of a polynomial with a satisfactory χ^2 is 4 and also (2/2) gives a satisfactory fit. Poles given by the both methods are:

[2/2] - $1.318 \pm 1.676i$

(2/2) - $1.831 \pm 1.357i$

They are not completely different but they differ considerably and in fact both of them are completely wrong - they even lie in the circle of convergence.

On the other hand in a situation, where results of both methods could be reconciled, they were not far from reality. We have seen also that results of the first method - direct fitting of a rational function - were definitely more exact in that situation. It was in a striking contradiction to what we have anticipated.

These all features of both methods appeared also in other numerical experiments I have done.

We can explain to some extent why the second method - ~~xxx xxxx~~ fitting of a polynomial to find coefficients of the Taylor series and construction of P.A. - can work unsatisfactorily. The reason could be that a discrepancy between a truncated

Taylor series and the best approximating polynomial, even on an interval symmetrical around the origin, could be more serious than we have expected. However, we find it very intriguing that the direct fitting of a rational function gives so good an information on poles and also on zeros of a function in the complex plane. We think it can be a feature of the least - square approximation, which, although not as convenient as the minimax approximation when only values on the real interval are interesting, can be advantageous for approximating functions in the complex plane, because it avoids taking modulus.

Bibliography:

1. G.A. Baker Jr., Essentials of Padè Approximants, Academic press 1975

2. J. Gilewicz, Approximants de Pade, p.306, Lecture Notes in Mathematics Springer-Verlag 1978

3. O. Dumbrajs, Rev. Roumanie de Physique 21 273 (1976)

Table I

Coefficients of polynomials fitted on (-1.5,1.5)

order	5	6	7	8
χ^2	669	316	284	283
a_0	.6363±.0001	.6371±.0001	.6371±.0001	.6371±.0001
a_1	1.228±.0002	1.230±.0002	1.229±.0003	1.229±.0002
a_2	1.020±.0003	1.016±.0004	1.014±.0005	1.013±.0003
a_3	.4832±.0002	.4740±.0005	.4757±.0006	.4739±.0001
a_5	.1393±.0002	.1383±.0002	.1431±.0008	.1437±.0003
a_6	.01996±.0001	.02520±.0003	.02611±.0003	.02870±.00006
a_7		.002208±.0001	.001252±.0004	.7443E-3±.8E-4
a_8			-.0008781±.0001	-.1766E-2±.5E-4
a_8				-.3764E-3±.1E-4

Table II

Coefficients of polynomials fitted on (-1.25,1.25)

order	5	6	7
χ^2	244	219	219
a_0	.6369±.0001	.6371±.0001	.6371±.0001
a_1	1.229±.0002	1.230±.0003	1.229±.0004
a_2	1.016±.0005	1.015±.0005	1.015±.0007
a_3	.4791±.0003	.4742±.0010	.4743±.0012
a_4	.1422±.0005	.1402±.0006	.1405±.0018
a_5	.02252±.0003	.02603±.0007	.02618±.0009
a_6		.002168±.0004	.001982±.0012
a_7			-.0001147±.0006

Table III

Zeros and poles of P.A. found by the method of O.Dumbrajs

M/N	7th ord. pol. on (-1.5,1.5)		6th ord. pol. on (-1.25,1.25)	
2/3	-1.558±.5167i (-1.743±1.693i)	1.531±2.244i (1.460±2.199i) 2.661 (2.568)	-1.618±5086i (-1.608±.5138i)	1.371±2.211i (1.401±2.222i) 2.495 2.528)
3/2	-1.939 (-1.460) -1.894±.7851i (-2.014±1.137i)	2.581±.847i (2.611±1.866i)	-1.678 (-1.720) -1.779±1.012i (-1.805±.9729i)	2.739±1.946i (2.707±1.927)
2/2	-1.402±.5475i (-1.471±.4360i)	1.888±1.379i (1.876±1.378i)	-1.397±.5502i (-1.390±.5508i)	1.871±1.384i (1.875±1.384i)
	zeros	poles	zeros	poles

Table IV

Minimal values of χ^2 obtained during fitting of rational functions

interval \ (m/n)	(2/2)	(2/3)	(3/2)	(2/4)	(3/3)	(4/2)
(-1.5,1.5)	—	3631	357	552	294	300
(-1.25,1.25)	5085	400	224	219	219	220

Table V

Poles and zeros of rational functions fitted on
(-1.5,1.5) and (-1.25,1.25)

interval	M/N	zeros	poles
(-1.5,1.5)	3/2	-1.840, -1.851±.8570i	2.779±1.870i
	3/3	-1.872, -1.940±.8752i	2.922±2.655i 6.093
	4/2	-1.866, -12.25 -1.911±.8759i	2.989±2.108i
(-1.25,1.25)	3/2	-1.802, -1.809±.8913i	2.840±1.905i
	2/4	-1.945±.3805i	2.217±.9404i .5145±2.363i

Some properties of rational methods for solving ordinary differential equations.

A. Wambecq

1. Introduction

We are concerned in solving the system of ordinary differential equations:

$$\frac{dy}{dx} = f(y), \tag{1.a}$$

with the initial conditions:

$$y(x_0) = y_0, \tag{1.b}$$

where y, y_0 and f are elements of $\mathbb{R}^n$.

Most numerical methods for solving (1) can be described by the rule:

$$\sum_{i=0}^{k} \alpha_i\, y_{n-i} = h\, \Phi_f(x_n;\, y_{n-k}, y_{n-k+1}, \ldots, y_n;\, h)$$

with $0 \leq n \leq N-k$

y_n, $n = 0, \ldots, N$ are approximations to

$y(x_n)$, $n = 0, \ldots, N$, $x_n = x_0 + nh$, and $Nh = x_N - x_0$.

Multistep methods, i.e. methods with $k > 1$, require also the starting values:

$$y_r = s_r(h) \qquad 0 \leq r < k.$$

In this paper, we will deal with rational expressions for Φ_f. This will mean that a definition for the product-quotient $\frac{a.b}{d}$ is needed first [5].

2. The product-quotient $\frac{a.b}{d}$.

We define the product-quotient of elements from $\mathbb{R}^n$ as:

$$\frac{a.b}{d} = \frac{a(b,d) + b(a,d) - d(a,b)}{(d,d)} \tag{2}$$

This definition defines a ternary operation, bilinear in a and b, symmetric in a and b, equal to a if $b = d$. It also transforms correctly under an orthogonal change of basis, and is undefined only when $d \equiv 0$.

As was illustrated in [5], the above definition is compatible with the Samelson inverse of a vector :

$$d^{-1} = \frac{d}{\|d\|^2}$$

Expressions of the form $\frac{a.b}{d}$ arise in formal computation of the second column in the Padé-table, when used to accelerate convergence (see [4] and [6]).

3. Rational onestep methods.

The rational Runge-Kutta methods treated in this section are of the form (2), with

$$\Phi_f = \frac{\sum_{I=1}^{\nu} \sum_{J=1}^{I} w_{IJ}\, g^I g^J}{\sum_{K=1}^{\nu} b_K\, g^K} \tag{3}$$

and $$g^I = f(y_0 + h \sum_{J=1}^{\nu} a_{IJ}\, g^J).$$

Following Butcher's approach for linear Runge-Kutta methods [1], one can find the necessary conditions [5]:

$$\sum_{I=1}^{\nu} \sum_{J=1}^{I} w_{IJ} (\Phi_i^I \Phi_j^J + \Phi_i^J \Phi_j^I) = \frac{\Phi_j}{\gamma_i} + \frac{\Phi_i}{\gamma_j} \tag{4}.$$

The Φ_i in these equations are called "elementary weights", polynomial functions of the parameters a_{IJ} and b_K, and Φ_i^I is defined as $\frac{\partial\Phi_i}{\partial b_I}$.

As Butcher showed [1], these elementary weights are closely related to the Taylor expansions of $y(x_1)$ and Φ_δ.

The γ_i are integer numbers corresponding to the elementary weights Φ_i.
The conditions (4) must be satisfied for :

$j \leq i$ and $r_i + r_j \leq r + 1$.

Then the rational Runge-Kutta method is of order r.
In table 1 the number of equations for rational and linear Runge-Kutta methods, for orders up to ten, are listed.

Table 1 : Number of equations

order	linear	rational
1	1	1
2	2	2
3	4	5
4	8	11
5	17	27
6	37	64
7	85	160
8	200	399
9	486	1021
10	1205	2628

Although the number of conditions in the rational case is considerably greater, a solution of order r with ν function evaluations certainly exists if a linear solution of the same order, with ν function evaluations exists.

Indeed, let $A = \sum_{I=1}^{\nu} d_I\, g^I$ produce a linear Runge-Kutta method of order r,

and let $B = \sum_{J=1}^{\nu} b_J\, g^J$ be an arbitrary nonzero linear combination of the g^J.

The function

$\psi = \frac{A.B}{B}$ produces a rational Runge-Kutta method, with

$w_{IJ} = (b_I d_J + b_J d_I)\,(1 - \frac{\delta_{IJ}}{2})$, and this method is of order r since

$\psi = \frac{A.B}{B} = A$ (if $B \not\equiv 0$).

The above also shows that linear Runge-Kutta methods can be found as special cases of the rational Runge-Kutta methods described by (3). But a more powerful link between rational and linear Runge-Kutta methods can be shown. Indeed, the linear Runge-Kutta methods can be characterized by the matrix notation:

$$\begin{array}{c|cccc} c_1 & a_{11} & \cdots & \cdots & a_{1\nu} \\ c_2 & \cdot & & & \cdot \\ \vdots & \vdots & & & \vdots \\ c_\nu & a_{\nu 1} & \cdots & \cdots & a_{\nu\nu} \\ \hline & b_1 & b_2 & \cdots & b_\nu \end{array} \;=\; \begin{array}{c|c} c & A_L \\ \hline & b^\tau \end{array}$$

In the same way, rational Runge-Kutta methods are characterized by:

$$\begin{array}{c|c} & W \\ \hline c & A_R \\ \hline & b^\tau \end{array} \qquad \text{with } W = \begin{bmatrix} w_{11} & & & 0 \\ w_{21} & w_{22} & & \\ \vdots & \vdots & \ddots & \\ w_{\nu 1} & w_{\nu 2} & \cdots & w_{\nu\nu} \end{bmatrix}$$

The following theorem establishes the link between these two characterisations.

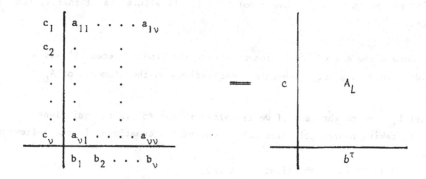

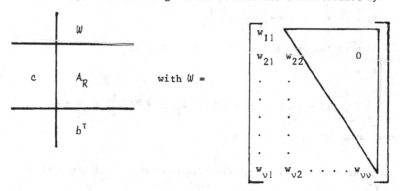

Theorem 1.

The elements of A_L and A_R must satify the same conditions.

Proof : Following Butcher's notation, all parameters of the linear Runge-Kutta methods can be found from the nonlinear system of equations:

$$\Phi_i = \frac{1}{\gamma_i} \qquad r_i \leqslant r$$

Regarding this as a linear system in the b_I, one can explicitly write:

$$\sum_{I=1}^{\nu} b_I \, \Phi_i^I = \frac{1}{\gamma_i} \quad r_i \leqslant r \tag{5}$$

This linear system can only have a solution if all eliminants of that system are zero.

Let L_α denote the set of conditions such that the linear system (5) has a solution. Only this set L_α contains restrictions on the elements of A_L.

This set L_α remains the same if we transform system (5) in an equivalent system by taking nonzero linear combinations of the equations, in the following way:

$\Phi_0 = 1$ (this is the first equation)

and for $j \leqslant i$, $r_i + r_j \leqslant r + 1$

$$\Phi_i \cdot \frac{1}{\gamma_j} + \Phi_j \cdot \frac{1}{\gamma_i} = \frac{1}{\gamma_i\gamma_j} + \frac{1}{\gamma_j\gamma_i}$$

This system is the same as:

$$\left[\begin{array}{l} \Phi_0 = 1 \\ \sum_{I=1}^{\nu} \sum_{J=1}^{I} \left(1 - \frac{\delta_{IJ}}{2}\right) (b_I b_J + b_J b_I) (\Phi_i^I \Phi_j^J + \Phi_i^J \Phi_j^I) = \frac{\Phi_j}{\gamma_i} + \frac{\Phi_i}{\gamma_j} \end{array}\right.$$

since we used the equations from (5)

In its turn, this system of equations is equivalent with the system:

$$\left[\begin{array}{ll} \sum b_I = 1 & (\Phi_0 = 1) \\ w_{IJ} = (2-\delta_{IJ})\, b_I b_J & \\ \sum_{I=1}^{\nu} \sum_{J=1}^{I} w_{IJ} (\Phi_i^I \Phi_j^J + \Phi_j^I \Phi_i^J) = \frac{\Phi_j}{\gamma_i} + \frac{\Phi_i}{\gamma_j} & \end{array}\right. \tag{6}$$

where we introduced the additional variables w_{IJ}.

Now let R_α denote the set of conditions that must be satisfied, so that the linear system in w_{IJ} and d_I:

$$\sum_{I=1}^{\nu} \sum_{J=1}^{I} w_{IJ} (\phi_i^I \phi_j^J + \phi_j^I \phi_i^J) = \sum_{I=1}^{\nu} d_I \left(\frac{\phi_j^I}{\gamma_i} + \frac{\phi_i^I}{\gamma_j}\right)$$

has a nonzero solution. (A nonzero solution for the denominator means

$$\sum_{I=1}^{\nu} d_I = p \neq 0)$$

From system (6) we see that L_α is the union of the set R_α, and the set of conditions

$$\begin{cases} \sum b_I = 1 \\ w_{IJ} = (2-\delta_{IJ})\, b_I b_J. \end{cases} \tag{7}$$

Now let us take the rational method:

$$\frac{\left(\sum_{I=1}^{\nu} b_I g^I\right)\left(\sum_{J=1}^{\nu} d_J g^J\right)}{\left(\sum_{J=1}^{\nu} d_J g^J\right)}$$

that reduces to the linear method of order r:

$$\sum b_I g^I.$$

Since this rational method must also satisfy the conditions for a linear method, the following set of conditions must be satisfied.

$$\begin{cases} \sum b_I = 1 \\ \sum w_{IJ} (\phi_i^I \phi_j^J + \phi_j^I \phi_i^J) = \frac{\phi_i}{\gamma_j} + \frac{\phi_j}{\gamma_i} \\ \sum d_I = p \neq 0 \text{ (nonzero solution)} \\ w_{IJ} = (1-\frac{\delta_{IJ}}{2})\,(b_I d_J + b_J d_I) \end{cases}$$

This is the same as the set L_α, and also is the union of R_α with

$$\sum b_I = 1$$
$$\sum d_I = p \qquad (8)$$
$$w_{IJ} = (1-\frac{\delta_{IJ}}{2})\,(b_I d_J + b_J d_I)$$

Now we can choose p in infinitely many ways so that $b_I \neq d_I$ for all I, but the sets (7) and (8) must always be the same.

From this we can conclude that both sets are empty, so that $L_\alpha = R_\alpha$, which proves our assumption.

The above theorem means that for a given order, and a given number of function evaluations, the a_{IJ} in both matrices A_L and A_R are restricted by the same, or equivalent, conditions.

This enables us to extend known results from linear Runge-Kutta processes to rational ones, such as, for example, order considerations. It is for example not possible to derive a fifth order methods with only five function evaluations, despite of the many free parameters in the rational formula. Another example is the fourth order case, where it is necessary for both linear and rational methods to have $c_4 = 1$.

So the only advantage of rational methods will be the greater freedom in choosing w_{IJ} and b_K, and the resulting numerical aspects of particular choices.

But apart from this, most properties of linear methods can be extended to rational ones, as the following theorem illustrates.

Theorem 1.

if (i) f satisfies a Lipschitz condition, i.e.

$$\|f(y) - f(y^*)\| \leq L\,\|y - y^*\|,$$

(ii) $\sum_{K=1}^{\nu} b_i = 1 = \sum_{I=1}^{\nu} \sum_{J=1}^{I} w_{IJ}$. (consistency condition)

(iii) The system (1.a) includes the equation:

$y_k' = f_k(y)$ for some k, such that $f_k(y) = \sqrt{\alpha} > 0$

or $f_k(y) = -\sqrt{\alpha} < 0$

(iv) $\|\mathfrak{f}(y)\| \leq M$,

then method (3) is convergent.

proof:

By a theorem found in Henrici [3], it is sufficient to show that

(i') $\Phi_{\mathfrak{f}}$ satisfies a Lipschitz condition and

(ii') $\Phi_{\mathfrak{f}}(y, 0) = \mathfrak{f}(y)$.

(ii'): follows from:

$$\Phi_{\mathfrak{f}}(y,0) = \frac{\sum_{I=1}^{\nu} \sum_{J=1}^{I} w_{IJ} \, (\mathfrak{f}(y))^2}{\sum_{K=1}^{\nu} b_K \, \mathfrak{f}(y)} = \frac{\sum_{I=1}^{\nu} \sum_{J=1}^{I} w_{IJ}}{\sum_{K=1}^{\nu} b_K} \cdot \frac{(\mathfrak{f}(y))^2}{\mathfrak{f}(y)}$$

$$= \mathfrak{f}(y)$$

by the consistency condition and a simplification property of (2).

(i'): $\|\Phi(y^\star,h) - \Phi(y,h)\| = P(y^\star,y) =$

$$\left\| \frac{\sum_i \sum_j \sum_l w_{ij} \, b_l [g_\star^i(g_\star^j,g_\star^\ell) + g_\star^j(g_\star^i,g_\star^\ell) - g_\star^\ell(g_\star^i,g_\star^j)]}{\sum_m \sum_n b_m b_n (g_\star^m,g_\star^n)} - \frac{\sum_i \sum_j \sum_l w_{ij} \, b_l [g^i(g^j,g^\ell) + g^j(g^i,g^\ell) - g^\ell(g^i,g^j)]}{\sum_m \sum_n b_m b_n (g^m,g^n)} \right\|$$

Since by assumption (iii) $\sum_m \sum_n b_m b_n (g_\star^m,g_\star^n) > \alpha$

we have

$$P(y^\star,y) \leq \alpha^2 \left\| \sum_{i,j,l,m,n} \alpha_{ijlmn} \, [g_\star^i(g_\star^j,g_\star^\ell)(g^m,g^n) - g^i(g^j,g^\ell)(g_\star^m,g_\star^n)] \right\|.$$

Then $P(y^\star,y) \leq K. \ \max(\|g_\star^i(g_\star^j,g_\star^\ell)(g^m,g^n) - g^i(g^j,g^\ell)(g_\star^m,g_\star^n)\|)$, and the right-hand side of this inequality can be shown to satisfy a Lipschitz condition. The triangle inequality and condition (iv) is needed for this purpose.

This result is an analogue of that for the linear Runge-Kutta methods, where consistency is necessary and sufficient for convergence.

The clue in the above proof is the one equation with the property (iii). This equation has an impact on the whole system of equations, since it keeps the denominator away from zero. If the rational method would be applied conponentwise the method remains consistent, it even keeps its order, but the convergence can be lost.

One further remark on the condition (iii): this condition can always be satisfied by introducing the extra equation :

$$y'_{n+1} = 1,$$

if this equation was not yet present to bring the set of differential equations in form (1.a).

One possibility to make advantage of the free parameters in rational methods, is to make the stability region of the method as great as possible. This region can be found by applying the rational method to the test equation:

$$y' = \lambda y \qquad \lambda \in \mathbb{C},$$

and is the set of points in the complex plane where $\frac{\|y_{n+1}\|}{\|y_n\|} < 1$.

In this manner, it is possible to derive explicit A- and L- stable methods in the rational case, whilst there are only implicit linear methods that are A- or L- stable.

An example would make this clear: consider a second order rational formula with two function evaluations.

The nonlinear system to determine the parameters is:

$$\begin{cases} w_{11} + w_{21} + w_{22} = b_1+b_2 \\ w_{21}c_2 + w_{22}(2c_2) = b_1c_2 + \dfrac{b_1+b_2}{2} \\ b_1 + b_2 = 1 \end{cases}$$

Together with the additional conditions:

$$\begin{cases} w_{21} = 0 \\ b_2 = -\dfrac{1}{2c_2} \end{cases}$$

this gives rise to the following method:

$$
\begin{array}{c|cc}
 & 1 & \\
 & 0 & 0 \\
\hline
0 & 0 & \\
c_2 & c_2 & 0 \\
\hline
 & 1+\frac{1}{2c_2} & -\frac{1}{2c_2}
\end{array}
$$

or with $c_2 = \frac{1}{2}$:

$$y_{n+1} = y_n + h \frac{(g^1)^2}{2(g^1)-g^2} \qquad (9)$$

$$g^1 = f(y_n)$$

$$g^2 = f(y_n + c_2 h g^1)$$

The stability region of this method is

$$\left| \frac{1+\frac{1}{2}z}{1-\frac{1}{2}z} \right| \leq 1 \text{ for all } \mathrm{Re}\,(z) \leq 0$$

since this is a Padé-approximant on the main diagonal, of e^z.

Another example is an L-stable 4-th order method:

$$
\begin{array}{c|cccc}
 & \frac{19}{15} & & & \\
 & -\frac{10}{3} & \frac{44}{15} & & \\
 & \frac{64}{15} & -\frac{88}{15} & \frac{8}{5} & \\
 & \frac{8}{15} & -\frac{2}{5} & 0 & 0 \\
\hline
0 & & & & \\
\frac{1}{4} & \frac{1}{4} & & 0 & \\
\frac{1}{2} & 0 & \frac{1}{2} & & \\
1 & 1 & -2 & 2 & \\
\hline
 & \frac{23}{6} & -\frac{56}{6} & \frac{14}{15} & -\frac{1}{30}
\end{array}
$$

with stability region:

$$\left| \frac{1 + \frac{2}{5}z + \frac{1}{20}z^2}{1 - \frac{3}{5}z + \frac{3}{20}z^2 - \frac{1}{60}z^3} \right| \leq 1$$

and this is also a Padé-approximant of e^z.

The use of nonlinear methods has some drawbacks, as can be seen by the following example, put under my attention by Prof. Curtis.

Considering the single differential equation:

$$\frac{dy}{dx} = \lambda y + g'(x) - \lambda g(x)$$

with solution $y = e^{\lambda x} + g(x)$,
where $g(x)$ is a smooth function and λ is large and negative. At the moment the stiff conponent died out, we can find the following expression for the truncation error of method (9) at y_{n+1}:

$$T = \frac{-\frac{\lambda h^3}{8} g_n' g_n'' - \frac{h^3}{24} g_n' g_n''' + \frac{h^3}{4} g_n''^2 + \theta(h^4)}{g'_n - \frac{h}{2} g_n'' + \frac{\lambda h^2}{8} g_n'' - \frac{h^2}{8} g_n''' + \theta(h^3)} .$$

This truncation error is $\theta(h^3)$ as it should be, but when λ is very large, the truncation error is only $\theta(h)$. Of course, this will not happen in most numerical computations, carried out in double precision, and $|\lambda| = \theta(10^6)$ or so.

Another fact that is worth mentioning is that polynomials, even of degree 0, are not integrated exactly, as linear methods do. And even uncoupled systems will not have uncoupled results, since the equations are not handled componentwise.

4. Rational multistep methods.

We will restrict ourselves to rational Adams-Bashforth or Adams-Moulton methods, of the form:

$$y_n - y_{n-1} = h \frac{\sum_{I=0}^{\nu} \sum_{J=0}^{I} \beta_{IJ} \, \delta_{n-I} \, \delta_{n-J}}{\sum_{K=0}^{\nu} \alpha_K \, \delta_{n-K}} .$$

Following an analogous way as with the rational Runge-Kutta methods, the following necessary order conditions are derived:

$$\sum_{I=0}^{\nu} \sum_{J=0}^{I} \beta_{IJ}(I^i J^j + I^j J^i) = \sum \alpha_K \left(\frac{K^i}{j+1} + \frac{K^j}{i+1}\right)$$

These conditions must be satisfied for $j \leq i$, $i+j \leq r-1$ and $r \leq \nu$ or $r \leq \nu+1$, depending on the method being explicit or implicit.

Let us take for example an Adams-Bashforth method with stepnumber 2:

$$y_n - y_{n-1} = h \frac{\beta_{11} f_{n-1}^2 + (1-\beta_{11}-\beta_{22}) f_{n-1} f_{n-2} + \beta_{22} f_{n-2}^2}{\frac{2\beta_{11}-2\beta_{22}-1}{2} f_{n-1} + \frac{3-2\beta_{11}+2\beta_{22}}{2} f_{n-2}} \tag{5}$$

that we can find by solving the linear system

$$\alpha_1 + \alpha_2 = \beta_{11} + \beta_{21} + \beta_{22}$$

$$3\alpha_1 + 5\alpha_2 = 4\beta_{11} + 6\beta_{21} + 8\beta_{22}$$

together with the scaling equation

$$\alpha_1 + \alpha_2 = 1.$$

For the particular choice $\beta_{11} = \frac{3+6\beta_{22}}{2}$, this reduces to the linear Adams-Bashforth method

$$y_n - y_{n-1} = \frac{h}{2}(3 f_{n-1} - f_{n-2}).$$

The notion of convergence and consistency can also be introduced for nonlinear multistep methods. For this purpose, one can define the functions:

$$\rho(\zeta) = \zeta - 1$$

$$\text{and } \sigma(\zeta) = \frac{\sum_{I=0}^{\nu} \sum_{J=0}^{I} \beta_{IJ} \zeta^{\nu-I} \zeta^{\nu-J}}{\sum_{K=0}^{\nu} \alpha_K \zeta^{\nu-K+1}}$$

A method is consistent if

$$\rho(1) = 0$$

$$\rho'(1) = \sigma(1)$$

For the proposed methods, this means

$$\sum_{I=0}^{\nu} \sum_{J=0}^{I} \beta_{IJ} = \sum_{K=0}^{\nu} \alpha_K \neq 0$$

Convergence follows from consistency, and $\rho(\zeta)$ having all its roots in the unit disk.

One can also try to derive explicit A-stable multistep methods. Taking $\beta_{11} = \frac{1}{2}$, and $\beta_{22} = 0$, the method (5) reduces to:

$$y_n - y_{n-1} = \frac{h}{2}\left(\frac{f_{n-1}^2 + f_{n-1}\, f_{n-2}}{f_{n-2}}\right)$$

Let us seek a solution of the form $y_n = x^n$ for this nonlinear difference equation, for the scalar differential equation $y' = \lambda y$, $\mathrm{Re}(\lambda) < 0$. We find:

$$x^n - x^{n-1} = \frac{\lambda h}{2}\left(\frac{x^{2n-2} + x^{2n-3}}{x^{n-2}}\right)$$

$$\text{or } (x - 1) = \frac{\lambda h}{2}(x + 1)$$

where from:

$$x = \frac{1 + \frac{\lambda h}{2}}{1 - \frac{\lambda h}{2}} \text{ as the only solution.}$$

Now x lies within the unit disk, what was the intention to obtain an A-stable method.

This stability analysis can also be done by introducing the stability function.

$$\pi(\zeta,h) = \rho(\zeta) - \lambda h\sigma(\zeta),$$

and determinning all λh so that π has all its roots within the unit disk.

5. Conclusion

We introduced rational methods for solving ordinary differential equations of the form:

$$y_n - y_{n-1} = \frac{\sum_{I=1}^{\nu}\sum_{J=1}^{I} w_{IJ}\, g^I g^J}{\sum_{K=1}^{\nu} b_K\, g^K}$$

and

$$y_n - y_{n-1} = \frac{\sum_{I=0}^{\nu}\sum_{J=0}^{\nu} \beta_{IJ}\, f_{n-I}\, f_{n-J}}{\sum_{K=0}^{\nu} \alpha_K\, f_{n-K}}.$$

These methods have some properties that cannot be found in linear methods, such as stability properties with explicitness. However, the linearity of the approximation is lost, what could be a drawback in numerical computation.

6. References.

[1] Butcher J.C.: Coefficients for the study of Runge-Kutta integration processes.
J. Aust. Math. Soc. 3, 185-201 (1963)

[2] Curtis A.R.: private communication.

[3] Henrici P.: Discrete variable methods in ordinary differential equations.
John Wiley & Sons Inc. New York 1968.

[4] Wambecq A.: Nonlinear methods in solving ordinary differential equations.
J. Comp. Appl. Math. 2, 27-33 (1976)

[5] Wambecq A.: Rational Runge-Kutta Methods for Solving Systems of Ordinary Differential Equations.
Computing 20, 333-342 (1978)

[6] Wambecq A.: Nonlinear methods in solving ordinary differential equations.
Lecture in: "Colloque sur les approximants de Padé"
Lille, 28-30 mars 1978

A. Wambecq
K.U.Leuven
Applied Mathematics and Programming Division
Celestijnenlaan 200 A

B-3030 Heverlee

Belgium

RECENT REFERENCES ON SEQUENCES

AND

SERIES TRANSFORMATIONS

This list of references is a complement to the one published in Cabannes' book and to the two internal reports of the University of Lille I recently produced. It only contains items published in 1978 and 1979 and which are not included in the previous lists.

Incomplete references denote preprints or papers to be published in a journal.

I shall be grateful for any correction, addition or comment about this bibliography.

Claude BREZINSKI
Université de Lille I
U.E.R. d'IEEA - Informatique B.P. 36
F - 59650 VILLENEUVE d'ASCQ (FRANCE)

1 - O. ABERTH
A method for exact computation with rational numbers. J. Comp. Appl. Math., 4 (1978), 285-288.

2 - P. ACHUTHAN, T. CHANDRAMOHAN, K. VENKATESAN
Continued fractions and Padé approximants - A historical perspective. Third Annual congress of the Indian Association for the History and Philosophy of Science, Ahmedabad, 1978.

3 - Z.P. ANISINOVA
Convergence of recurrent processes of Robbins-Monro type in a scheme of series (in Russian). Dokl. Akad. Nauk. Ukrain SSR, ser. A, 9 (1978), 835-838.

4 - R.A. ASKEY, M.E.H. ISMAIL
Recurrence relations, continued fractions and orthogonal polynomials.

5 - G.A. BAKER jr, P.R. GRAVES MORRIS
Padé approximants. In "Encyclopedie of Mathematics and its applications", chapter 1, Addison-Wesley, 1978.

6 - G.A. BAKER jr, P.R. GRAVES - MORRIS
Padé approximants. Addison-Wesley.

7 - D. BESSIS
A new method in the combinatorics of the topological expansion. Comm. Math. Phys. .

8 - C. BREZINSKI
Convergence acceleration of some sequences by the ε-algorithm. Numer. Math., 29 (1978), 173-177.

9 - C. BREZINSKI
Algorithmes d'accélération de la convergence. Etude numérique. Technip, Paris, 1978.

10 - C. BREZINSKI
Survey on convergence acceleration methods in numerical analysis. Mathematics Student, 46 (1978), 28-41.

11 - C. BREZINSKI
Programmes Fortran pour transformations de suites et de séries. Publication ANO-3, Laboratoire de Calcul, Université de Lille, 1979.

12 - C. BREZINSKI
Padé-type approximation and general orthogonal polynomials. Birkhäuser-Verlag, Basel, 1979.

13 - C. BREZINSKI
Limiting relationships and comparison theorems for sequences. Rend. Circ. Mat. Palermo.

14 - C. BREZINSKI
Rational approximation to formal power series. J. Approx. theory,

15 - C. BREZINSKI
Padé-type approximants for double power series. J. Indian Math. Soc..

16 - M.G. DE BRUIN
Some special generalized c-fractions. Intern. Congress of Math. Helsinki, 1978.

17 - A. BULTHEEL
Recursive algorithms for non normal Padé tables. Report TW 40, KU Leuven, Belgium, July 1978.

18 - A. BULTHEEL
Division algorithms for continued fractions and the Padé table. Report TW 41, KU Leuven, Belgium, August 1978.

19 - A. BULTHEEL
Fast algorithms for the factorization of Hankel and Toeplitz matrices and the Padé approximation problem. Report TW 42, KU Leuven, Belgium, oct. 1978.

20 - A. BULTHEEL
Remark on " A new look at the Padé table and different methods for computing its elements". J. Comp. Appl. Math., 5 (1979), 67.

21 - R.C. BUSSY, W. FAIR
Convergence of "periodic in the limit" operator continued fraction.

22 - R. CHAUVAUX, A. MAGNUS
Separation of variables for a potential problem of a hemisphere and the degenerate kernel method. Contact Group on applied mathematics FNRS-NFWO, the numerical solution of integral equations, Louvain La Neuve, nov. 17, 1978.

23 - R. CHAUVAUX, A. MEESEN
Multiple oscillations in small hemispherical particles. Thin Solid Films.

24 - T.S. CHIHARA
An introduction to orthogonal polynomials. Gordon and Break, New-York 1978.

25 - J.S.R. CHISHOLM
Multivariate approximants with branch points. I. Diagonal approximants. Proc. Roy. Soc. London, A 358 (1978), 351-366.

26 - J.S.R. CHISHOLM
Multivariate approximants with branch points. II. Off-diagonal approximants. Proc. R. Soc. London, A 362 (1978), 43-56.

27 - C. CHLOUBER
Comput. Phys. 15 (1978), 153.

28 - F. CORDELLIER
Une classe de transformation non linéaires de suites basée sur l'approximation au sens des moindres carrés. Colloque d'Analyse Numérique, Giens, mai 1978.

29 - V.R. DELEGARD
Linear and nonlinear techniques for accelerating convergence in sequences and series. D.A., Univ. of Northern Colorado, 1977.

30 - J. DELLA DORA
Approximation topologique - Approximation algébrique. Colloque AFCET-SMF, IRIA, septembre 1978.

31 - P. DEWILDE, A. VIEIRA, Th. KAILATH
On a generalized Szego-Levinson realization algorithm for optimal linear predictors based on a network synthesis approach. IEEA Trans. CAS.

32 - B.W. DICKINSON
Efficient solution of linear equations with banded Toeplitz matrices. IEEE Trans. ASSP.

33 - R.P. EDDY
Extrapolating to the limit of a vector sequence. Proceedings of the first annual workshop on the information linkage between applied mathematics and industry, US Naval Postgraduate School, Monteren, Feb. 1978, Academic Press, P.C.C. Wang ed., 1979.

34 - A. EDREI
Angular distribution of the zeros of Padé polynomials. J. Approx. Theory, 24 (1978), 251-265.

35 - G. FEDERICI
Zeros and poles of Padé approximants. Thesis, Syracuse University, New-York, 1978.

36 - P. FLAJOLET
Analyse d'algorithme de manipulation de fichiers. Rapport 321, IRIA, 1978.

37 - P. FLAJOLET
Histoires de fichiers (II) : aspects algébriques. Colloque sur les arbres en algèbre et en programmation, Lille, 22-24 février, 1979.

38 - P. FLAJOLET, J. FRANCON, J. VUILLEMIN
Computing integrated costs of sequences of operations with application to dictionaries. 11th Conference on Theory of Computing, 1979.

39 - E. FRANK
Continued fraction expansions for kth roots. Intern. Congress of Math., Helsinki, 1978.

40 - B. FRIEDLANDER, M. MORF, T. KAILATH, L. LJUNG
New inversion formules for matrices classified in terms of their distance from Toeplitz matrices. SIAM J. Appl. Math.

41 - A. GALLI
Polynômes de Schur, approximants de Padé, interpolation. Séminaire d'Analyse Numérique, Granoble, 25 janvier 1979.

42 - B. GERMAIN BONNE
Ensembles de suites et de procédés associés pour l'accélération de la convergence. Séminaire d'Analyse Numérique, Lille, 14 décembre 1978.

43 - J. GILEWICZ
Approximants de Padé
Lecture Notes in Mathematics 667, Springer-Verlag, 1978.

44 - J. GILL
Modifying factors for sequences of linear fractional transformations
Kgl. Norske Vidensk. Selsk. Skr., 3 (1978).

45 - J. GILL
A generalization of certain corresponding continued fractions. Bull. Calcutta Math. Soc., 69 (1978).

46 - A.A. GONČAR, L.L. GIERMO
Markov's theorem for multipoint Padé approximants. Mat. Sb. (N.S.), 105 (147), (1978), 512-524, 639.

47 - P.R. GRAVES - MORRIS
A generalized q-d algorithm. Preprint, University of Kent, November 1978.

48 - P.R. GRAVES - MORRIS, T.R. HOPKINS
Reliable rational interpolation
Preprint, University of Kent, 1978.

49 - F.F. GRINSTEIN
Summation of partial wave expansions in the scattering by short range potential. Report, Instituto de Fisica, Niteroi, Brasil, feb. 1979.

50 - R.E. GRUNDY
On the solution of non linear Volterra integral equations using two point rational approximants. J. Inst. Maths Applics, 22 (1978), 317-320.

51 - J. GUNSON, P.H. NG
Approximation of special functions of complex variables. J. Inst. Math. Appl., 22 (1978), 43-52.

52 - S.A. GUSTAFSON
Convergence acceleration on a general class of power series. Computing, 21 (1978), 53-70.

53 - S.A. GUSTAFSON
Algorithm 38. Two computer codes for convergence acceleration. Computing, 21 (1978), 87-91.

54 - F.G. GUSTAVSON, D.Y.Y. YUN
Fast computation of Padé approximants and Toeplitz systems of equations via the extended euclidean algorithm. IBM, report RC 7551, 1979.

55 - F.G. GUSTAVSON, D.Y.Y. YUN
Fast computation for Toeplitz systems, Cauchy-Hermite-Padé approximants, and the extended euclidean algorithm. IBM, Report RC , 1979.

56 - A. ISERLES
On multivalued exponential approximations. Report NA 2/78, DAMTP, Univ. of Cambridge, England, 1978.

57 - A. ISERLES
A stable convoluted predictor-corrector method. Report NA 4/78, DAMTP, Univ. of Cambridge, England, 1978.

58 - A. ISERLES
A note on Padé approximations and generalized hypergeometric functions. Report NA 1/79, DAMTP, Univ. of Cambridge, England, 1979.

59 - A. ISERLES
On the A-acceptability of Padé approximations. SIAM J. Math. Anal.

60 - A. ISERLES
On the generalized Padé approximations to the exponential function. SIAM J. Num. Anal.

61 - A.K. JAIN
Fast inversion of banded Toeplitz matrices by circular decomposition. IEEE Trans. on Acoustics, speech, signal process (ASSP), 26 (1978), 121-126.

62 - M.J. JAMIESON, T.H. O'BEIRNE
A note on the generalisation of Aitken's δ^2 transformation. J. Phys. B : Atom. Molec. Phys., 11 (1978), L31-L35.

63 - K. JITTORNTRUM, M.R. OSBORNE
Trajectory analysis and extrapolation in barrier function methods. J. Austr. Math. Soc. .

64 - W.B. JONES, W.J. THRON
Sequences of meromorphic functions corresponding to a formal Laurent series. SIAM J. Math. Anal., 10 (1979), 1-17.

65 - W.B. JONES, W.J. THRON
Continued fractions : analytic theory and applications. Academic Press.

66 - W.B. JONES, W.J. THRON, H. WARDELAND
A strong Stieltjes moment problem.

67 - W.B. JONES, W.J. THRON, H. WAADELAND
Two point Padé tables for a class of confluent hypergeometric functions

68 - T. KULIKOWSKA

Nukleonika, 23 (1978), 691 (in Russian)

69 - D. LEVIN, A. SIDI
Two new classes of non linear transformations for accelerating the convergence of infinite integrals and series. J. Comp. Math. Appl. .

70 - G. LOPES
On the convergence of multipoint Padé approximants for functions of Stieltjes type. Soviet Math. Dokl., 19 (1978), 425-428.

71 - Y.L. LUKE
Some approximations and inequalities for Arc Tan and Ln. Rev. Tec. Ing. Univ. Zulia, 1 (1978), 55-58.

72 - J.H. Mc CABE
A further correspondance property of M fractions. Math. Comp., 32 (1978), 1303-1305.

73 - C. MOLER, C. VAN LOAN
Nineteen dubious ways to compute the exponential of a matrix. SIAM Rev., 20 (1978), 801-836.

74 - L.M. MURATOV
On improving the convergence of series. Izvestia VUZ. Matematika, 22 (1978), 53-57, Sov. Math., 22 (1978), 44-48.

75 - J. DE PILIS
Faster convergence for iterative solutions to systems via three-part splittings, SIAM J. Numer. Anal., 15 (1978), 888-911.

76 - N.M. REID
Uniform convergence and truncation error estimates of continued fractions $K(a_n/1)$. Ph. D., Univ. of Colorado, Boulder, 1978.

77 - D. ROSEN, J. SHALLIT
A continued fraction algorithm for approximating all real polynomial roots. Math. Mag., 51 (1978), 112-116.

78 - A. RUTTAN
Best complex rational approximations of real functions. Thesis, Kent State University, 1978.

79 - R.A. SACK
A numerical method for simultaneous convergence to the elements of a sequence. J. Comp. Appl. Math., 5 (1979), 29-35.

80 - R. SAIGAL, M.J. TODD
Efficient acceleration techniques for fixed point algorithms. SIAM J. Numer. Anal., 15 (1978), 997-1007.

81 - J. SANCHEZ - DEHESA
The spectrum of Jacobi matrices in terms of its associated weight function. J. Comp. Appl. Math., 4 (1978), 275-284.

82 - L. SHORT
The practical evaluation of multivariate approximants with branch points. Proc. R. Soc. London A, 362 (1978), 57-69.

83 - L. SHORT
The evaluation of Feymman integrals in the physical region using multi-valued approximants. J. Phys. G : Nucl. Phys., 5 (1979), 167-198.

84 - A. SIDI
Convergence properties of some non linear sequence transformations. Math. Comp., 33 (1979), 315-326.

85 - A. SIDI
Derivation of new numerical quadrature formulas by use of some non linear sequence transformations. Math. Comp. .

86 - A. SIDI
Analysis of convergence for some non linear convergence acceleration methods. Math. Comp. .

87 - A. SIDI
Some aspects of two-point Padé approximants. J. Comp. Appl. Math. .

88 - A. SIDI
Some properties of a generalization of the Richardson extrapolation process.

89 - A.C. SMITH
Asymptotic estimates of sums of series using difference equations. Utilitas Mathematica, 13 (1978), 249-269.

90 - D.A. SMITH, W.F. FORD
Acceleration of linear and logarithmic convergence. SIAM J. Numer. Anal., 16 (1979), 223-240.

91 - D.A. SMITH, W.F. FORD
Nonlinear acceleration of linear convergence.

92 - D.A. SMITH, W.F. FORD
Transformation of divergent series by convergence accelerators.

93 - Y. STARKAND
Explicit formulas for matrix-valued Padé approximants. J. Comp. Appl. Math., 5 (1979), 63-66.

94 - J. SULLIVAN
Padé approximants via the continued fraction approach. Am. J. Phys., 46 (5), (1978), 489-494.

95 - A.J. THAKKAR

Molec. Phys., 36 (1978), 887.

96 - J. THRASH
Approximation of improper integrals. 765th meeting of the AMS, New-York, April 19-20, 1979.

97 - H. WERNER, L. WUYTACK
Nonlinear quadrature rules in the presence of a singularity. Comp. and Maths with Appls., 4 (1978), 237-245.

- o -

COMMENTED BIBLIOGRAPHY

on techniques for computing Padé Approximants.

L. Wuytack
Department of Mathematics
University of Antwerp
Universiteitsplein 1
B-2610 Wilrijk (Belgium)

The papers and books in this bibliography are directly concerned with the computation of a solution to the following classical problem, as formulated by Padé in 1892. Let f be a given power series and R_n^m the class of rational functions $r=\frac{p}{q}$ where p is a polynomial of degree at most m and q a polynomial of degree at most n, such that $\frac{p}{q}$ is irreducible. The problem is to find an element $r=\frac{p}{q}$ from R_n^m satisfying

$$f.q-p=0\ (x^{m+n+1+j}) \tag{1}$$

where j is an integer which is as high as possible. This problem has a unique solution for all values of m and n. If we put

$$f=\sum_{i=0}^{\infty} c_i.x^i,\ p=\sum_{i=0}^{m} a_i.x^i,\ q=\sum_{i=0}^{n} b_i.x^i \tag{2}$$

then (1) implies that a_i and b_i must satisfy the following linear system

$$\sum_{i=0}^{n} c_{k-i}.b_i=a_k \quad \text{for } k=0, 1, \ldots, m \tag{3.a}$$

$$\text{and } \sum_{i=0}^{n} c_{m+k-i}.b_i=0 \quad \text{for } k=1, 2, \ldots, n \tag{3.b}$$

(3)

where $c_i=0$ if $i<0$ and (3.b) is empty if n=0. There is also a strong relation between the theory on Padé Approximation and the theory on continued fractions and orthogonal polynomials. This implies that algorithms for computing solutions of linear systems or for computing corresponding continued fractions or for computing orthogonal polynomials, can be used to compute Padé approximants. Papers on these topics are however only mentined in the list below if they are of classical importance (e.g. [1]-[5],[7],[8]) or consider the problem of Padé approximation explicitly as a special case. We tried to collect publications on original algorithms for computing Padé approximants, on variants of these algorithms and survey papers. Other publications can be found in C. Brezinski's bibliographies.

Several generalizations of the Padé approximation problem can be found in the literature, e.g. multi-variable Padé approximation and multi-point Padé approximation (also called rational Hermite interpolation). Algorithms for solving these more general problems are only mentioned in case they give new insight in the computation of classical Padé approximants.

Finally we remark that the comment for each paper refers in most cases only to the algorithmic aspects of the Padé approximation problem. Several papers are more general than the comment might indicate. The papers are ordered chronologically and marked with ^ if the computation of non-normal Padé approximants is included.

[1] VISKOVATOFF B. : *De la méthode générale pour réduire toutes sortes des quantités en fractions continues. Mémoires de l'Académie Impériale des Sciences de St. Petersburg 1 (1803-1806), pp. 226-247.*

A technique is proposed for transforming a quotient of two power series into a continued fraction expansion. The technique can be used to form the corresponding continued fraction of a given power series. (see also "KHOVANSKII A.N. : The Application of Continued Fractions and their Generalizations to problems in Approximation Theory. Noordhoff, Groningen, 1963")

[2] JACOBI C.G.J. : *Über die Darstellung einer Reihe gegebner Werthe durch eine gebrochne rationale Function. Journal für die Reine und Angewandte Mathematik 30 (1845), pp. 127-156.*

Explicit representations are given for the numerator and denominator of a Padé approximant in the form of determinants. Seven representations for the numerator and six representations for the denominator are derived (see §5, pp. 148-152) as special cases of the representation of interpolating rational functions. These representations are based on the property that the coefficients in the numerator and denominator have to satisfy the linear system of equations (3.a) and (3.b).

[3] KRONECKER L. : *Zur Theorie der Elimination einer Variabeln aus zwei algebraischen Gleichungen. Monatsberichte der Königlich Preussischen Akademie der Wissenschaften zu Berlin vom Jahre 1881, pp. 535-600.*

The following problem is considered : Let f and f_1 be given polynomials of degree n and $n-n_1$ respectively. Find polynomials p and q such that the degree of $(f_1.q - f.p).q$ is less than n. The problem of finding a rational function $r = \frac{p}{q}$ having the same derivatives (in a certain point) as a given function f_1 is treated (see pages 544 and 549) as a special case. (Note that this problem is equivalent with the Padé approximation problem).
Two basic techniques are indicated for solving the problem. The first is an Euclidean-type division algorithm for finding a continued fraction

representation for $\frac{f_1}{f}$. Polynomials g_1, f_2, g_2, f_3, ... are defined such that

$$f-g_1.f_1+f_2=0,\ f_1-g_2.f_2+f_3=0,\ f_2-g_3.f_3+f_4=0,\ \dots$$

From these relations follows

$$\frac{f_1}{f}=\frac{1|}{|g_1}-\frac{1|}{|g_2}-\frac{1|}{|g_3}-\dots$$

The convergents of this continued fraction represent certain solutions to the given problem. The second algorithm is by solving the set of equations which must be satisfied by the coefficients of p and q. These equations are derived from the condition that certain coefficients of the powers of x in $f_1.q-f.p$ must be zero.

[4] FROBENIUS G. : *Ueber Relationen zwischen den Näherungsbrüchen von Potenzreihen. Journal für die Reine und Angewandte Mathematik 90 (1881), pp. 1-17.*

Several relations between neighbouring elements in the Padé table are derived. These relations include the basic identities connecting separately the numerators and denominators of 3 neighbouring elements in the Padé table. The place of these elements is indicated in each of the following 4 configurations

	x	x	x	x	x	x	
x	x	x			x	x	x

The given identities form the basis of several recurrence algorithms for computing numerators and denominators of Padé approximants (see [18], [20], [34], [39]).

Some of the relations are connected with the determinantal representations in [2]. They also can be used to derive explicit formulas for the coefficients in a continued fraction representation of the form

$$k_0+\frac{x|}{|k_1}+\frac{x|}{|k_2}+\frac{x|}{|k_3}+\dots$$

for the elements on the principal diagonal of the Padé table. A recurrence relation for computing k_0, k_1, k_2, ... is given (page 16).

[5] *STIELTJES T.J. : Sur la Réduction en Fraction Continu d'une Série Précédant suivant les Puissances Descendants d'une Variable. Annales de la Faculté des Sciences de Toulouse 3 (1889), pp. 1-17.*

Relations are given between the coefficients $a_i^{(k)}$, $b_i^{(k)}$ in the continued fraction expansion

$$f_k(z) = \frac{c_k|}{|z-a^{(k)}} - \frac{b_0^{(k)}|}{|z-a_1^{(k)}} - \frac{b_1^{(k)}|}{|z-a_2^{(k)}} - \dots - \frac{b_{i-1}^{(k)}|}{|z-a_i^{(k)}} - \dots$$

which corresponds to the given power series

$$f_k(z) = \sum_{n=0}^{\infty} c_{k+n} . z^{-n-1} \quad \text{for } k \geqslant 0.$$

These relations can be used to define an algorithm for computing these coefficients.

[6] ^ PADE H. : *Sur la représentation approchée d'une fonction par des fractions rationnelles. Annales scientifiques de l'Ecole normale supérieure de Paris 9 (1892), pp. 1-93.*

This paper gives the first systematic study of the problem where a rational function $r = \frac{p}{q}$ of a certain degree approximates a given power series f such that $f.q-p=O(x^j)$, where j is an integer which is as high as possible. The rational functions of different degree satisfying this relation can be put into a table. For this table the "block structure" (which means that equal elements appear in square blocks) is proved and conditions for its normality (which means that all its elements are different from each other) are given.
Much attention is given to the relation between certain sequences of elements in the table and the theory on continued fractions. The case where $f(x)=e^x$ is studied in more detail.

[7] ^ PERRON O. : *Die Lehre von den Kettenbrüchen. Teubner, Stuttgart, 1929.*

This classical book on continued fractions also includes a chapter on Padé approximation. In this chapter the fundamental properties of Padé

approximants and the structure of the Padé table are given, following the lines of the paper of Padé [6].
The basic relations between the Padé table and continued fractions are given. Also the question of convergence of Padé approximants is treated. Although no algorithms for constructing Padé approximants are described the book contains the fundamental properties which form the basis of several of these algorithms.

[8] - WALL H.S. : *Analytic Theory of Continued Fractions*. *Van Nostrand, New York, 1948.*

The same comments as on Perron's book [7] also hold here.

[9] RUTISHAUSER H. : *Der Quotienten-Differenzen-Algorithmns. Zeitschrift für Angewandte Mathematik und Physik 5 (1954), pp. 233-251.*

A recursive algorithm is given for finding the coefficients $q_i^{(k)}$, $e_i^{(k)}$ in the continued fraction expansion

$$f_k(z)=\frac{c_k|}{|z}-\frac{q_1^{(k)}|}{|1}-\frac{e_1^{(k)}|}{|z}-\frac{q_2^{(k)}|}{|1}-\frac{e_2^{(k)}|}{|z}-\dots-\frac{q_i^{(k)}|}{|1}-\frac{e_i^{(k)}|}{|z}-\dots$$

which corresponds to the given power series

$$f_k(z)=\sum_{n=0}^{\infty} c_{n+k}.z^{-n-1} \quad \text{for } k\geq 0.$$

[10] BAUER F.L. : *The quotient-difference and epsilon algorithm. In "Numerical approximation" (LANGER R., ed.), University of Wisconsin Press, Madison, 1959, pp. 361-370.*

The ε-algorithm (see [12]) and a related η-algorithm are derived by means of the g-algorithm. This algorithm is similar to the qd-algorithm and computes the coefficients c and g_i in the following corresponding continued fraction

$$\frac{s_0|}{|z} + \sum_{i=0}^{\infty} \left(-\frac{g_i.(c-g_{i+1})|}{|1} - \frac{g_{i+1}.(1-g_{i+2})|}{|z}\right)$$

for the formal power series $\sum_{i=0}^{\infty} \frac{s_i}{i+1}$. The theoretical background for the g-algorithm is given in "BAUER F.L. : The g-algorithm. SIAM Journal 8 (1960), pp. 1-17."

[11] THACHER H.C.Jr and TUKEY J.W. : *Rational interpolation made easy by a recursive algorithm. Unpublished manuscript, 1960.*

A recursive algorithm is described which can be used for finding the coefficients a_i of the continued fraction

$$r_k(x) = \sum_{i=0}^{k} c_i.x^i + \frac{c_{k+1}.x^{k+1}|}{|1} + \sum_{i=0}^{\infty} \frac{x|}{|a_i}$$

whose convergents are the elements on a descending staircase in the Padé table for

$$f(x) = \sum_{i=0}^{\infty} c_i.x^i.$$

[12] WYNN P. : *The rational approximation of functions which are formally defined by a power series expansion. Mathematics of Computation 14 (1960), pp. 147-186.*

Several basic techniques are described for computing elements of the Padé table and of the E-array, which is similar to the Padé table but for a power series in $\frac{1}{z}$. These techniques are based on the use of recurrence relations between the numerators and denominators of 3 neighbouring elements in the Padé table (see [4]), the algorithm of Stieltjes (see [5]), the qd-algorithm (see [9]) and the ε-algorithm.
It is shown that the ε-algorithm (see "WYNN P. : On a device for computing the $e_m(S_n)$ transformation. Mathematics of Computation 10 (1956), pp. 91-96) can also be used to compute Padé approximants. This results in a recursive algorithm, based on the use of a nonlinear transformation, starting from the partial sums of a given power series.

The basic techniques are combined to form algorithms which can be used to find either a particular element or a sequence of elements or a complete array of elements from either the Padé table or the E-array. To compare them, in terms of computational effort, further distinction is made between the processes of diriving explicit formulas for the rational functions and that of computing the values of these functions for some prescribed values of the argument. It is remarked that some of the algorithms can be considered as codes for computing the solution of the linear system, which must be satisfied by the coefficients of the numerator and denominator of a Padé approximant or of an element of the E-array.

[13] - MAGNUS A. : *Certain continued fractions associated with the Padé table. Mathematische Zeitschrift 78 (1962), pp. 361-374.*

Two methods are described (see page 362 and page 368, theorem 5) for computing the denominators b_i (which are polynomials in $\frac{1}{z}$) of the principal part expansion

$$b_0 + \sum_{i=1}^{\infty} \frac{1|}{|b_i}$$

of a given power series

$$f(z) = \sum_{n=-N}^{\infty} a_n . z^n$$

The algorithms are recursive in nature and involve either the inversion of power series or the solution of systems of linear equations. Due to the connection with the theory on Padé approximation (see page 370, theorem 6) the algorithms can also be used to construct Padé approximants in the non-normal case.

[14] - MAGNUS A. : *Expansion of power series into P-fractions. Mathematische Zeitschrift 80 (1962), pp. 209-216.*

A third method is described for the problem considered in [13]. This method involves the recursive calculation of a sequence of determinants. The numerators and denominators of the convergents of the principal part expansion of the given power series are expressed in terms of these determinants.

[15] DONNELLY J.D.P. : *The Padé Table. In "Methods of Numerical Approximation" (HANDSCOMB D.C., ed.) Pergamon Press, Oxford, 1966, pp. 125-130.*

A short survey is given of the definition and structure of the Padé table together with its relation to the theory on continued fractions. The qd - (see [9]) and ε-algorithm (see [12]) are indicated as possible algorithms for computing the elements of the table or their values for certain values of the argument.

[16] WYNN P. : *Upon systems of recursions which obtain among the quotients of the Padé table. Numerische Mathematik 8 (1966), pp. 264-269.*

Starting from the ε-algorithm the following relation between 5 adjacent elements in the Padé table is derived.

$$(C-N)^{-1}+(C-S)^{-1}=(C-W)^{-1}+(C-E)^{-1}$$

The elements C, N, E, S, W correspond to the following configuration in the Padé table

$$\begin{array}{ccc} & N & \\ W & C & E \\ & S & \end{array}$$

If 2 consecutive rows of columns in the Padé table are given than the given relation can be used to find the other rows or columns.

[17] MASSEY J.L. : *Shift-register synthesis and BCH decoding. IEEE Transactions on Information Theory IT-15 (1969), pp. 122-127.*

An algorithm is given for constructing the denominator of a Padé approximant. This algorithm is based on an iterative technique for solving the set of equations (3.b) and is a particular application of an algorithm of Berlekamp (see "BERLEKAMP E.R. : Algebraic coding theory. McGraw-Hill, New York, 1968").

[18] BAKER G.A. Jr : *The Padé approximant methods and some related generalizations. In "The Padé Approximant in Theoretical Physics" (BAKER G.A. Jr and GAMMEL J.L., eds.), Academic Press, New York, 1970, pp. 1-39.*

A survey is given of the basic results concerning the Padé table and its applications. Some techniques are indicated for computing Padé approximants, e.g. evaluation of explicit expressions for the Padé approximant, solving the system (3.a) and (3.b), the ε-algorithm (see [12]), the use of recurrence relations (of Frobenius type) for numerator and denominator.

[19] HOUSEHOLDER A.S. : *The Padé Table, the Frobenius Identities, and the qd Algorithm. Linear Algebra and Its Applications 4 (1971), pp. 161-174.*

A survey is given of some properties of Padé approximants, the structure of the Padé table and the relations between its elements.
The identities underlying the qd-algorithm (see [9]) as well as the basic identities of Frobenius (see [4]) are reproved. This is done in terms of bigradient determinants and Hankel determinants.

[20] LONGMAN I.M. : *Computation of the Padé table. International Journal of Computer Mathematics 3 (1971), pp. 53-64.*

An algorithm is described for computing the coefficients in the numerator p and denominator q of a Padé approximant. To compute the coefficients of p a recurrence relation (of Frobenius-type) between the numerators of three neighbouring elements in the Padé table is used. A similar relation for the denominators is used to compute the coefficients of q.

[21] RISSANEN J. : *Recursive identification of linear systems. SIAM Journal on Control 9 (1971), pp. 420-430.*

An algorithm is given for finding the solution of a linear system, that is equivalent with the system (3.b). To solve the system an iterative technique is applied which uses a factorization algorithm in each step.

A similar algorithm is given in "RISSANEN J. : Solution of linear equations with Hankel and Toeplitz Matrices. Numerische Mathematik 22 (1974), pp. 361-366.
The algorithm has been extended to the case of matrix functions in "RISSANEN J. : Recursive evaluation of Padé Approximants for matrix sequences. IBM Journal of Research and Development 16 (1972), pp. 401-406."

[22] GRAGG W.B. : *The Padé table and its relation to certain algorithms of Numerical Analysis*. SIAM Review 14 (1972), pp. 1-62.

A survey is given of the theory on Padé approximation and most of its basic results are reproved. This survey includes the structure of the Padé table, the normality conditions, the identities of Frobenius (see [4]), Wynn's identity (see [16]) and connections with the theory on continued fractions.
Several algorithms related to the Padé table are given, e.g. the ε-algorithm (see [12]), the η-algorithm (see [10]) and the qd-algorithm (see [9]). A new algorithm of qd-type is given for computing the coefficients $a_i^{(k)}$, $b_i^{(k)}$ in the following continued fraction

$$r_k(x) = \sum_{i=0}^{k} c_i . x^i - \frac{c_k . x^k|}{|1} - \frac{a_1^{(k)}|}{|x} - \frac{b_1^{(k)}|}{|1} - \frac{a_2^{(k)}|}{|x} - \frac{b_2^{(k)}|}{|1} - \ldots - \frac{a_k^{(k)}|}{|x}$$

whose convergents form the elements on an ascending staircase in the Padé table for $f(x) = \sum_{i=0}^{\infty} c_i . x^i$.

[23] BAKER G.A. Jr. : *Recursive calculation of Padé approximants*. In "*Padé Approximants and their Applications*" (ed. GRAVES-MORRIS P.R., Academic Press, London, 1973), pp. 83-91.

A survey is given of some methods for computing Padé approximants. The following methods are discussed

(a) direct computation by solving the linear system (3.a) and (3.b),
(b) recursive calculation by using the ε-algorithm (see [12]), Baker's algorithm (see [18]), the qd-algorithm (see [9]) and Gragg's algorithm (see [22]).

[24] WATSON P.J.S. : *Algorithms for differentiation and integration.* In "*Padé Approximants and their Applications*". (GRAVES-MORRIS P.R., ed.) Academic Press, London, 1973, pp. 93-97.

The paper includes an algorithm for computing the coefficients $a_i^{(k)}$, $b_i^{(k)}$ in the following continued fraction

$$r_k(x) = \sum_{i=0}^{k} c_i . x^i + \frac{c_{k+1}.x^{k+1}|}{|1} + \sum_{i=1}^{\infty} \left(\frac{a_i^{(k)}.x|}{|1} + \frac{b_i^{(k)}.x|}{|1} \right)$$

whose convergents form the elements on an descending staircase in the Padé table. These coefficients are computed using the coefficients of the denominators of two preceding convergents of the continued fraction.

[25] ^ GRAGG W.B. : *Matrix interpretations and applications of the continued fraction algorithm.* Rocky Mountain Journal of Mathematics 4 (1974), pp. 213-225.

An interpretation in terms of matrices is given for an algorithm that can be used to find recursively the denominators (and numerators) of a Padé approximant. It is shown that the algorithm corresponds to the Gauss-Banachiewicz factorization of a symmetric Hankel matrix. A variant of the algorithm is related to the Lanczos algorithm for tridiagonalization.

It is also indicated how the algorithm can be modified in case the Padé table is not normal.

[26] PATRY J. and GUPTA S. : *Computing analytical functions by means of power series or continued fractions.* In "*International Computing Symposium 1973*" (A. GÜNTHER et al.,eds.) North-Holland, New York, 1974, pp. 323-329.

An algorithm is included for converting a given series $\sum_{n=1}^{\infty} c_n . x^{-2n+1}$ into a corresponding continued fraction of the form

$$\sum_{i=1}^{\infty} \frac{a_i|}{|x}.$$

Numerical experiments indicate that the algorithm is more stable than the qd-algorithm.

[27]^ WARNER D.D. : *Hermite interpolation with rational functions*. Ph.D.thesis, University of California, 1974.

A survey of algorithms for computing interpolatory rational functions of Hermite-type is included. The following algorithms are discussed : Kronecker's Division Algorithm, Thiele's continued fraction algorithm, the algorithms of Thacher and Tukey, Stoer, Larkin. A new algorithm, based on a generalized Wynn identity, is given.

[28]^ BAKER G.A. Jr. : *Essentials of Padé Approximants*. Academic Press, New York, 1975.

This book on Padé Approximation and its applications also contains a discussion of the basic techniques for computing Padé approximants. Several techniques are considered in more detail, e.g. the ε-algorithm (see [12]), Baker's algorithm (see [18]), Watson's algorithm (see [24]), Gragg's variant (see [22]) of the qd-algorithm (see [9]).

[29] CLAESSENS G. : *A new look at the Padé table and the different methods for computing its elements*. Journal of Computational and Applied Mathematics 1 (1975), pp. 141-152.

A survey is given of the algorithms for computing Padé approximants. The following methods and their relations are discussed : The algorithms of Baker (see [18]), Longman (see [20]), Gragg (see [22]), Watson (see [24]), Thacher and Tukey (see [11]), Rutishauser (or qd-algorithm, see [9]) and Wynn (or ε-algorithm, see [12]). A new method of Watson-type is given for computing the coefficients $a_i^{(k)}$, $b_i^{(k)}$ in the following continued fraction

$$r_k(x) = \sum_{i=0}^{k} c_i . x^i - \frac{c_k . x^k|}{|1} - \frac{a_1^{(k)}|}{|x} - \frac{b_1^{(k)}|}{|1} - \frac{a_2^{(k)}|}{|x} - \frac{b_2^{(k)}|}{|1} - \dots - \frac{a_k^{(k)}|}{|x}$$

whose convergents form the elements on an ascending staircase in the Padé table. A correction to the basic formulas for this method is given in "BULTHEEL A. : Remark on "A new look at the Padé table and different methods for computing its elements". Journal of Computational and Applied Mathematics 5 (1979), p. 67."

Reference is made to the algorithms of Bauer (see [10]).
A comparison between these algorithms, in terms of computational effort, is given.

[30] MILLS W.H. : *Continued fractions and linear recurrences. Mathematics of Computation 29 (1975), pp. 173-180.*

An algorithm, related to Berlekamp's algorithm (see [17]), is given to produce a solution of a linear system with a Hankel matrix. The algorithm is of recursive type and can be used to compute a diagonal in the Padé table.

[31] BREZINSKI C. : *Computation of Padé approximants and continued fractions. Journal of Computational and Applied Mathematics 2 (1976), pp. 113-123.*

The algorithm of Trench (see TRENCH W.F. : An algorithm for the inversion of finite Hankel matrices. SIAM Journal 13 (1965), pp. 1102-1107) is used to derive two methods for computing Padé approximants. Both methods recursively compute the elements on a sub-diagonal in the Padé table. The first method is based on a relation between two consecutive elements on a sub-diagonal in the Padé table. The second method computes seperately the coefficients of numerator and denominator.
Modifications of these algorithms can be used to compute a particular diagonal in the qd-array (this is the array built up by the qd-algorithm) and in the ε-array (this is the array built up by the ε-algorithm). Extensions of the algorithm to more general cases are indicated.

[32] CLAESSENS G. : *Some aspects of the rational Hermite interpolation table and its applications. Ph.D.thesis, University of Antwerp, 1976.*

Several algorithms are given for computing rational Hermite interpolating functions. These algorithms reduce to the qd-algorithm (see [9]), the tg-algorithm (see [33]), the algorithm of Gragg (see [22]) and a modification of Kronecker's algorithm (see [3]). The connection between these algorithms is shown.

[33] CLAESSENS G. : *A new algorithm for osculatory rational interpolation.* *Numerische Mathematik* 27 (1976), pp. 77-83.

An algorithm, similar to the qd-algorithm, is given for computing the coefficients $t_i^{(k)}$, $g_i^{(k)}$ in the continued fraction

$$r_k(x) = \sum_{i=0}^{k} c_i . x^i + \frac{c_{k+1}.x^{k+1}|}{|1} + \sum_{i=1}^{\infty} \left(\frac{z|}{|t_i^{(k)}} + \frac{(1-g_i^{(k)}).z|}{|g_i^{(k)}} \right)$$

whose convergents form the elements on a descending staircase in the Padé table.

[34] PINDOR M. : *A simplified algorithm for calculting the Padé table derived from Baker and Longman schemes. Journal of Computational and Applied Mathematics* 2 (1976), pp. 255-258.

An algorithm is given, based on a Frobenius-type recurrence relation, for computing recursively the numerator and denominator of a Padé approximant. It is more effective, in terms of computational effort, than Baker's (see [18]) and Longman's (see [20]) algorithm.

[35]^ MURPHY J.A. and D'DONOHOE M.R. : *A class of algorithms for obtaining, rational approximants to functions which are defined by power series. Journal of Applied Mathematics and Physics* (ZAMP) 28 (1977), pp. 1121-1131.

A general technique is given to find a generalized corresponding continued fraction to a given power series. This continued fraction is of the form

$$\frac{p_0}{q_0(z)} + \sum_{i=1}^{\infty} \frac{p_i.z^{n_i}|}{|q_i(z)}$$

where $\{n_i\}$ is a sequence of positive integers, p_i are constants for $i \geq 0$ and $q_i(z)$ are polynomials of a certain degree. The technique is similar to the algorithm of Viskovatoff (see [1]) and based on a recurrence relation. Several classical algorithms for computing C-fractions, S-fractions, J-fractions (see [8]) and P-fractions (see [13]) are shown to be special cases. Other applications of the algorithm are given in "DREW D.M. and

MURPHY J.A. : Branch points, M-Fractions and Rational approximants generated by Linear Equations. Journal of the Institute of Mathematics and Its Applications 19 (1977), pp. 169-185".

[36] - BULTHEEL A. : *Recursive algorithms for non normal Padé tables. Report TW 40, Applied Mathematics and Programming Division, University of Leuven, 1978.*

It is shown how the Berlekamp-Massey algorithm (see [17]) can be modified to compute the elements of a non-normal Padé table. Some variants of this modified algorithm reduce to the algorithms of Brezinski (see [31]), of Watson (see [24]), of Thacher and Tukey (see [11]). It is remarked that there is some duality with Cordellier's algorithms (see [40]).

[37] - BULTHEEL A. : *Division algorithms for continued fractions and the Padé table. Report TW 41, Applied Mathematics and Programming Division, University of Leuven, 1978.*

It is shown that the algorithm of Viskovatoff (see [1]) can also be used for constructing non-normal Padé approximants, lying on a descending staircase in the Padé table. This generalization is seen to be equivalent with the Berlekamp-Massey algorithm (see [17]).
A similar division algorithm, for constructing associated continued fractions, is also generalized to compute (reducible) Padé approximants lying on a diagonal in the Padé table. This algorithm is seen to coincide with Brezinki's algorithm (see [31]) in the normal case.
Variants of both division algorithms can be used to find the elements on other paths in the Padé table.

[38] - BULTHEEL A. : *Fast algorithms for the factorization of Hankel and Toeplitz matrices and the Padé approximation problem. Report TW 42, Applied Mathematics and Programming Division, University of Leuven, 1978.*

Matrix interpretations are given for several recursive algorithms to compute Padé approximants in the normal as well as in the non-normal case.

They are based on techniques for solving a system of linear equations with a Hankel or Toeplitz matrix, such as the algorithms of Trench (see [31]) and Rissanen (see [21]). The algorithms of Brezinski (see [31]), Stieltjes (see [5]), Watson (see [24]), Baker (see [18]) and variants of these are seen to be special cases. Also the "continued fraction algorithms" (see [25], [35], [37], [38]) can be interpreted in a similar way.

[39] BUSSONAIS D. : *"Tous" les algorithmes de calcul par recurrence des approximants de Padé d'une série. Construction des fractions continues correspondantes. Talk at the "Colloque sur les approximants de Padé", Lille, March 1978.*

The identities of Frobenius (see [4]) are reviewed. It is shown how they can be used to compute any sequence of consecutive elements in the Padé table. The algorithms of Baker (see [18]), of Longman (see [20]), of Pindor (see [34]), of Watson (see [24]), of Thacher and Tukey (see [11]) and of Brezinski (see [31]) are indicated as special cases of a general technique. A comparison between these algorithms, in terms of computational effort, is made.

[40]^ CORDELLIER F. : *Deux algorithmes de calcul récursif des éléments d'une table de Padé non normale. Talk at the "Colloque sur les approximants de Padé", Lille, March 1978.*

Two recursive techniques are given for finding the elements in a Padé table. The first computes the (reducible form of) elements on an ascending staircase, while the second computes the (reducible form of) elements on an ascending diagonal in the Padé table. Both algorithms are Euclidean-type division algorithms (see [3]).

[41]^ GILEWICZ J. : *Approximants de Padé. Springer-Verlag (Lecture Notes in Mathematics 667), Berlin, 1978.*

This book on the mathematical aspects of Padé approximation contains a discussion on several computational aspects of the problem of Padé approximation. Some algorithms are considered in more detail, e.g. the ε-algorithm (see [12]), Viskovatoff-type algorithms (see [1]), the algorithms of Baker (see [18]), Longman (see [20]), Pindor (see [34]) and Wynn (see [16]).

[42]* McELIECE R.J. and SHEARER J.B. : *A property of Euclid's algorithm and an application to Padé Approximation. SIAM Journal on Applied Mathematics 34 (1978), pp. 611-615.*

It is shown that Euclid's algorithm, for finding the greatest divisor of two polynomials, can be used to compute (the reducible forms of) Padé approximants. See also [3] and [32].

[43] BREZINSKI C. : *Sur le calcul de certains rapports de déterminants. This paper is published in these Proceedings.*

Several algorithms are described for computing e.g. Shanks transformation (see SHANKS D. : Nonlinear transformations of divergent and slowly convergent sequences. Journal of Mathematics and Physics 34 (1955), pp. 1-42) of a given series. Due to the connection between Shanks transformation and the ε-algorithm (see [12]), these algorithms can be used to compute Padé approximants.

[44]* BULTHEEL A. : *Recursive algorithms for the Padé table : two approaches. This paper is published in these Proceedings.*

A matrix interpretation of several algorithms for computing sequences of Padé approximants is given. Continued fraction algorithms (of Viskovatoff-type) and recursive algorithms are linked together within this framework.

[45]^ CLAESSENS G. and WUYTACK L. : *On the computation of non-normal Padé approximants.* To appear in *Journal of Computational and Applied Mathematics* 5 (1979).

Techniques are described for modifying some classical algorithms in case the Padé table is not normal. A modification of the qd- and ε-algorithm is given. The modified ε-algorithm is based on a generalization of Wynn's identity (see [16]) to the non-normal case.

[46]^ GEDDES K.O. : *Symbolic computation of Padé approximants.* *ACM Transactions on Mathematical Software* 5 (1979), pp. 218-233.

A fraction-free variant of Gaussian elimination is used to solve the system of linear equations (3.b) for the coefficients of the denominators of a Padé approximant. Kahan's version for symmetric triangularization is used as basic algorithm in performing the elimination procedure. The block structure of the Padé table is exploited in case of non-normality. The result is an algebraic manipulation algorithm which may be applied to a power series over an arbitrary integral domain.

[47] GRAVES-MORRIS P.R. : *The numerical calculation of Padé approximants. This paper is published in these Proceedings.*

A survey is given of the different techniques for computing Padé approximants. These techniques are compared in terms of e.g. reliability, stability, efficiency. It is illustrated that the problem of determining the coefficients of a Padé approximant is ill-conditioned.

[48] STARKAND Y. : *Explicit formulas for matrix-valued Padé approximants.* *Journal of Computational and Applied Mathematics* 5 (1979), pp. 63-66.

Gauss elimination is used to solve the linear system for the coefficients of the denominator of a matrix-valued Padé approximant.

Vol. 609: General Topology and Its Relations to Modern Analysis and Algebra IV. Proceedings 1976. Edited by J. Novák. XVIII, 225 pages. 1977.

Vol. 610: G. Jensen, Higher Order Contact of Submanifolds of Homogeneous Spaces. XII, 154 pages. 1977.

Vol. 611: M. Makkai and G. E. Reyes, First Order Categorical Logic. VIII, 301 pages. 1977.

Vol. 612: E. M. Kleinberg, Infinitary Combinatorics and the Axiom of Determinateness. VIII, 150 pages. 1977.

Vol. 613: E. Behrends et al., L^p-Structure in Real Banach Spaces. X, 108 pages. 1977.

Vol. 614: H. Yanagihara, Theory of Hopf Algebras Attached to Group Schemes. VIII, 308 pages. 1977.

Vol. 615: Turbulence Seminar, Proceedings 1976/77. Edited by P. Bernard and T. Ratiu. VI, 155 pages. 1977.

Vol. 616: Abelian Group Theory, 2nd New Mexico State University Conference, 1976. Proceedings. Edited by D. Arnold, R. Hunter and E. Walker. X, 423 pages. 1977.

Vol. 617: K. J. Devlin, The Axiom of Constructibility: A Guide for the Mathematician. VIII, 96 pages. 1977.

Vol. 618: I. I. Hirschman, Jr. and D. E. Hughes, Extreme Eigen Values of Toeplitz Operators. VI, 145 pages. 1977.

Vol. 619: Set Theory and Hierarchy Theory V, Bierutowice 1976. Edited by A. Lachlan, M. Srebrny, and A. Zarach. VIII, 358 pages. 1977.

Vol. 620: H. Popp, Moduli Theory and Classification Theory of Algebraic Varieties. VIII, 189 pages. 1977.

Vol. 621: Kauffman et al., The Deficiency Index Problem. VI, 112 pages. 1977.

Vol. 622: Combinatorial Mathematics V, Melbourne 1976. Proceedings. Edited by C. Little. VIII, 213 pages. 1977.

Vol. 623: I. Erdelyi and R. Lange, Spectral Decompositions on Banach Spaces. VIII, 122 pages. 1977.

Vol. 624: Y. Guivarc'h et al., Marches Aléatoires sur les Groupes de Lie. VIII, 292 pages. 1977.

Vol. 625: J. P. Alexander et al., Odd Order Group Actions and Witt Classification of Innerproducts. IV, 202 pages. 1977.

Vol. 626: Number Theory Day, New York 1976. Proceedings. Edited by M. B. Nathanson. VI, 241 pages. 1977.

Vol. 627: Modular Functions of One Variable VI, Bonn 1976. Proceedings. Edited by J.-P. Serre and D. B. Zagier. VI, 339 pages. 1977.

Vol. 628: H. J. Baues, Obstruction Theory on the Homotopy Classification of Maps. XII, 387 pages. 1977.

Vol. 629: W.A. Coppel, Dichotomies in Stability Theory. VI, 98 pages. 1978.

Vol. 630: Numerical Analysis, Proceedings, Biennial Conference, Dundee 1977. Edited by G. A. Watson. XII, 199 pages. 1978.

Vol. 631: Numerical Treatment of Differential Equations. Proceedings 1976. Edited by R. Bulirsch, R. D. Grigorieff, and J. Schröder. X, 219 pages. 1978.

Vol. 632: J.-F. Boutot, Schéma de Picard Local. X, 165 pages. 1978.

Vol. 633: N. R. Coleff and M. E. Herrera, Les Courants Résiduels Associés à une Forme Méromorphe. X, 211 pages. 1978.

Vol. 634: H. Kurke et al., Die Approximationseigenschaft lokaler Ringe. IV, 204 Seiten. 1978.

Vol. 635: T. Y. Lam, Serre's Conjecture. XVI, 227 pages. 1978.

Vol. 636: Journées de Statistique des Processus Stochastiques, Grenoble 1977, Proceedings. Edité par Didier Dacunha-Castelle et Bernard Van Cutsem. VII, 202 pages. 1978.

Vol. 637: W. B. Jurkat, Meromorphe Differentialgleichungen. VII, 194 Seiten. 1978.

Vol. 638: P. Shanahan, The Atiyah-Singer Index Theorem, An Introduction. V, 224 pages. 1978.

Vol. 639: N. Adasch et al., Topological Vector Spaces. V, 125 pages. 1978.

Vol. 640: J. L. Dupont, Curvature and Characteristic Classes. X, 175 pages. 1978.

Vol. 641: Séminaire d'Algèbre Paul Dubreil, Proceedings Paris 1976–1977. Edité par M. P. Malliavin. IV, 367 pages. 1978.

Vol. 642: Theory and Applications of Graphs, Proceedings, Michigan 1976. Edited by Y. Alavi and D. R. Lick. XIV, 635 pages. 1978.

Vol. 643: M. Davis, Multiaxial Actions on Manifolds. VI, 141 pages. 1978.

Vol. 644: Vector Space Measures and Applications I, Proceedings 1977. Edited by R. M. Aron and S. Dineen. VIII, 451 pages. 1978.

Vol. 645: Vector Space Measures and Applications II, Proceedings 1977. Edited by R. M. Aron and S. Dineen. VIII, 218 pages. 1978.

Vol. 646: O. Tammi, Extremum Problems for Bounded Univalent Functions. VIII, 313 pages. 1978.

Vol. 647: L. J. Ratliff, Jr., Chain Conjectures in Ring Theory. VIII, 133 pages. 1978.

Vol. 648: Nonlinear Partial Differential Equations and Applications, Proceedings, Indiana 1976–1977. Edited by J. M. Chadam. VI, 206 pages. 1978.

Vol. 649: Séminaire de Probabilités XII, Proceedings, Strasbourg, 1976–1977. Edité par C. Dellacherie, P. A. Meyer et M. Weil. VIII, 805 pages. 1978.

Vol. 650: C*-Algebras and Applications to Physics. Proceedings 1977. Edited by H. Araki and R. V. Kadison. V, 192 pages. 1978.

Vol. 651: P. W. Michor, Functors and Categories of Banach Spaces. VI, 99 pages. 1978.

Vol. 652: Differential Topology, Foliations and Gelfand-Fuks-Cohomology, Proceedings 1976. Edited by P. A. Schweitzer. XIV, 252 pages. 1978.

Vol. 653: Locally Interacting Systems and Their Application in Biology. Proceedings, 1976. Edited by R. L. Dobrushin, V. I. Kryukov and A. L. Toom. XI, 202 pages. 1978.

Vol. 654: J. P. Buhler, Icosahedral Golois Representations. III, 143 pages. 1978.

Vol. 655: R. Baeza, Quadratic Forms Over Semilocal Rings. VI, 199 pages. 1978.

Vol. 656: Probability Theory on Vector Spaces. Proceedings, 1977. Edited by A. Weron. VIII, 274 pages. 1978.

Vol. 657: Geometric Applications of Homotopy Theory I, Proceedings 1977. Edited by M. G. Barratt and M. E. Mahowald. VIII, 459 pages. 1978.

Vol. 658: Geometric Applications of Homotopy Theory II, Proceedings 1977. Edited by M. G. Barratt and M. E. Mahowald. VIII, 487 pages. 1978.

Vol. 659: Bruckner, Differentiation of Real Functions. X, 247 pages. 1978.

Vol. 660: Equations aux Dérivée Partielles. Proceedings, 1977. Edité par Pham The Lai. VI, 216 pages. 1978.

Vol. 661: P. T. Johnstone, R. Paré, R. D. Rosebrugh, D. Schumacher, R. J. Wood, and G. C. Wraith, Indexed Categories and Their Applications. VII, 260 pages. 1978.

Vol. 662: Akin, The Metric Theory of Banach Manifolds. XIX, 306 pages. 1978.

Vol. 663: J. F. Berglund, H. D. Junghenn, P. Milnes, Compact Right Topological Semigroups and Generalizations of Almost Periodicity. X, 243 pages. 1978.

Vol. 664: Algebraic and Geometric Topology, Proceedings, 1977. Edited by K. C. Millett. XI, 240 pages. 1978.

Vol. 665: Journées d'Analyse Non Linéaire. Proceedings, 1977. Edité par P. Bénilan et J. Robert. VIII, 256 pages. 1978.

Vol. 666: B. Beauzamy, Espaces d'Interpolation Réels: Topologie et Géometrie. X, 104 pages. 1978.

Vol. 667: J. Gilewicz, Approximants de Padé. XIV, 511 pages. 1978.

Vol. 668: The Structure of Attractors in Dynamical Systems. Proceedings, 1977. Edited by J. C. Martin, N. G. Markley and W. Perrizo. VI, 264 pages. 1978.

Vol. 669: Higher Set Theory. Proceedings, 1977. Edited by G. H. Müller and D. S. Scott. XII, 476 pages. 1978.

Vol. 670: Fonctions de Plusieurs Variables Complexes III, Proceedings, 1977. Edité par F. Norguet. XII, 394 pages. 1978.

Vol. 671: R. T. Smythe and J. C. Wierman, First-Passage Perculation on the Square Lattice. VIII, 196 pages. 1978.

Vol. 672: R. L. Taylor, Stochastic Convergence of Weighted Sums of Random Elements in Linear Spaces. VII, 216 pages. 1978.

Vol. 673: Algebraic Topology, Proceedings 1977. Edited by P. Hoffman, R. Piccinini and D. Sjerve. VI, 278 pages. 1978.

Vol. 674: Z. Fiedorowicz and S. Priddy, Homology of Classical Groups Over Finite Fields and Their Associated Infinite Loop Spaces. VI, 434 pages. 1978.

Vol. 675: J. Galambos and S. Kotz, Characterizations of Probability Distributions. VIII, 169 pages. 1978.

Vol. 676: Differential Geometrical Methods in Mathematical Physics II, Proceedings, 1977. Edited by K. Bleuler, H. R. Petry and A. Reetz. VI, 626 pages. 1978.

Vol. 677: Séminaire Bourbaki, vol. 1976/77, Exposés 489–506. IV, 264 pages. 1978.

Vol. 678: D. Dacunha-Castelle, H. Heyer et B. Roynette. Ecole d'Eté de Probabilités de Saint-Flour. VII-1977. Edité par P. L. Hennequin. IX, 379 pages. 1978.

Vol. 679: Numerical Treatment of Differential Equations in Applications, Proceedings, 1977. Edited by R. Ansorge and W. Törnig. IX, 163 pages. 1978.

Vol. 680: Mathematical Control Theory, Proceedings, 1977. Edited by W. A. Coppel. IX, 257 pages. 1978.

Vol. 681: Séminaire de Théorie du Potentiel Paris, No. 3, Directeurs: M. Brelot, G. Choquet et J. Deny. Rédacteurs: F. Hirsch et G. Mokobodzki. VII, 294 pages. 1978.

Vol. 682: G. D. James, The Representation Theory of the Symmetric Groups. V, 156 pages. 1978.

Vol. 683: Variétés Analytiques Compactes, Proceedings, 1977. Edité par Y. Hervier et A. Hirschowitz. V, 248 pages. 1978.

Vol. 684: E. E. Rosinger, Distributions and Nonlinear Partial Differential Equations. XI, 146 pages. 1978.

Vol. 685: Knot Theory, Proceedings, 1977. Edited by J. C. Hausmann. VII, 311 pages. 1978.

Vol. 686: Combinatorial Mathematics, Proceedings, 1977. Edited by D. A. Holton and J. Seberry. IX, 353 pages. 1978.

Vol. 687: Algebraic Geometry, Proceedings, 1977. Edited by L. D. Olson. V, 244 pages. 1978.

Vol. 688: J. Dydak and J. Segal, Shape Theory. VI, 150 pages. 1978.

Vol. 689: Cabal Seminar 76–77, Proceedings, 1976–77. Edited by A.S. Kechris and Y. N. Moschovakis. V, 282 pages. 1978.

Vol. 690: W. J. J. Rey, Robust Statistical Methods. VI, 128 pages. 1978.

Vol. 691: G. Viennot, Algèbres de Lie Libres et Monoïdes Libres. III, 124 pages. 1978.

Vol. 692: T. Husain and S. M. Khaleelulla, Barrelledness in Topological and Ordered Vector Spaces. IX, 258 pages. 1978.

Vol. 693: Hilbert Space Operators, Proceedings, 1977. Edited by J. M. Bachar Jr. and D. W. Hadwin. VIII, 184 pages. 1978.

Vol. 694: Séminaire Pierre Lelong – Henri Skoda (Analyse) Année 1976/77. VII, 334 pages. 1978.

Vol. 695: Measure Theory Applications to Stochastic Analysis, Proceedings, 1977. Edited by G. Kallianpur and D. Kölzow. XII, 261 pages. 1978.

Vol. 696: P. J. Feinsilver, Special Functions, Probability Semigroups, and Hamiltonian Flows. VI, 112 pages. 1978.

Vol. 697: Topics in Algebra, Proceedings, 1978. Edited by M. F. Newman. XI, 229 pages. 1978.

Vol. 698: E. Grosswald, Bessel Polynomials. XIV, 182 pages. 1978.

Vol. 699: R. E. Greene and H.-H. Wu, Function Theory on Manifolds Which Possess a Pole. III, 215 pages. 1979.

Vol. 700: Module Theory, Proceedings, 1977. Edited by C. Faith and S. Wiegand. X, 239 pages. 1979.

Vol. 701: Functional Analysis Methods in Numerical Analysis, Proceedings, 1977. Edited by M. Zuhair Nashed. VII, 333 pages. 1979.

Vol. 702: Yuri N. Bibikov, Local Theory of Nonlinear Analytic Ordinary Differential Equations. IX, 147 pages. 1979.

Vol. 703: Equadiff IV, Proceedings, 1977. Edited by J. Fábera. XIX, 441 pages. 1979.

Vol. 704: Computing Methods in Applied Sciences and Engineering, 1977, I. Proceedings, 1977. Edited by R. Glowinski and J. L. Lions. VI, 391 pages. 1979.

Vol. 705: O. Forster und K. Knorr, Konstruktion verseller Familien kompakter komplexer Räume. VII, 141 Seiten. 1979.

Vol. 706: Probability Measures on Groups, Proceedings, 1978. Edited by H. Heyer. XIII, 348 pages. 1979.

Vol. 707: R. Zielke, Discontinuous Čebyšev Systems. VI, 111 pages. 1979.

Vol. 708: J. P. Jouanolou, Equations de Pfaff algébriques. V, 255 pages. 1979.

Vol. 709: Probability in Banach Spaces II. Proceedings, 1978. Edited by A. Beck. V, 205 pages. 1979.

Vol. 710: Séminaire Bourbaki vol. 1977/78, Exposés 507–524. IV, 328 pages. 1979.

Vol. 711: Asymptotic Analysis. Edited by F. Verhulst. V, 240 pages. 1979.

Vol. 712: Equations Différentielles et Systèmes de Pfaff dans le Champ Complexe. Edité par R. Gérard et J.-P. Ramis. V, 364 pages. 1979.

Vol. 713: Séminaire de Théorie du Potentiel, Paris No. 4. Edité par F. Hirsch et G. Mokobodzki. VII, 281 pages. 1979.

Vol. 714: J. Jacod, Calcul Stochastique et Problèmes de Martingales. X, 539 pages. 1979.

Vol. 715: Inder Bir S. Passi, Group Rings and Their Augmentation Ideals. VI, 137 pages. 1979.

Vol. 716: M. A. Scheunert, The Theory of Lie Superalgebras. X, 271 pages. 1979.

Vol. 717: Grosser, Bidualräume und Vervollständigungen von Banachmoduln. III, 209 pages. 1979.

Vol. 718: J. Ferrante and C. W. Rackoff, The Computational Complexity of Logical Theories. X, 243 pages. 1979.

Vol. 719: Categorial Topology, Proceedings, 1978. Edited by H. Herrlich and G. Preuß. XII, 420 pages. 1979.

Vol. 720: E. Dubinsky, The Structure of Nuclear Fréchet Spaces. V, 187 pages. 1979.

Vol. 721: Séminaire de Probabilités XIII. Proceedings, Strasbourg, 1977/78. Edité par C. Dellacherie, P. A. Meyer et M. Weil. VII, 647 pages. 1979.

Vol. 722: Topology of Low-Dimensional Manifolds. Proceedings, 1977. Edited by R. Fenn. VI, 154 pages. 1979.

Vol. 723: W. Brandal, Commutative Rings whose Finitely Generated Modules Decompose. II, 116 pages. 1979.

Vol. 724: D. Griffeath, Additive and Cancellative Interacting Particle Systems. V, 108 pages. 1979.

Vol. 725: Algèbres d'Opérateurs. Proceedings, 1978. Edité par P. de la Harpe. VII, 309 pages. 1979.

Vol. 726: Y.-C. Wong, Schwartz Spaces, Nuclear Spaces and Tensor Products. VI, 418 pages. 1979.

Vol. 727: Y. Saito, Spectral Representations for Schrödinger Operators With Long-Range Potentials. V, 149 pages. 1979.

Vol. 728: Non-Commutative Harmonic Analysis. Proceedings, 1978. Edited by J. Carmona and M. Vergne. V, 244 pages. 1979.